INSECT-PLANT INTERACTIONS

Volume 1

Influence of Biotic and Abiotic Factors

Insect-Plant Interactions, 2-volume set:
Volume 1: Influence of Biotic and Abiotic Factors
Volume 2: Mechanisms and Strategies
ISBN: 978-1-779644-26-8 (hbk)

Insect-Plant Interactions: Volume 1: Influence of Biotic and Abiotic Factors
ISBN: 978-1-779644-27-5 (hbk)
ISBN: 978-1-779644-28-2 (ebk)

Insect-Plant Interactions: Volume 2: Mechanisms and Strategies
ISBN: 978-1-779644-29-9 (hbk)
ISBN: 978-1-779644-30-5 (ebk)

INSECT-PLANT INTERACTIONS

Volume 1

Influence of Biotic and Abiotic Factors

Jamin Ali, PhD
Adil Tonğa, PhD
Adnan Shakeel, PhD
Nasheeman Ashraf, PhD
Editors

AAP | APPLE ACADEMIC PRESS

First edition published 2026

Apple Academic Press Inc.
1265 Goldenrod Circle, NE,
Palm Bay, FL 32905 USA

760 Laurentian Drive, Unit 19,
Burlington, ON L7N 0A4, Canada

CRC Press
2385 NW Executive Center Drive,
Suite 320, Boca Raton FL 33431

4 Park Square, Milton Park,
Abingdon, Oxon, OX14 4RN, UK

Library and Archives Canada Cataloguing in Publication

CIP data on file with Canada Library and Archives

Library of Congress Cataloging-in-Publication Data

CIP data on file with US Library of Congress

ISBN: 978-1-77964-427-5 (hbk)
ISBN: 978-1-77964-428-2 (ebk)

DOI: 10.1201/9781779644282

About the Editors

Jamin Ali, PhD
Foreign Research Expert, Jilin Agricultural University, China

Jamin Ali, PhD, is a Foreign Research Expert at Jilin Agricultural University in China, where he focuses on applied chemical ecology and sustainable pest management solutions. He earned his PhD in Chemical Ecology from Keele University, United Kingdom, with a thesis exploring the interactions between the aphid Myzus persicae and its natural enemies. His academic journey includes a Master of Science in Zoology from Aligarh Muslim University, India, and advanced training in Integrated Crop Management at the University of Neuchâtel, Switzerland. Dr. Ali's career includes international research and teaching roles, such as a Visiting Scientist position at Ben-Gurion University, Israel, and a Teaching Assistant role at Keele University, UK. His work emphasizes using semiochemicals to develop eco-friendly pest control strategies, contributing to integrated pest management (IPM) practices globally. He has been recognized with prestigious awards, including the 2024 Outstanding Postdoctoral Award in China and scholarships from Türkiye and the Indian government. Dr. Ali has authored multiple influential books, such as *The Peach Potato Aphid (Myzus persicae): Ecology and Management and Chemical Ecology: Insect-Plant Interactions*. His research articles in high-impact journals explore plant defenses and sustainable agricultural practices. He has presented his research at conferences and workshops in the United Kingdom, Kenya, the Netherlands, Malaysia, and India. Currently, Dr. Ali advances IPM through collaborations, focusing on phytohormones, plant-insect interactions, and innovative pest control technologies. His work connects scientific research with practical solutions to improve agricultural sustainability worldwide.

Adil Tonğa, PhD
Researcher, Diyarbakir Plant Protection Research Institute, Diyarbakir, Türkiye

Adil Tonğa, PhD, currently serves as a dedicated researcher at the Diyarbakir Plant Protection Research Institute, Diyarbakir, Türkiye. His academic journey includes earning MSc and PhD degrees from Dicle University in Diyarbakir, Türkiye. He also maintains a strong interest in both national and international scientific events, reflecting his commitment to staying updated on advancements in entomology. His enthusiasm is evident in his efforts to build collaborations with researchers globally to advance collective knowledge and address agricultural challenges. His expertise spans various dimensions of the field of entomology, with a particular focus on exploring intricate interactions within trophic levels and investigating relationships among plants, pests, and their natural enemies. Inspired by the chemical communication underpinning ecological interactions, he explores insect behavior. Beyond theoretical exploration, he devotes significant effort to the practical application of this knowledge in devising effective pest management tactics. A notable focus of his research involves projects on selective insecticides that not only contribute to understanding the ecotoxicology of egg parasitoids but also play a pivotal role in assessing the compatibility of these insecticides within integrated pest management strategies (IPMs). His holistic approach underscores his commitment to advancing both theoretical understanding and practical solutions within the field of plant protection.

Adnan Shakeel, PhD
Department of Botany, Central University of Kashmir, Ganderbal, Jammu and Kashmir, India

Adnan Shakeel, PhD, is a molecular plant–nematode researcher, currently working as INSPIRE FACULTY at the Dept. of Botany, Central University of Kashmir, Ganderbal. Dr. Shakeel has undertaken postdoctoral training at Guangxi University (China) and CSIR–IIIM (India). He completed his PhD in Plant Science from Aligarh Muslim University, India, with a thesis titled Evaluation of Fly Ash and Nitrogen Fertilizer for the Management of Root-Knot Nematodes. Through seven years of research experience on plant–nematode dialogue, Dr. Shakeel has identified the major parasitic nematodes in the Aligarh and Kashmir regions of India, estimated their effect on root vegetable crops like carrot and beetroot, and on an important medicinal crop, saffron. Dr. Adnan Shakeel has published more than 30 SCI articles on plant–nematode/ microbe interactions, along with five book chapters and five conference abstracts. He has been awarded the Inspire Faculty Award by DST Govt. of India and Young Scientist Associate Award in Plant nematology by AETDS-Uttarakhand, India. He has also reviewed articles for several globally reputed journals published by Elsevier and Springer Nature.

Nasheeman Ashraf, PhD

Professor, Jilin Agricultural University, China

Nasheeman Ashraf, PhD, is a plant molecular biologist, currently working on modern genomic approaches to understand molecular plant–microbe interactions at CSIR-IIIM, Srinagar. Dr. Nasheeman completed her PhD from NIPGR, JNU, New Delhi, working on plant–microbe molecular talks. She is currently leading the saffron biology group with five active PhD students and two postdocs. Dr. Nasheeman has been mainly working on the metabolite and apocarotenoid pathways of saffron plants in response to biotic and abiotic factors, including fungi, bacteria, nematodes, and endophytes. Her contribution has significantly boosted crop production in the Kashmir region. She has published more than 30 SCI research articles in reputable scientific journals and has edited one book so far. She is actively reviewing for various international journals. She has also received two international/national patents for her research. She has been honored with several international awards, including an EMBO Short-Term Fellowship to work as a visiting scientist at the Botanical Institute, University of Castilla-La Mancha, Albacete, Spain, and a CSIR-Raman Research Fellowship to work as a visiting scientist at the University of Kentucky, USA (2015–2016).

Contents

Contributors

Arzlan Abbas
College of Plant Protection, Jilin Agricultural University, Changchun, Jilin, China

Sohail Abbas
School of Agricultural and Pest Control, College of Plant Protection, Jilin Agricultural University, Changchun, Jilin, China

Gholamreza Abdi
Department of Biotechnology, Persian Gulf Research Institute, Persian Gulf University, Bushehr, Iran

Zeeshan Ahmad
Department of Zoology, Hazara University, Mansehra, Khyber Pakhtunkhwa, Pakistan

Mobolade Dele Akinbuluma
Department of Crop Protection and Environmental Biology, University of Ibadan, Nigeria; Department of Evolutionary and Population Biology, University of Amsterdam, The Netherlands

Aleena Alam
School of Agricultural and Pest Control, College of Plant Protection, Jilin Agricultural University, Changchun, Jilin, China

Nooral Amin
Plant Biotechnology Center, College of Agronomy, Jilin Agricultural University, Changchun, Jilin, China

Achyuta Basak
Department of Genetics and Plant Breeding, Uttar Banga Krishi Viswavidyalaya, Pundibari, Cooch Behar, West Bengal, India

Aashaq Hussain Bhat
Department of Biosciences, Chandigarh University, Mohali, Punjab, India; University Center for Research and Development, Chandigarh University, Mohali, Punjab, India

Roland Bocco
Texas A&M AgriLife Research and Extension Center, Aggie Dr., Beaumont, TX, USA

Muhammad Zulqar Nain Dara
College of Plant Protection, Jilin Agricultural University, Changchun, Jilin, China

Sozan E. El-Abeid
Nanotechnology and Advanced Nano-Materials Laboratory (NANML), Plant Pathology Research Institute, Agricultural Research Center, Giza, Egypt; Department of Mycology and Disease Survey Research, Plant Pathology Research Institute, Agricultural Research Center, Giza, Egypt

Gozde Busra Eroglu
Department of Molecular Biology and Genetics, Erzurum Technical University, Erzurum, Türkiye

Shimaa Ghareeb
Plant Protection Research Institute, Agricultural Research Center, Giza, Egypt; Faculty of Agriculture, Kyushu University, 744 Motooka, Fukuoka, Japan

Hossam E. Harb
Department of Biotechnology, Faculty of Science, Cairo University, Cairo, Egypt

Rana Hussien
Plant Protection Research Institute, Agricultural Research Center, Giza, Egypt

Dina S. S. Ibrahim
Department of Nematode Diseases and Central Laboratory of Biotechnology, Plant Pathology Research Institute, Agricultural Research Center, Giza, Egypt

Mohsin Ikram
School of Agriculture, Uttaranchal University, Dehradun, Uttarakhand, India

Javad Karimi
Department of Plant Protection, School of Agriculture, Ferdowsi University of Mashhad, Mashhad, Iran

Amandeep Kaur
Department of Entomology, Punjab Agricultural University, Ludhiana, Punjab, India

Azhar Abbas Khan
Department of Entomology, College of Agriculture, University of Layyah, Layyah, Pakistan

Vijay Kumar
Department of Entomology, Punjab Agricultural University, Ludhiana, Punjab, India

Lei Li
Jilin Agricultural University, Changchun, Jilin, China

Mubasher Ahmad Malik
Department of Entomology, College of Agriculture, University of Layyah, Layyah, Pakistan

Yomna A. Mohamed
Department of Biotechnology, Faculty of Science, Cairo University, Cairo, Egypt

Md. Sabir Ahmed Mondol
Department of Agricultural Biochemistry, Bidhan Chandra Krishi Viswavidyalaya, Mohanpur, Nadia, West Bengal, India

Usama Nawaz
Department of Zoology, Hazara University, Mansehra, Khyber Pakhtunkhwa, Pakistan

James Adebayo Ojo
Department of Crop Production, Kwara State University, Malete, Kwara State, Nigeria

Liu Pengwei
QiLu ZhongKe Academy of Modern Microbiology Technology – Crop Precision Breeding Center, Jinan, Shandong, China

Jorge Poveda
Recognized Research Group AGROBIOTECH, UIC-370 (JCyL), Department of Plant Production and Forest Resources, Higher Technical School of Agricultural Engineering of Palencia, University Institute for Research in Sustainable Forest Management (iuFOR), University of Valladolid, Palencia, Spain

Rajendra Prasad
School of Agriculture, Uttaranchal University, Dehradun, Uttarakhand, India

Vaseem Raja
University Center for Research and Development, Chandigarh University, Mohali, Punjab, India

Ali Rajabpour
Department of Plant Protection, Agricultural Sciences and Natural Resources University of Khuzestan, Mollasani, Khuzestan, Iran

Amisha Rani
Department of Forest Products, Dr. Yashwant Singh Parmar University of Horticulture and Forestry, Nauni, Solan, Himachal Pradesh, India

Sergio Rasmann
Functional Ecology Laboratory, Institute of Biology, University of Neuchâtel, Neuchâtel, Switzerland

Sabahat Saeed
College of Animal Sciences and Technology, Jilin Agricultural University, Changchun, Jilin, China; Department of Zoology, Hazara University, Mansehra, Khyber Pakhtunkhwa, Pakistan

Mansi Sahu
School of Applied and Life Sciences, Uttaranchal University, Dehradun, Uttarakhand, India

Sudipta Samanta
Department of Genetics and Plant Breeding, Indian Agricultural Research Institute, Kolkata Hub, Barrackpore, Kolkata, West Bengal, India

Komal Shoukat
Government College University, Faisalabad, Punjab, Pakistan

Mahipal Singh
School of Agriculture, Uttaranchal University, Dehradun, Uttarakhand, India

Aroge Temitope
College of Plant Protection, Jilin Agricultural University, Changchun, Jilin, China

Abou Togola
International Maize and Wheat Improvement Center (CIMMYT), United Nations Avenue, Gigiri, Nairobi, Kenya

Adil Tonğa
Department of Entomology, Diyarbakır Plant Protection Research Institute, Diyarbakır, Türkiye

Nadia K. Toukem
Department of Zoology and Entomology, University of Pretoria, Pretoria, South Africa

Noman Waheed
The United Graduate School of Agricultural Sciences, Kagoshima University, Kagoshima, Japan; Department of Zoology, Hazara University Mansehra KPK, Pakistan

Fatemeh Yarahmadi
Department of Plant Protection, Agricultural Sciences and Natural Resources University of Khuzestan, Mollasani, Khuzestan, Iran

Abdullahi Yusuf
Department of Zoology and Entomology, University of Pretoria, Pretoria, South Africa

Abbreviations

12-OPDA	12-oxophytodienoic acid
AAP	acquisition access period
ABA	abscisic acid
AcMNPV	*Autographa californica* multiple nucleopolyhedrovirus
AG	above ground
AlpMV	alpinia mosaic virus
AMF	arbuscular mycorrhizal fungi
AMV	alfalfa mosaic virus
AnFBMV	angelonia flower break mosaic virus
AnYV	anthriscus yellows virus
AOX	alternative oxidase
APX	ascorbate peroxidase
BABA	β-aminobutyric acid
BBTV	banana bunchy tops virus
BBWV-1	broad bean wilt virus-1
BCA	biocontrol agent
BCRV	black currant reversion virus
BCTV	beet curly top virus
BG	below ground
BlSHV	blueberry shock virus
BLCV	beet leaf curl virus
BMMV	barley mild mosaic virus
BMV	brome mosaic virus
BNYVV	beet necrotic yellow vein virus
BYDV	barley yellow dwarf virus
BYSMV	barley yellow striate mosaic virus
CaMV	cauliflower mosaic virus
CBDV	colocasia bobone disease virus
CCYV	cucurbit chlorotic yellows virus
CDPK	calcium-dependent protein kinases
CEC	cation-exchange capacity
CenMV	centrosema mosaic virus
CFDV	coconut foliar decay virus
CFMMV	cocksfoot mild mottle virus
CGCMV	cucumber green chlorotic mottle virus
CiLV-C	citrus leprosis virus-c

CLas	candidatus liberibacter asiaticus
CMMV	cowpea mild mottle virus
CMOLV	cherry mottle leaf virus
CMoV	carrot mottle virus
CMV	cucumber mosaic virus
COI1	coronatin-insensitive 1
CRW	clover root weevil
CSSV	cacao swollen shoot virus
CTV	citrus tristeza virus
CYDV	cereal yellow dwarf virus
CYMBV	citrus yellow mosaic badnavirus
CymMV	cymbidium mosaic virus
CYMV	citrus yellow mosaic virus
CYP450	cytochrome p450 monooxygenase
CYSDV	cucurbit yellow stunting disorder virus
CYV	cucumber yellows virus
DALYs	disability adjusted life years
DAMPs	damage-associated molecular patterns
DMADP	dimethylallyl diphosphate
EC	ecosystem conditions
eDNA	environmental DNA
EFN	extrafloral nectar
EPNs	entomopathogenic nematodes
ES	ecosystem service
ET	ethylene
FACs	fatty acids-amino acid conjugates
FAO	food and agriculture organization
FBNSV	faba bean necrotic stunt virus
FDP	farnesyl diphosphate
GAPDH	glyceraldehyde 3-phosphate dehydrogenase
GC	gas chromatography
GC-MS	gas chromatography–mass spectrometry
GDP	geranyl diphosphate
GGDP	geranylgeranyl diphosphate
GHG	greenhouse gas
GI	green infrastructure
GLVs	green leaf volatiles
GOX	glucose oxidase
GR	glutathione reductase
GSL	glucosinolate
GSL-ITC	glucosinolate-isothiocyanate
GSSG	oxidized glutathione

GST	glutathione-s-transferase
GV-A	grapevine virus A
GV-B	grapevine virus B
GVLRaV	grapevine leaf roll associated viruses
HAEs	herbivore-associated stimulants
HAMPs	herbivore-associated molecular patterns
HIPVs	herbivore-induced plant volatiles
HLB	huanglongbing disease
HMG	high-mobility group
HO2	hydroperoxide
HR	hypersensitive response
IAP	inoculation access period
IDP	isopentenyl diphosphate
IJ	infective juvenile
IPDM	integrated pest and disease management
IPM	integrated pest management
IPMS	integrated pest management strategies
ISR	induced systemic resistance
IYSV	iris yellow spot virus
JA	jasmonic acid
LBVV	lettuce big vein virus
LNYV	lettuce necrotic yellows virus
LOXs	lipoxygenases
LP	latent period
LSV	lily symptomless virus
MAMPs	microbe-associated molecular patterns
MAPK	mitogen-activated protein kinase
MCMV	maize chlorotic mottle virus
MEP	methylerythritol pathway
MeSA	methyl salicylate
MMV	maize mosaic virus
MNSV	melon necrotic spot virus
MRFV	maize rayado fino virus
MRLStV	maize red leaf stripe virus
MSV	maize streak virus
MVA	mevalonate route
MVOC	microbial volatile organic compound
NGS	next-generation sequencing
OFV	orchid fleck virus
OG	oligogalacturonide
OPR3	opda reductase 3
ORNs	olfactory receptor neurons

OSV	oat sterile virus
PCD	programmed cell death
PEMV-1	pea enation mosaic virus-1
PepMMoV	pepino mild mottle virus
PG	polygalacturonases
PGPR	plant growth-promoting rhizobacteria
PI	protein inhibitor
PID	photoionization detector
PLRV	potato leaf roll virus
PLV	plum latent virus
PMTV	potato mop top virus
PMV-M	potato mosaic virus-m
PNRSV	prunus necrotic ring spot virus
POD	peroxidase
PPO	polyphenol oxidase
PPP	plant protection product
PR	pathogenesis-related
PRRs	pattern recognition receptors
PTR-MS	proton transfer reaction mass spectrometry
PVS	potato virus S
PVX	potato virus X
PVY	potato virus Y
PYFV	parsnip yellows fleck virus
PYVMVD	pumpkin yellow vein mosaic viral disease
RBSDV	rice black-streaked dwarf virus
RDV	rice dwarf virus
RGMV	ryegrass mosaic virus
RGSV	rice grassy stunt virus
RHIPVs	root herbivore-induced plant volatiles
RKN	root-knot nematodes
ROS	reactive oxygen species
RpLV	raspberry latent virus
RRSV	rice ragged stunt virus
RRV	rose rosette virus
RSV	rice stripe virus
RTSV	rice tungro spherical virus
RYMV	rice yellow mottle virus
SA	salicylic acid
SAG	salicylic acid glucoside
SAR	systemic acquired resistance
SAW	surface acoustic wave
SCLRSV	strawberry crinkle latent ring spot virus

SCSV	subterranean clover stunt virus
SCTV	sugar beet curly top virus
SIFT-MS	selected ion flow tube-mass spectrometry
SLCCNV	squash leaf curl china virus
SODs	superoxide dismutases
SoMV	sowbane mosaic virus
SPMMV	sweet potato mild mottle virus
TBSV	tomato bushy stunt virus
TCV	turnip crinkle virus
TLCV	tomato leaf curl virus
TNV	tobacco necrosis virus
TOCV	tomato chlorosis virus
ToLCNDV	tomato leaf curl new delhi virus
ToTV	tomato torrado virus
TPCTV	tomato pseudo curly top virus
TPS	terpene synthases
TSWV	tomato spotted wilt virus
TYLCV	tomato yellow leaf curl virus
TYMV	turnip yellow mosaic virus
VOC	volatile organic compound
WASP	wound-activated surface potential
WStMV	wheat streak mosaic virus

Preface

Insect–plant interactions have long been at the heart of ecological and evolutionary studies, shaping biodiversity, agricultural landscapes, and natural ecosystems. These interactions, ranging from mutualistic partnerships to antagonistic conflicts, govern crucial processes such as pollination, herbivory, and plant defense strategies. Understanding the intricate relationship between insects and plants has never been more important, particularly in the face of global challenges like climate change, urbanization, and food security threats. This book, *Insect-Plant Interactions: Influence of Biotic and Abiotic Factors,* offers an in-depth exploration of these complex relationships by integrating perspectives from ecology, entomology, plant physiology, and agroecosystem management. It brings together cutting-edge research and comprehensive reviews to highlight both well-established concepts and emerging trends in insect–plant dynamics.

The book begins by examining mutualistic relationships (Chapter 1), shedding light on how insects and plants collaborate for mutual benefit. It then transitions to the darker side of these interactions, addressing insect-mediated plant pathogen transmission (Chapter 2) and the complex web of interactions among sucking pests, herbivores, and natural enemies (Chapter 3). The functional ecology of insectivorous foraging in agroecosystems (Chapter 4) further emphasizes the role of predators in maintaining ecological balance. A major focus of the book is the evolutionary arms race between plants and insects (Chapter 5), including the biochemical and structural defense mechanisms plants employ to deter herbivores. Induced plant defenses and their applications in pest management (Chapters 6 and 11) are explored in detail, bridging laboratory discoveries with practical field applications. The role of nematodes in insect–plant interactions (Chapter 7) presents an often-overlooked yet crucial component of belowground ecological dynamics. Expanding the scope beyond traditional agricultural systems, this book delves into insect–plant interactions in urban environments (Chapter 8), illustrating how urban stressors influence ecological relationships. Climate change and environmental shifts are examined in Chapters 9 and 10, offering insights into how plant-insect cascades may be altered and the broader implications for food security. Novel approaches to pest management are discussed in Chapters 12 and 13, with a particular emphasis on the use of endophytes and their volatile organic compounds, as well as biocontrol agents for sustainable plant protection.

By synthesizing diverse perspectives and recent advancements, this book serves as an invaluable resource for researchers, students, and professionals in entomology, plant sciences, and agricultural ecology. We extend our deepest appreciation to all contributors for their insightful work and dedication. As we navigate the ever-changing landscape of ecological and agricultural challenges, we hope this volume will inspire further research and innovation in the field of insect–plant interactions.

CHAPTER 1

Friend or Foe: The Intricacies of Mutualistic Plant-Insect Partnerships

JAMES ADEBAYO OJO,[1,*] ABOU TOGOLA,[2] ROLAND BOCCO,[3] and MOBOLADE DELE AKINBULUMA[4,5]

[1]*Department of Crop Production, Kwara State University, Malete, Kwara State, Nigeria*

[2]*International Maize and Wheat Improvement Center (CIMMYT), United Nations Avenue, Gigiri, Nairobi, Kenya*

[3]*Texas A&M AgriLife Research and Extension Center, Aggie Dr., Beaumont, TX, USA*

[4]*Department of Crop Protection and Environmental Biology, University of Ibadan, Nigeria*

[5]*Department of Evolutionary and Population Biology, University of Amsterdam, The Netherlands*

*Corresponding author

ABSTRACT

The interaction between plants and insects represents a dynamic and coevolved system central to ecosystem stability. Insects, whether beneficial or pests, engage in interdependent relationships with plants, fostering mutualism that shapes biodiversity. This chapter explores the complexities of plant-insect partnerships, focusing on mutualisms that enhance nutrition, defense, pollination, seed dispersal, and other forms of symbiosis. It discusses the importance of these alliances in ecosystem function and biodiversity preservation. The co-evolutionary dynamics between plants

Insect-Plant Interactions, Volume 1: Influence of Biotic and Abiotic Factors.
Jamin Ali, Adil Tonğa, Adnan Shakeel, and Nasheeman Ashraf (Eds.)

DOI: 10.1201/9781779644282-1

and insects are examined, highlighting adaptations like floral structures and chemical signaling that influence phenotypic traits in both partners. Additionally, the chapter explores insect morphological traits, species abundance, and trait-mediated interactions within these partnerships. It emphasizes reciprocity, discussing the benefits, trade-offs, and potential costs of these relationships. The chapter also examines the environmental impacts of mutualistic interactions, including their influence on plant populations, insect communities, and the potential disruption of these systems by climate change or invasive species. The role of human activities in disrupting these natural balances is addressed, along with the need for conservation efforts. The chapter concludes by identifying emerging research areas in plant-insect interactions and future directions for studying their mutualistic roles in ecosystems.

1.1 INTRODUCTION

Plant-insect interactions are essential elements of terrestrial ecosystems, serving a variety of vital functions in ecological processes. Pollination, herbivory, mutualism, seed distribution, ecosystem services (ES), trophic cascades, biodiversity, and the effects of climate change are all examples of the importance of these interactions in forming ecosystems (Gurevitch et al., 2002; Neuschulz et al., 2016). In order to ensure the reproduction of several plant species, bees, butterflies, and other insects aid in the movement of pollen from one flower to another. Fruit and seed production, as well as the preservation of plant diversity, depend on this process. Herbivorous insects consume plant tissues, which affects the survival and growth of the plants. Although herbivory can be harmful to individual plants, it also promotes the development of herbivore adaptations and plant defenses (McCall and Irwin, 2006). This dynamic interplay keeps plant populations in ecosystems in balance. Certain insects establish mutualistic connections with plants. For example, ants benefit both plants and herbivores by offering protection against the former in exchange for food or shelter. By gathering and moving seeds to new areas, insects like ants and beetles help spread seeds (Ben-Zvi et al., 2021). This increases the likelihood of plants surviving and aids in their colonization of new places. Interactions between plants and insects deliver essential ecological services. For instance, pollinators support both the growth of crops and the preservation of natural plant ecosystems. Both biodiversity and food security depend on these services. Plant-insect interactions can start trophic cascades. For example, the extinction of apex predators such as birds may result in a rise in herbivorous insects, which may have adverse effects on plant ecosystems. In terms of biodiversity, there are connections between the diversity found in plants and insects. Insects can find a wide range of resources in a diversified plant community, which in turn facilitates seed distribution and plant pollination (Ben-Zvi et al., 2021; Teixeira et al., 2023). Climate change has an impact on the timing of insect activity and plant blossoming, which may disrupt the synchronicity of these interactions and jeopardize the stability of ecosystems. Interactions between plants and insects are essential to ecosystem function. The health of ecosystems, biodiversity, and human societies depend on our ability to comprehend and preserve these relationships.

Mutualism is a type of symbiotic relationship between two species in which both organisms benefit from their interaction (Ferrière et al., 2004). Mutualism is characterized by advantages for each species, as opposed to parasitism or predation. These relationships can be either facultative, signifying that the association is advantageous but not necessary, or obligate, in which the species are interdependent for their survival (Bronstein, 2015). Mutualism is essential for ecological stability, nutrient cycling, biodiversity, and succession. Mutualistic relationships encourage species cooperation, which supports environmental stability (Hoeksema and Bruna, 2015). Mycorrhizal relationships between fungi and plants, for example, improve nutrient absorption, increasing the availability of important nutrients for plants and promoting overall ecosystem productivity (Enebe and Erasmus, 2023). The interdependence of species promotes the diversity of creatures within ecosystems, which enhances biodiversity. Mutualistic interactions, which influence ecological succession, aid in the establishment and growth of pioneer species. Thus, as ecosystems evolve over time, the makeup of entire ecosystems may be influenced. Ecological concepts such as mutualism are fundamental to ecosystem stability, nutrient cycling, and biodiversity (Chamberlain et al., 2014). It highlights the complex web of interactions that keeps life on Earth functioning and serves as an example of the harmonious relationships found in nature (Ferrière et al., 2004).

1.2 OVERVIEW OF PLANT-INSECT MUTUALISM IN ECOLOGICAL CONTEXTS

Plant-insect mutualism is a symbiotic relationship between plants and insects. In this complex ecological relationship, insects aid in the reproductive and defense systems of plants by providing them with nectar, pollen, or shelter, and plants reciprocate by offering protection and other services like pollination. This mutualistic relationship affects ecosystem stability and biodiversity, and it is essential to ecosystem dynamics. Comprehending the intricacies of mutualism between plants and insects is crucial in order to fully understand ecological interconnectivity and sustain biodiversity within heterogeneous ecosystems.

1.2.1 ROLE OF INSECTS IN POLLINATION AND SEED DISPERSAL FOR PLANTS

Insects play pivotal roles in the survival and proliferation of numerous plant species through pollination and seed dispersal, fostering plant diversification, ecological succession, and ES (Chamberlain et al., 2014). Most flowering plants rely on insects, primarily bees, butterflies, moths, beetles, and flies, for sexual reproduction. During nectar-seeking visits, these insects inadvertently transfer pollen from one flower's anthers to another's stigma, ensuring fertilization (Kenneth and Jayashankar, 2021). This process is vital for fruit and seed production, including many agricultural crops. Bees are especially effective pollinators due to their pollen and nectar-gathering behaviors, enhancing the resilience and redundancy of pollination services (Sponsler et al., 2023). While beetles and flies ingest pollen and nectar on a regular basis, butterflies and moths mostly visit flowers for their nectar. The diversity of insect pollinators improves the resilience and redundancy of pollination services (Barbero and Maffei, 2023). Interactions between pollinators and plants have frequently resulted in coevolution. Pollinators acquire specialized mouthparts and behaviors to obtain nectar and pollen supplies, whereas plants create specific adaptations, such as floral structures, colors, and odors, in order to attract specific pollinators (Rader et al., 2016). This mutual adaptation, an illustration of coevolution, highlights the close ecological and evolutionary relationships that exist between plants and insects.

Insects also play a crucial role in seed distribution for various plant species. Ants and beetles, for example, are known for carrying seeds, with ants often depositing seeds in their nests, providing an ideal environment for germination (Giladi, 2006). Beetles and other insects disperse seeds as they consume fruits (Arachchige et al., 2023). These ecological functions, including pollination and seed dispersal, enhance biodiversity, support crop productivity, and preserve wild plant populations (Neuschulz et al., 2016). The decline in insect pollinator populations, driven by factors such as habitat loss, pesticide use, and climate change, poses risks to ecosystem health and food availability (Doi et al., 2008; Kudo et al., 2004). In summary, insects are indispensable to terrestrial ecosystems, influencing plant community organization, biodiversity, and plant development.

1.2.2 INSECTS AS DEFENDERS AGAINST HERBIVORES OR PATHOGENS FOR PLANTS

Insects, despite being mostly recognized for their herbivorous behavior, also play a fascinating and little-known role in safeguarding plants from diseases and other herbivores. Many bug species are herbivorous, which means they feed on plants and hunt other herbivorous insects. Ladybugs, lacewings, and praying mantises, among other insects, devour aphids, caterpillars, and other insects that consume plants (Egan et al., 2021; Ramos and Schiestl, 2020). They reduce plant damage by assisting in the management of herbivore populations. Parasitoid insects are another class of insects that indirectly protect plants. Their larvae consume their hosts from within, killing the herbivorous insects on which they lay their eggs (De-la-Cruz et al., 2022; Togola et al., 2023). This organic biocontrol reduces plant damage and herbivore populations. Certain insects have developed the capacity to trigger plant chemical defense mechanisms. These insects cause plants to emit chemicals that either repel or damage herbivores when they eat them. As a result of this interaction, plants develop a defense mechanism that deters potential herbivores (De-la-Cruz et al., 2022; Mithöfer and Boland, 2012). A variety of insects develop mutualistic connections with plants, offering protection in return for nourishment or a place to reside. In exchange for sugary nectar or cover from plant structures, ants, for example, actively repel or eradicate herbivores from plants. Beneficial insects are widely acknowledged in agriculture for their importance in organic pest management (Togola et al., 2023). Integrated pest management (IPM) techniques use these defenders to reduce the demand for chemical pesticides, promoting ecologically responsible and sustainable agricultural methods (Agunbiade et al., 2018; Togola et al., 2023).

Insects' ability to defend themselves against diseases or herbivores may be hampered by a variety of factors. Pest control services might become less available as a result of chemicals' negative effects on beneficial insects, habitat loss, and climate change. To preserve and strengthen the functions of these protectors, conservation measures are essential (Drossart and Gérard, 2020). Among the strategies are reducing pesticide use, establishing pollinator-friendly gardens, and building habitat corridors. The natural balance that these insects provide can be preserved by fostering biodiversity and health. Insect defenders support plant populations in natural situations to be more resilient and healthier (Giannetti et al., 2019). By reducing herbivore pressures, these insects contribute to preserving plant diversity and the coexistence of different species within ecosystems.

1.2.3 INSECTS AIDING PLANTS WITH NUTRIENT ACQUISITION

Ecosystem interactions between plants and insects go beyond herbivory and pollination, with certain insects aiding nutrient absorption in plants, especially in nutrient-deficient soils (De-la-Cruz et al., 2022). Insects provide plants access to diverse nutrient sources, including micronutrients, phosphorus, and nitrogen, which are crucial for nutrient cycling (Chamberlain et al., 2014). Some ants, including *Allomerus* spp. and *Pseudomyrmex* spp., work together with nitrogen-fixing bacteria to turn atmospheric nitrogen into forms that plants can use (Farrell, 2017; Hölldobler and Kwapich, 2022). Carnivorous plants, such as sundews, pitcher plants, and Venus's flytraps, capture insects to obtain essential micronutrients such as phosphorus and nitrogen, allowing them to thrive in nutrient-poor conditions. Dung beetles, carrion beetles, and burying beetles enrich the soil by consuming decomposing organic matter and recycling vital nutrients (Fagundes et al., 2005).

Enhancing ecosystem resilience necessitates insects to mediate the nutrient acquisition process. By guaranteeing that plants obtain vital nutrients even in nutrient-deficient surroundings, it enhances ecosystems' resilience and stability. Insect conservation that aids in the acquisition of nutrients is essential to maintaining the health of ecosystems (Drossart and Gérard, 2020). An essential first step in the conservation of insects is to acknowledge the ecological significance of these interactions (Rico-Gray and Oliveira, 2008). The intricate network of ecological interactions that supports the health of ecosystems is best illustrated by insects helping plants absorb nutrients. These interactions, which can involve carnivorous plants, nitrogen-fixing ants, or nutrient-recycling insects, improve the resilience and overall health of natural systems in addition to aiding in the cycling of nutrients (Herrera and Pellmyr, 2009). Understanding the ecological relevance of these interactions is essential to our comprehension of biodiversity conservation and ecosystem dynamics.

1.3 CO-EVOLUTIONARY DYNAMICS

One of the classic co-evolutionary dynamics is the relationship between plants and insects, particularly herbivorous insects. In these co-evolutionary dynamics, plants and insects adapt to each other's traits over time.

1.3.1 HOW PLANTS AND INSECTS ADAPT TO EACH OTHER'S TRAITS

Plants produce a wide range of secondary metabolites, which perform various ecological functions but are not directly involved in their primary metabolic activities as part of their defensive system. Alkaloids, tannins, coumarin, phenolic compounds, terpenoids, and a wide range of other secondary metabolites are poisonous or deterring. These compounds can be unpalatable to herbivores. Two broad plant defense mechanisms use these chemical compounds. The constitutive defense system is always present in the plant genetic system, while the induced defense system is activated in response to herbivore attack. These chemicals diversify or concentrate as plants evolve. Plants repel herbivores via chemical compounds and physical defenses like thorns, spines, or tough, leathery leaves. These structures block herbivores from eating plant tissues. For instance, trichomes are hair-like structures that can poison herbivores or distract them, preventing the pest from feeding (Nandi et al., 2022; War et al., 2020). Some plants mimic unpleasant or toxic species to deter herbivores (Divekar et al., 2022; Gebreziher, 2018). Other plants develop camouflage to blend into their surroundings (Dettner and Liepert, 1994; Lev-Yadun, 2016) or respond to herbivore damage by increasing defensive compound production as a "call for help" by attracting predators or parasites. In response to herbivore feeding on plant tissues, these plants emit signaling compounds such as jasmonic acid (JA) or salicylic acid (Schweiger et al., 2014).

1.3.2 ROLE OF GENETIC AND ECOLOGICAL FACTORS IN SHAPING INSECT-PLANT RELATIONSHIPS

In response to plant defense systems, herbivorous insects evolve adaptation mechanisms, including genetic changes or detoxification mechanisms, that allow them to metabolize or excrete plant toxins, making them less susceptible to the plant's chemical defenses. Insect genetic changes are a fundamental aspect of co-evolutionary dynamics, particularly in interactions with other organisms like plants, predators, parasites, and competitors. Co-evolution often drives genetic changes in insects, allowing them to adapt to and exploit specific ecological niches. Over time, specific genetic traits associated with detoxification enzymes, behavioral adaptations, or host plant specialization may become more prevalent in insect populations. This can result in the co-evolution of plant-insect interactions and improve their ability to evade or resist predation, involving the development of protective structures or behaviors. Insects within a population possess genetic diversity that serves as the foundation for adaptation. These are several ways in which insect genetic changes occur in co-evolutionary dynamics. On the other hand, herbivore specializations play a key role in the co-evolutionary dynamics between plants and herbivores. These specializations encompass the development of specific traits, behaviors, and adaptations that enable herbivorous organisms to exploit particular plant species or families. Many herbivores exhibit specialization in feeding and often target a narrow range of host plants or even a single plant species. This specialization is often a product of the co-evolutionary relationship between the herbivore and its host plant. These adaptations include specialized mouthparts tailored to their preferred host plants, enhancing their ability to efficiently extract resources from specific plant tissues.

For example, long proboscises are adapted for nectar feeding, while mandibles are adapted for leaf-chewing (Bauder and Karolyi, 2019; Krenn et al., 2021). Some herbivores have also developed ways to neutralize or sequester plant toxins. These include detoxification enzymes or protective structures that store toxic compounds (Erb and Robert, 2016; War et al., 2020). Herbivores can exhibit specific behaviors to increase their efficiency in exploiting host plants, such as blending in with their surroundings for predator avoidance through camouflage, synchronized life cycle patterns with their host plants' growth and phenology to access fresh and nutritious plant tissue, or even evolving mimicry strategies to deceive their host plants. Additionally, some herbivores have developed heightened senses, like specialized olfactory receptors, to detect chemical cues from their host plants, enabling them to locate suitable host plants more effectively (Anton and Cortesero, 2022).

1.4 COMMUNICATION AND SIGNALING

Complex systems like signaling and communication enable mutualistic interactions and ecological balance. Plants use a variety of chemical, visual, and tactile cues to lure beneficial insects for pollination or to protect themselves against herbivorous animals. In reaction, insects change their behavior, emit pheromones, or use visual signals to communicate information about the availability of resources or possible threats. The coordination of processes including symbiotic relationships, herbivore deterrence, and pollination depends on this dynamic signal exchange. Comprehending the subtleties of communication in thc rclationships between plants and insects not only enhances our comprehension of ecological networks but also has consequences for agricultural activities, conservation, and the health of ecosystems.

1.4.1 CHEMICAL CUES USED BY PLANTS TO ATTRACT BENEFICIAL INSECTS

Over time, insects may become more specialized in their interactions, shapes, behaviors, morphology, and life history traits. These interactions foster mutual benefits between insects and various other organisms, including plants and fungi, while playing a key role in co-evolutionary dynamics and influencing the ecology and evolution of the entire system.

For instance, insects seeking nectar and pollen for sustenance visit flowers, inadvertently facilitating plant reproduction by transferring pollen from one flower to another (Krenn et al., 2021; Kumar and Khan, 2022). This mutualism benefits both the plants, ensuring pollination and reproduction, and the insects, providing a food source. Some plants have co-evolved with specific pollinators, attracting insects with rewards like nectar to aid in pollination. Certain plants form mutualistic bonds with ants, offering shelter and food (typically nectar) in exchange for protection against herbivores as well as habitat maintenance (Beattie, 1985; Beattie and Hughes, 2002).

Myrmecophytes, which are plants with specialized structures to house ant colonies, receive protection from herbivores and contribute to nutrient cycling through ant waste deposition (Dejean et al., 2020; Mayer et al., 2014). Parasitoid wasps lay their eggs on other insects, like caterpillars, with their larvae feeding on the host insect. In return, they protect the host's body from scavengers and pathogens, influencing host defenses. Lichens form mutualistic partnerships between fungi and photosynthetic algae or cyanobacteria, with insects, such as scale insects, contributing essential nutrients to the photosynthetic partner (Calcott et al., 2018; Pereira and Kjellberg, 2021). Also, some herbivores form mutualistic relationships with microorganisms that help detoxify plant compounds or aid in digestion, making them more capable of exploiting a particular host plant. For instance, insects like termites and wood-feeding cockroaches have mutualistic relationships with gut microbes that aid in digesting complex plant material. Another example is aphid-ant mutualism, in which the ants protect the herbivores from predators and sometimes even help them access plant resources in exchange for sugary secretions or other rewards from aphids (Ivens and Kronauer, 2022). In certain situations, herbivores have developed unique ways to rely on plants as their source of sustenance. Examples include gall-forming insects, which manipulate plant tissues to create structures that provide both protection and a food source (Miller and Raman, 2019).

1.4.2 INSECT PHEROMONES AND THEIR ROLE IN FACILITATING PARTNERSHIPS

Chemical signaling is a predominant mode of communication among insects. Pheromone-based communication is the most common form of chemical signaling that holds fundamental importance in insect biology and ecology (mate attraction, foraging, territorial demarcation, threat or danger signaling, etc.). Insects utilize pheromones to convey information to conspecific individuals. For instance, female moths release sex pheromones to entice males, while ants employ pheromones to delineate foraging trails and establish their nest territories (Yahner, 2012). Pheromones are widely recognized for their important role in mating, with female insects emitting sex pheromones to allure males and males reciprocating with aggregation pheromones to attract females. These chemical cues facilitate the identification and location of potential mates. Social insects, like bees, use aggregation pheromones to convene colony members at a common location, as observed in the recruitment of workers to a food source (Czaczkes et al., 2015; Hubbell and Johnson, 1978). Ants and termites employ trail pheromones to mark paths to food sources and nesting sites, serving as navigational cues for colony members, thus ensuring efficient foraging and colony organization (Czaczkes et al., 2015; Yahner, 2012). Insects also employ pheromones to label objects, conspecifics, or specific areas for future recognition. The majority of social insects utilize maternal pheromones for communication with their offspring, as seen in the case of the honeybee queen mandibular pheromone, which regulates worker bee behavior and development, ensuring colony cohesion (Czaczkes et al., 2015). When insects perceive threats or sustain injuries, they release alarm pheromones that act as cautionary signals to conspecifics, alerting them to potential danger. Honeybees, for example, release alarm pheromones in response to threats from predators, rallying other bees to protect the hive (Mattila et al., 2021; Tan et al., 2016). In some insect colonies, pheromones play a role in maintaining social cohesion by orchestrating tasks among colony members and conveying critical information regarding the colony's requirements, such as the need for additional foragers, builders, or guards. Certain insects mimic the pheromones of other species to deceive or manipulate their behavior, such as parasitic wasps mimicking the sex pheromones of their hosts to facilitate oviposition (Nufio and Papaj, 2001). In social insect colonies, the queen has the ability to produce a specific pheromone that suppresses the reproductive capabilities of worker females (Keller and Nonacs, 1993).

1.4.3 VISUAL SIGNALING

Visual signaling is another important form of communication in insect communities. Insects have evolved a variety of visual signals and behaviors that serve multiple ecological functions. Many insects use visual signals to attract potential mates. These lights are species-specific and often coordinated to maximize the chances of successful mating. Fireflies (lightning bugs), for instance, use bioluminescent flashes to attract mates (Owens, 2022). These displays can include aerial acrobatics, perching, or aggressive behaviors aimed at deterring intruders. Insects often perform intricate courtship rituals involving visual displays. Male peacock spiders, for example, perform complex dances and display colorful abdominal flaps to court females (Girard et al., 2015). These displays help females assess the quality of potential mates. Warning coloration, as seen in many toxic or unpalatable insects, serves as a visual signal to potential predators to prevent attacking these insects.

In aposematism, some insects possess bright and conspicuous coloration to signal to predators that they are unpalatable, toxic, or harmful. For example, some toxic or stinging insects, such as wasps and certain butterflies, exhibit warning coloration to deter potential predators (Lev-Yadun, 2021; Lindstedt et al., 2019). In contrast to warning coloration, some insects have evolved visual adaptations that allow them to blend into their surroundings. This can involve color patterns, body shape, and behaviors that make them less conspicuous to predators (Lev-Yadun, 2021). Insect species engage in various visual displays to assert dominance, especially in competitive situations. For instance, male stag beetles engage in fierce battles that involve locking mandibles to establish dominance and access to females (Goyens et al., 2015). Some insects mimic the appearance of potential predators to deter their own predators. For example, certain moths have evolved patterns on their wings that resemble the eyes of owls, which can discourage attacks from predators (Stevens, 2005).

1.4.4 ACOUSTIC COMMUNICATION

Acoustic communication is a common form of signaling in which insects evolve the ability to produce sounds or to detect sounds produced by others as a means of communication. The chirping of crickets, the buzzing of bees, and the trilling of grasshoppers are examples of acoustic communication (Neil and Holderied, 2021). These sounds serve various functions, including mating, territory establishment, predator avoidance, and warning of danger. The ability to produce and interpret these sounds has evolved in response to the ecological and social challenges insects face in their environments. In many insect species, males produce characteristic sounds or calls to attract females for mating. These calls are often species-specific and can be produced through various mechanisms, including wing vibrations, stridulation (rubbing body parts together), or sound production organs. Acoustic signals are used for species identification, especially in environments with multiple species. Each species may have a unique acoustic signature, allowing individuals to recognize and locate members of their own species for mating.

In social insects, alarm calls or sounds are used to alert colony members to the presence of danger, such as predators or intruders. The alarm signals trigger defensive responses and can include both sound and vibration cues. Acoustic signals can be associated with aggressive behaviors among conspecifics. For example, male stag beetles create stridulatory sounds during combat to establish dominance and access to mates (Goyens et al., 2015). Some insects use acoustic signals to establish and defend territories. Male cicadas, for example, produce loud calls to claim and protect their territory (Heller and Hemp, 2020). Insects may produce sounds to deter potential predators as well. For instance, certain ants produce stridulatory sounds to alert other colony members to the presence of a predator (Masoni et al., 2021). In social insects, such as ants and bees, acoustic signals may be used for parent-offspring communication. For example, worker honeybees' "piping" behaviorally communicates their intent to swarm and establish a new colony site (Mattila et al., 2021). Insects like bees and ants use acoustic signals to coordinate group activities, such as swarming, foraging, and nest maintenance.

1.4.5 TACTILE COMMUNICATION

Tactile communication is a vital mode of interaction in various insect species, serving multiple functions in conveying information. Social insects like ants and bees employ tactile behaviors, such as trophallaxis (food exchange) and grooming, to share resources and maintain colony cohesion (Hölldobler and Kwapich, 2022; Meurville and LeBoeuf, 2021). Trophallaxis involves mouth-to-mouth or mouth-to-anus fluid exchange and plays a key role in resource sharing within colonies (Meurville and LeBoeuf, 2021). Grooming not only maintains hygiene but also facilitates social bonding. Tactile cues are crucial in distinguishing nestmates from non-nestmates in social insects, with antennal and body contact used to assess cuticular chemical cues (Jungnickel et al., 2004). Tactile courtship behaviors, such as antennal stroking and leg tapping, precede mating in many species (Jungnickel et al., 2004). Tactile communication is evident in worker bees caring for the queen and larvae, where larvae signal hunger through head-thrusting (Smiseth and Moore, 2007). Intraspecific communication involves establishing social bonds, cooperation, division of labor, and, in competitive situations, conveying aggression and dominance. Tactile contact is used for sharing resources, caste recognition, nest construction, and nest maintenance. Besides tactile communication, various other signaling forms regulate insect biology and behavior, including vibration, mimicry, sexual dimorphism, aggressive displays, and chemical defense signaling.

1.5 MECHANISMS OF RECIPROCITY

The reciprocity mechanism in plant-insect interactions is based on the two species exchanging resources and services in a way that is mutually advantageous. In order to obtain vital services like pollination and protection, plants lure insects with the promise of nectar, pollen, or shelter. In exchange, insects help plants reproduce by supporting pollination and strengthening their resistance to herbivores. Both sides maximize their fitness through cooperation in this reciprocal connection, which is regulated by a delicate balance. Comprehending the nuances of the reciprocity mechanism in these exchanges is essential to clarifying the coevolutionary processes forming ecosystems and preserving biodiversity.

1.5.1 HOW BOTH PLANTS AND INSECTS GAIN ADVANTAGES FROM THEIR INTERACTIONS

This reciprocity relationship is known as a mutualistic one since both species involved gain something from their interaction. The fact that the insect gains from helping the plant is interesting enough to be pointed out. According to Adroit et al. (2020), the equilibrium of forest ecosystems depends on the ongoing interactions between plants and insects. Gurevitch et al. (2002) and Sauvion et al. (2017) found that both insects and plants depend on each other for pollination. Gurevitch et al. (2002) reported that interactions between the two groups of organisms can either be advantageous to both, as in the case of entomophilous pollination, or they can be advantageous solely to the insect. Between partners that react differently to changes, evolutionary interactions between insects and plants are dynamic processes that take place over time and space. A male flower's pollen needs to touch a female flower's egg for pollination to take place. Pollen is transported from blossom to blossom by pollinating insects. With the help of reproduction, a seed is created, which is subsequently spread to produce another individual. Insects use plants not just as a source of food but also as a place to hide, lay eggs, and reproduce. The attraction implications previously discussed in Sections 1.2 and 1.4 are what typically determine which host plant is chosen. Frequently, the adult female decides which host plant to use for egg-laying (Mosquera et al., 2023; Petrén et al., 2021). To enhance the reproductive success of its progeny, it generally favors a location where both the number and quality of resources are crucial.

1.5.2 TRADE-OFFS AND COSTS: EXAMINING POTENTIAL DRAWBACKS OF MUTUALISM FOR EACH PARTNER

The interactions between the two groups of organisms might be advantageous to both (mutualistic), as in the scenario of entomophilia pollination, or they can be advantageous to the insect alone, as in the case of herbivory (Adroit et al., 2020). Insects that eat only plants are known as phytophagous or herbivorous insects. It is possible that all plant parts (stem, leaf, root, seed, fruit, sap, nectar, and pollen) will be ingested. Consumers gain a lot from this relationship because they get their energy from the plant's tissues, unlike the plant, which loses energy and needs to find a means to make up for it (Adroit et al., 2020).

A commensalism is known as a relationship between individuals of two species in which one species obtains food or other benefits from the other without either harming or benefiting the other (Capinera, 2008). The host organism is essentially unchanged by the interaction, whereas the commensal species may show great morphological adaptation. This relationship can be contrasted with mutualism, in which both species benefit. But antagonistic interactions are ones in which one species gains while the other suffers. This relationship encompasses both insectivory by plants (such as carnivorous plants) and phytophagy by insects (such as insect pests). When insects consume a plant's parts, it becomes less able to compete with its neighbors and create progeny. This is an example of an antagonistic interaction between plants and insects. One of the main pests that harm horticultural and agricultural crops are insects (Teixeira et al., 2023). In addition to spreading bacterial, viral, or fungal illnesses, they can directly harm plants (Franco et al., 2021; Wielkopolan et al., 2021). Aphids, leafhoppers, and scales are sucking insects that drain the sap or contents of plants' cells, weakening them.

1.6 HUMAN INTERVENTIONS AND APPLICATIONS

Numerous applications of human intervention in plant-insect interactions are intended to improve ES, optimize practices in agriculture, and lessen pest-related difficulties. By means of selective breeding, genetic alteration, and pesticide application, humans modify plant characteristics to either attract pollinating insects or deter damaging pests. Furthermore, cutting-edge technologies that maximize the balance between plants and insects for sustainable agricultural yields, such as integrated pest control systems and precision agriculture, take advantage of scientific developments. Comprehending and managing these interplays is crucial to tackling worldwide food security issues, advancing biological diversity, and formulating ecologically responsible methods for managing pests in diversified ecosystems.

1.6.1 AGRICULTURAL PRACTICES LEVERAGING POLLINATOR SERVICES

Food production and human livelihoods depend on pollinators, who are crucial to the ecosystem's sustainability. Insect pollination is required for the great majority of flowering plant species to transfer pollen from the anthers to the stigmas of their blooms, which is the way these species make seeds. Without this support, a number of interdependent species and ecological processes would cease to exist (Shrestha et al., 2018). Agricultural activities have the potential to either exacerbate the decline of pollinators or offer ways to assist pollinator ecosystems. Farmers created and implemented management techniques and agroforestry to preserve and control wild pollinator populations. In addition to improving pollination ES, these methods also enhance crop diversity (biodiversity), soil health, and the need for fewer pesticides, to name a few (Woodcock et al., 2019). Building capacity by providing training to a variety of target clients (such as farmers, farmer groups, extension agents, and educational institutions) offers opportunities to learn about various facets of this issue, enhance local technical capacity (such as taxonomic knowledge), and establish expert networks (Porto et al., 2020).

1.6.2 UTILIZATION OF BENEFICIAL INSECTS FOR PEST CONTROL

A safe and effective natural pest control method for farms and gardens is the use of beneficial insects. They are frequently employed as a component of an IPM program, which attempts to minimize the use of pesticides, enhance the health of crops, and increase harvest yields (Anderson et al., 2019; Deguine et al., 2021). Predators or parasites are the two basic categories in which beneficial insects might be placed (Pijnakker et al., 2020). Insects known as predators consume pests to eradicate them. Predator insects frequently include ladybugs, praying mantises, and green lacewing larvae. The most prevalent beneficial insects are ladybugs, lacewings, earwigs, assassin bugs, ground beetles, and tachinid flies. These insects can either be drawn naturally or artificially to a garden habitat (Pijnakker et al., 2020).

A parasitoid is an organism that coexists with its host in close proximity at the host's expense and eventually kills the host, according to evolutionary ecology (Korenko et al., 2022; Pijnakker et al., 2020). To minimize the number of pests, parasitoids are utilized in biological pest control. The immature stages of parasitoids can be found inside or outside of other insects, or "hosts," where they grow. The host that parasitoids feed on eventually dies, as opposed to parasites like fleas and ticks that normally feed on hosts without killing them.

1.7 UNRAVELING THE COMPLEXITY

Investigating the complicated network of interactions between these organisms is necessary to unravel the mystery of plant-insect cooperation.

Their alliances, meanwhile, are complex and impacted by species-specific characteristics, coevolutionary processes, and environmental variables. To understand the complex nuances of these relationships, there are interdisciplinary approaches that combine behavior research, genetics, and ecology. Understanding the intricacy of plant-insect relationships has implications for conservation, agriculture, and the sustainable management of natural resources, in addition to improving our knowledge of ecosystem dynamics.

1.7.1 CHALLENGES IN STUDYING AND UNDERSTANDING INTRICATE PLANT-INSECT PARTNERSHIPS

Numerous factors play a role in the intricate and dynamic relationships between plants and herbivores. These include evolutionary processes, communication, signaling, and their contributions to the survival of both plants and insects. Understanding this complex interplay remains a challenge due to the influence of various global forces, such as crop pest pressure and climate change (Giron et al., 2018). Phytochemistry, a significant facet of chemical ecology, primarily focuses on plant-derived compounds that deter herbivore damage. Progress in this field has been hindered by a limited focus on a few active compounds, neglecting advances in organic chemistry, and insufficient collaboration between ecologists and chemists (Dyer et al., 2018). Characterizing the complete phytochemical profile of individual plants is challenging but essential for advancing plant-insect interactions. Omics techniques like genomics, proteomics, and transcriptomics offer insights into plant adaptation to stress and the development of resistant and tolerant plant species. However, the high diversity of plant secondary metabolites and unresolved biosynthetic pathways pose challenges, necessitating a combination of traditional approaches, structural determination, bioactivity assessment of pure compounds, and characterization of mixtures with numerous unknown compounds (Dyer et al., 2018). Climate change and extreme weather events, both natural and anthropogenic, affect insects and plants globally. These changes disrupt ecological and evolutionary processes, impacting species interactions at various organizational levels. Examples include mismatches between plants and pollinators and between food resources and herbivores, ultimately affecting species fitness (Both et al., 2006; Doi et al., 2008; Karthik et al., 2021; Kudo et al., 2004; Post and Forchhammer, 2008; Tamura et al., 2022).

1.8 EMERGING RESEARCH AND FUTURE DIRECTIONS

Volatile organic compounds (VOCs) serve as crucial chemical signals in plant-insect interactions (Barbero and Maffei, 2023). Plants produce VOCs, attractive to insect predators and parasitoids, as a form of indirect defense against herbivores (Yadav et al., 2023). Additionally, some plants employ direct defenses, releasing compounds detrimental to insect growth or possessing structures that deter oviposition and feeding (De-la-Cruz et al., 2022; Mithöfer and Boland, 2012). Some authors have highlighted insights into the mechanisms involved. For example, Turlings et al. (2000) reported that the compound "volicitin" in the saliva of a caterpillar (the larva of a moth species) is triggered when the caterpillar eats maize leaf. The released compound then stimulates receptors on the leaf, which later respond by engaging the plant's production of some aromatic compounds, terpenoids, and other VOCs. These molecules attract a group of Hymenopterans, the wasps, which implant their eggs into the caterpillar's body. The wasp larva then eats the harmful insect pest from the inside. A similar observation was reported in the storage environment against the cowpea cosmopolitan chrysomelid (*Callosobruchus maculatus*). In an effort to evaluate the efficacy of the pteromalid parasitic wasp *Dinarmus basalis* Rondani (Hymenoptera) in reducing infestations by *C. maculatus* (L.), Akinbuluma and Chinaka (2023) reported that the adult *D. basalis* caused a reduced emergence of *C. maculatus* and reduced damage to the cowpea seeds. Olfactory cues were reported to play a crucial role in the host-searching efforts of *D. basalis*, agreeing with the observation of Kumazaki et al. (2000) that *D. basalis* utilizes the oviposition-marking pheromone of *Callosobruchus* species as a host-recognizing kairomone.

In a nutshell, these models have become sources of novel approaches to the biological control of insect pests. The system is worthwhile since it enables sustainable management of insect pests with a sound holistic approach. With this understanding, crop varieties that produce compounds that are most attractive or repellent to beneficial insects are carefully selected and used, as the case may be. Another emerging concept is the synthesis of the odorous molecules of the plant in order to attract predators of harmful species. In addition to herbivores and their natural enemies like parasitoids and predators, plant defensive traits also attract or repel pollinators (De-la-Cruz et al., 2022; Egan et al., 2021;

Galen and Cuba, 2001; Poelman et al., 2008; Ramos and Schiestl, 2020). It was recently pointed out that interactions between plants, herbivores, and pollinators can lead to evolutionary change (Egan et al., 2021). However, the roles that pollinators play as selective agents in direct and indirect plant defense have not been fully explored. As only a few works have been done in this direction, further investigations could provide greater insight into the eco-evolutionary dynamics of plant-herbivore-pollinator interactions.

The nonliving environment also shapes co-evolutionary dynamics, which can result in different species and functional traits (Hamann et al., 2021; Jander, 2014; Karageorgi et al., 2019). Understanding how environmental factors like drought, increased CO_2 concentrations, rising temperatures, and salinity affect plant-arthropod interactions is crucial (Hamann et al., 2021). To comprehensively address these interactions, a multidisciplinary approach is essential. Integrating omics techniques, including genomics, transcriptomics, and proteomics, offers insights into the molecular basis of these interactions. Tools like CRISPR-Cas9 genome editing technology present exciting prospects for targeted manipulation (Barbero and Maffei, 2023). Collaborations among molecular biologists, ecologists, entomologists, chemists, and agronomists can drive innovative research, contributing to sustainable crop protection, ecosystem function, and environmental conservation.

1.9 CONCLUSION

Mutualistic interactions are important for ecology and evolution, but they are not impervious to dangers. Issues like invasive species, habitat degradation, and climate change can upset these intricate relationships. Conservation efforts require an understanding of the significance of these linkages. Mutualistic relationships, a complex web of interactions between plants and insects, shape ecosystem dynamics. These partnerships serve as excellent examples of the diversity of ecological relationships due to their mechanisms, adaptations, and coevolutionary processes. It is essential for conservation, ecological management, and our general understanding of the interdependence of life on Earth to comprehend the complexities of these mutualisms.

KEYWORDS

- **biodiversity**
- **coevolution**
- **ecosystem**
- **parasitoids**
- **plant-insect interactions**
- **spatio-temporal dynamics**

REFERENCES

Adroit, B., Zhuang, X., Wappler, T., Terral, J. F., & Wang, B. (2020). A case of long-term herbivory: Specialized feeding trace on *Parrotia* (Hamamelidaceae) plant species. *Royal Society Open Science, 7*(201449).

Agunbiade, T., Sun, W., Coates, B., Traore, F., Ojo, J., Lutomia, A., Bello-Bravo, J., Miresmailli, S., Huesing, J., & Tamo, M. (2018). Cowpea field insect pests and integrated pest management techniques for cowpea cultivation in West Africa. In *Book Chapter* (Vol. *1*, pp. 297–320).

Akinbuluma, M. D., & Chinaka, O. P. (2023). Efficacy of the parasitic wasp, *Dinarmus basalis* Rondani (Hymenoptera: Pteromalidae), in reducing infestations by the cowpea beetle, *Callosobruchus maculatus* (L.) (Coleoptera: Chrysomelidae: Bruchinae). *Egyptian Journal of Biological Pest Control, 33*, 43.

Anderson, J. A., Ellsworth, P. C., Faria, J. C., Head, G. P., Owen, M. D., Pilcher, C. D., Shelton, A. M., & Meissle, M. (2019). Genetically engineered crops: Importance of diversified integrated pest management for agricultural sustainability. *Frontiers in Bioengineering and Biotechnology, 7*, 24.

Anton, S., & Cortesero, A. M. (2022). Plasticity in chemical host plant recognition in herbivorous insects and its implication for pest control. *Biology, 11*, 1842.

Arachchige, E. C. S., Evans, L. J., Campbell, J. W., Delaplane, K. S., Rice, E. S., Cutting, B. T., Kendall, L. K., Samnegård, U., & Rader, R. (2023). A global assessment of the species composition and effectiveness of watermelon pollinators and the management strategies to inform effective pollination service delivery. *Basic and Applied Ecology, 66*, 50–62.

Barbero, F., & Maffei, M. E. (2023). Recent advances in plant-insect interactions. *International Journal of Molecular Sciences, 24*, 11338.

Bauder, J. A. S., & Karolyi, F. (2019). Superlong proboscises as co-adaptations to flowers. In *Insect Mouthparts: Form, Function, Development, and Performance* (pp. 479–527).

Beattie, A. J. (1985). *The Evolutionary Ecology of Ant-Plant Mutualisms.* Cambridge University Press.

Beattie, A. J., & Hughes, L. (2002). Ant–plant interactions. In *Plant–Animal Interactions: An Evolutionary Approach* (pp. 211–235).

Ben-Zvi, G., Seifan, M., & Giladi, I. (2021). Ant guild identity determines seed fate at the post-removal seed dispersal stages of a desert perennial. *Insects, 12*, 147.

Both, C., Bouwhuis, S., Lessells, C., & Visser, M. E. (2006). Climate change and population declines in a long-distance migratory bird. *Nature, 441*, 81–83.

Bronstein, J. L. (2015). *Mutualism.* Oxford University Press.

Calcott, M. J., Ackerley, D. F., Knight, A., Keyzers, R. A., & Owen, J. G. (2018). Secondary metabolism in the lichen symbiosis. *Chemical Society Reviews, 47*, 1730–1760.

Capinera, J. (2008). Native American culture and insects. In *Encyclopedia of Entomology* (pp. 2546–2555).

Chamberlain, S. A., Bronstein, J. L., & Rudgers, J. A. (2014). How context dependent are species interactions? *Ecology Letters, 17*, 881–890.

Czaczkes, T. J., Grüter, C., & Ratnieks, F. L. W. (2015). Trail pheromones: An integrative view of their role in social insect colony organization. *Annual Review of Entomology, 60*, 581–599.

Deguine, J. P., Aubertot, J. N., Flor, R. J., Lescourret, F., Wyckhuys, K. A., & Ratnadass, A. (2021). Integrated pest management: Good intentions, hard realities. A review. *Agronomy for Sustainable Development, 41*, 38.

Dejean, A., Petitclerc, F., & Azémar, F. (2020). Seasonality influences ant-mediated nutrient acquisition (myrmecotrophy) by a Neotropical myrmecophyte. *Evolutionary Ecology, 34*, 645–657.

De-la-Cruz, I. M., Batsleer, F., Bonte, D., Diller, C., Hytönen, T., Muola, A., Osorio, S., Posé, D., Vandegehuchte, M. L., & Stenberg, J. A. (2022). Evolutionary ecology of plant-arthropod interactions in light of the "omics" sciences: A broad guide. *Frontiers in Plant Science, 13*, 808427.

Dettner, K., & Liepert, C. (1994). Chemical mimicry and camouflage. *Annual Review of Entomology, 39*, 129–154.

Divekar, P. A., Narayana, S., Divekar, B. A., Kumar, R., Gadratagi, B. G., Ray, A., Singh, A. K., Rani, V., Singh, V., & Singh, A. K. (2022). Plant secondary metabolites as defense tools against herbivores for sustainable crop protection. *International Journal of Molecular Sciences, 23*, 2690.

Doi, H., Gordo, O., & Katano, I. (2008). Heterogeneous intra-annual climatic changes drive different phenological responses at two trophic levels. *Climate Research, 36*, 181–190.

Drossart, M., & Gérard, M. (2020). Beyond the decline of wild bees: Optimizing conservation measures and bringing together the actors. *Insects, 11*, 649.

Dyer, L. A., Philbin, C. S., Ochsenrider, K. M., Richards, L. A., Massad, T. J., Smilanich, A. M., Forister, M. L., Parchman, T. L., Galland, L. M., Hurtado, P. J., Espeset, A. E., Glassmire, A. E., Harrison, J. G., Mo, C., Yoon, S., Pardikes, N. A., Muchoney, N. D., Jahner, J. P., Slinn, H. L., Shelef, O., Dodson, C. D., Kato, M. J., Yamaguchi, L. F., & Jeffrey, C. S. (2018). Modern approaches to study plant-insect interactions in chemical ecology. *Nature Reviews Chemistry, 2*, 50–64.

Egan, P. A., Muola, A., Parachnowitsch, A. L., & Stenberg, J. A. (2021). Pollinators and herbivores interactively shape selection on strawberry defence and attraction. *Evolution Letters, 5*, 636–643.

Enebe, M. C., & Erasmus, M. (2023). Susceptibility and plant immune control—A case of mycorrhizal strategy for plant colonization, symbiosis, and plant immune suppression. *Frontiers in Microbiology, 14*, 1178258.

Erb, M., & Robert, C. A. (2016). Sequestration of plant secondary metabolites by insect herbivores: Molecular mechanisms and ecological consequences. *Current Opinion in Insect Science, 14*, 8–11.

Fagundes, M., Neves, F. S., & Fernandes, G. W. (2005). Direct and indirect interactions involving ants, insect herbivores, parasitoids, and the host plant *Baccharis dracunculifolia* (Asteraceae). *Ecological Entomology, 30*, 28–35.

Farrell, A. (2017). *The Combined Effect of Harvester Ant (Messor ebeninus) Nests and Desert Shrubs on Surrounding Soil and Annual Plant Community* [Master's thesis].

Ferrière, R., Dieckmann, U., & Couvet, D. (2004). *Evolutionary Conservation Biology.* Cambridge University Press.

Franco, F. P., Túler, A. C., Gallan, D. Z., Gonçalves, F. G., Favaris, A. P., Peñaflor, M. F. G., Leal, W. S., Moura, D. S., Bento, J. M. S., & Silva-Filho, M. C. (2021). Fungal phytopathogen modulates plant and insect responses to promote its dissemination. *The ISME Journal, 15*, 3522–3533.

Galen, C., & Cuba, J. (2001). Down the tube: Pollinators, predators, and the evolution of flower shape in the alpine sky pilot, *Polemonium viscosum. Evolution, 55*, 1963–1971.

Gebreziher, H. G. (2018). The role of herbivore-induced plant volatiles (HIPVs) as an indirect plant defense mechanism in a diverse plant and herbivore species: A review. *International Journal of Agriculture, Environment and Food Sciences, 2*, 139–147.

Giannetti, D., Castracani, C., Spotti, F., Mori, A., & Grasso, A. (2019). Gall-colonizing ants and their role as plant defenders: From "bad job" to "useful service." *Insects, 10*(11).

Giladi, I. (2006). Choosing benefits or partners: A review of the evidence for the evolution of myrmecochory. *Oikos, 112*, 481–492.

Girard, M. B., Elias, D. O., & Kasumovic, M. M. (2015). Female preference for multi-modal courtship: Multiple signals are important for male mating success in peacock spiders. *Proceedings of the Royal Society B: Biological Sciences, 282*, 20152222.

Giron, D., Dubreuil, G., Bennett, A., Dedeine, F., Dicke, M., Dyer, L. A., Erb, M., Harris, M. O., Huguet, E., & Kaloshian, I. (2018). Promises and challenges in insect–plant interactions. *Entomologic Experimentalism et Applicate, 166*, 319–343.

Goyens, J., Dirckx, J., & Aerts, P. (2015). Stag beetle battle behavior and its associated anatomical adaptations. *Journal of Insect Behavior, 28*, 227–244.

Gurevitch, J., Scheiner, S. M., & Fox, G. A. (2002). *The Ecology of Plants.* Sinauer Associates.

Hamann, E., Blevins, C., Franks, S. J., Jameel, M. I., & Anderson, J. T. (2021). Climate change alters plant–herbivore interactions. *New Phytologist, 229*, 1894–1910.

Heller, K. G., & Hemp, C. (2020). Hyperdiverse songs, duetting, and the roles of intra- and intersexual selection in the acoustic communication of the genus *Eurycorypha* (Orthoptera: Tettigonioidea, Phaneropterinae). *Organisms, Diversity & Evolution, 20*, 597–617.

Herrera, C. M., & Pellmyr, O. (2009). *Plant-Animal Interactions: An Evolutionary Approach.* John Wiley & Sons.

Hoeksema, J. D., & Bruna, E. M. (2015). Context-dependent outcomes of mutualistic interactions. *Mutualism, 10*, 181–202.

Hölldobler, B., & Kwapich, C. L. (2022). *The Guests of Ants: How Myrmecophiles Interact with their Hosts.* Harvard University Press.

Hubbell, S. P., & Johnson, L. K. (1978). Comparative foraging behavior of six stingless bee species exploiting a standardized resource. *Ecology, 59*, 1123–1136.

Ivens, A. B., & Kronauer, D. J. (2022). Aphid-farming ants. *Current Biology, 32*, R813–R817.

Jander, G. (2014). Revisiting plant-herbivore co-evolution in the molecular biology era. In *Annual Plant Reviews: Insect-Plant Interactions* (Vol. *47*, pp. 361–384).

Jungnickel, H., Da Costa, A., Tentschert, J., Patricio, E. F. L., Imperatriz-Fonseca, V. L., Drijfhout, F., & Morgan, E. (2004). Chemical basis for inter-colonial aggression in the stingless bee *Scaptotrigona bipunctata* (Hymenoptera: Apidae). *Journal of Insect Physiology, 50*, 761–766.

Karageorgi, M., Groen, S. C., Sumbul, F., Pelaez, J. N., Verster, K. I., Aguilar, J. M., Hastings, A. P., Bernstein, S. L., Matsunaga, T., & Astourian, M. (2019). Genome editing retraces the evolution of toxin resistance in the monarch butterfly. *Nature, 574*, 409–412.

Karthik, S., Reddy, M. S., & Yashaswini, G. (2021). Climate change and its potential impacts on insect-plant interactions. In *Natural Causes, Effects, and Mitigation of Climate Change on the Environment* (Vol. *10*).

Keller, L., & Nonacs, P. (1993). The role of queen pheromones in social insects: Queen control or queen signal? *Animal Behavior, 45*, 787–794.

Kenneth, A., & Jayashankar, M. (2021). Insects and extra-floral nectaries: Sugar-coated strategy more than anatomical anomaly.

Korenko, S., Sýkora, J., Kostro-Ambroziak, A., & Pekár, S. (2022). Two lines of defense in the pupas of ichneumonid wasp parasitoids associated with spider hosts. *Frontiers in Ecology and Evolution, 10*, 1061700.

Krenn, H. W., Karolyi, F., Lampert, P., Melin, A., & Colville, J. F. (2021). Nectar uptake of a long-proboscid *Prosoeca* fly (Nemestrinidae)—Proboscis morphology and flower shape. *Insects, 12*, 371.

Kudo, G., Nishikawa, Y., Kasagi, T., & Kosuge, S. (2004). Does seed production of spring ephemerals decrease when spring comes early? *Ecological Research, 19*, 255–259.

Kumar, G., & Khan, M. S. (2022). Insects as crop pollinators. In *Insects as Service Providers* (pp. 37–64). Springer.

Kumazaki, M., Matsuyama, S., Suzuki, T., Kuwahara, Y., & Fujii, K. (2000). Parasitic wasp, *Dinarmus basalis*, utilizes oviposition-marking pheromone of host azuki bean weevils as host-recognizing kairomone. *Journal of Chemical Ecology, 26*, 2677–2695.

Lev-Yadun, S. (2016). *Defensive (Anti-Herbivory) Coloration in Land Plants.* Springer.

Lev-Yadun, S. (2021). Avoiding rather than resisting herbivore attacks is often the first line of plant defence. *Biological Journal of the Linnean Society, 134*, 775–802.

Lindstedt, C., Murphy, L., & Mappes, J. (2019). Antipredator strategies of pupae: How to avoid predation in an immobile life stage? *Philosophical Transactions of the Royal Society B, 374*, 20190069.

Masoni, A., Frizzi, F., Nieri, R., Casacci, L. P., Mazzoni, V., Turillazzi, S., & Santini, G. (2021). Ants modulate stridulatory signals depending on the behavioral context. *Scientific Reports, 11*, 5933.

Mattila, H. R., Kernen, H. G., Otis, G. W., Nguyen, L. T. P., Pham, H. D., Knight, O. M., & Phan, N. T. (2021). Giant hornet (*Vespa soror*) attacks trigger frenetic antipredator signaling in honeybee (*Apis cerana*) colonies. *Royal Society Open Science, 8*, 211215.

Mayer, V. E., Frederickson, M. E., McKey, D., & Blatrix, R. (2014). Current issues in the evolutionary ecology of ant–plant symbioses. *New Phytologist, 202*, 749–764.

McCall, A. C., & Irwin, R. E. (2006). Florivory: The intersection of pollination and herbivory. *Ecology Letters, 9*, 1351–1365.

Meurville, M. P., & LeBoeuf, A. C. (2021). Trophallaxis: The functions and evolution of social fluid exchange in ant colonies (Hymenoptera: Formicidae). *Myrmecological News, 31*.

Miller, D. G., & Raman, A. (2019). Host–plant relations of gall-inducing insects. *Annals of the Entomological Society of America, 112*, 1–19.

Mithöfer, A., & Boland, W. (2012). Plant defense against herbivores: Chemical aspects. *Annual Review of Plant Biology, 63*, 431–450.

Mosquera, K. D., Martínez Villegas, L. E., Rocha Fernandes, G., Rocha David, M., Maciel-de-Freitas, R., Moreira, L. A., & Lorenzo, M. G. (2023). Egg-laying by female *Aedes aegypti* shapes the bacterial communities of breeding sites. *BMC Biology, 21*, 97.

Nandi, D., Chakraborty, A., Biswas, T., Meher, D., & Singh, A. P. (2022). Role of trichomes in plant defence—A crop-specific review. *Crop Research, 57*, 460–475.

Neil, T. R., & Holderied, M. W. (2021). Sound production and hearing in insects. In *Advances in Insect Physiology* (pp. 101–139). Elsevier. https://doi.org/10.1016/bs.aiip.2021.10.001.

Neuschulz, E. L., Mueller, T., Schleuning, M., & Böhning-Gaese, K. (2016). Pollination and seed dispersal are the most threatened processes of plant regeneration. *Scientific Reports, 6*, 29839. https://doi.org/10.1038/srep29839.

Nufio, C. R., & Papaj, D. R. (2001). Host marking behavior in phytophagous insects and parasitoids. *Entomologia Experimentalis et Applicata, 99*, 273–293. https://doi.org/10.1046/j.1570-7458.2001.00827.x.

Owens, A. C. (2022). *The Lightning Bug and the Lightening: Firefly Conservation Under Artificial Illumination.*

Pereira, R. A. S., & Kjellberg, F. (2021). Mutualism as a source of evolutionary innovation: Insights from insect-plant interactions. In *Plant-Animal Interactions: Source of Biodiversity* (pp. 307–332). Springer.

Petrén, H., Gloder, G., Posledovich, D., Wiklund, C., & Friberg, M. (2021). Innate preference hierarchies coupled with adult experience, rather than larval imprinting or transgenerational acclimation, determine host plant use in *Pieris rapae*. *Ecology and Evolution, 11*, 242–251. https://doi.org/10.1002/ece3.7013.

Pijnakker, J., Vangansbeke, D., Duarte, M., Moerkens, R., & Wäckers, F. L. (2020). Predators and parasitoids-in-first: From inundative releases to preventative biological control in greenhouse crops. *Frontiers in Sustainable Food Systems, 4*, 595630. https://doi.org/10.3389/fsufs.2020.595630.

Poelman, E. H., van Loon, J. J., & Dicke, M. (2008). Consequences of variation in plant defense for biodiversity at higher trophic levels. *Trends in Plant Science, 13*, 534–541. https://doi.org/10.1016/j.tplants.2008.06.003.

Porto, R. G., de Almeida, R. F., Cruz-Neto, O., Tabarelli, M., Viana, B. F., Peres, C. A., & Lopes, A. V. (2020). Pollination ecosystem services: A comprehensive review of economic values, research funding, and policy actions. *Food Security, 12*, 1425–1442. https://doi.org/10.1007/s12571-020-01071-3.

Post, E., & Forchhammer, M. C. (2008). Climate change reduces reproductive success of an Arctic herbivore through trophic mismatch. *Philosophical Transactions of the Royal Society B: Biological Sciences, 363*, 2367–2373. https://doi.org/10.1098/rstb.2007.2207.

Rader, R., Bartomeus, I., Garibaldi, L. A., Garratt, M. P. D., Howlett, B. G., Winfree, R., Cunningham, S. A., Mayfield, M. M., Arthur, A. D., Andersson, G. K. S., Bommarco, R., Brittain, C., Carvalheiro, L. G., Chacoff, N. P., Entling, M. H., Foully, B., Freitas, B. M., Gemmill-Herren, B., Ghazoul, J., Griffin, S. R., Gross, C. L., Herbertsson, L., Herzog, F., Hipólito, J., Jaggar, S., Jauker, F., Klein, A. M., Kleijn, D., Krishnan, S., Lemos, C. Q., Lindström, S. A. M., Mandelik, Y., Monteiro, V. M., Nelson, W., Nilsson, L., Pattemore, D. E., De O. Pereira, N., Pisanty, G., Potts, S. G., Reemer, M., Rundlöf, M., Sheffield, C. S., Scheper, J., Schüepp, C., Smith, H. G., Stanley, D. A., Stout, J. C., Szentgyörgyi, H., Taki, H., Vergara, C. H., Viana, B. F., & Woyciechowski, M. (2016). Non-bee insects are important contributors to global crop pollination. *Proceedings of the National Academy of Sciences, 113*, 146–151. https://doi.org/10.1073/pnas.1517092112.

Ramos, S. E., & Schiestl, F. P. (2020). Herbivory and pollination impact on the evolution of herbivore-induced plasticity in defense and floral traits. *Evolution Letters, 4*, 556–569.

Rico-Gray, V., & Oliveira, P. S. (2008). *The Ecology and Evolution of Ant-Plant Interactions*. University of Chicago Press.

Sauvion, N., Thiery, D., & Calatayud, P. A. (2017). *Insect-Plant Interactions in a Crop Protection Perspective*. Academic Press.

Schweiger, R., Heise, A., Persicke, M., & Müller, C. (2014). Interactions between the jasmone and salicylic acid pathway modulate the plant metabolome and affect herbivores of different feeding types. *Plant, Cell & Environment, 37*, 1574–1585.

Shrestha, M., Garcia, J. E., Bukovac, Z., Dorin, A., & Dyer, A. G. (2018). Pollination in a new climate: Assessing the potential influence of flower temperature variation on insect pollinator behavior. *PLOS ONE, 13*, e0200549. https://doi.org/10.1371/journal.pone.0200549.

Smiseth, P. T., & Moore, A. J. (2007). Signalling of hunger by senior and junior larvae in asynchronous broods of a burying beetle. *Animal Behavior, 74*, 699–705.

Sponsler, D., Iverson, A., & Steffan-Dewenter, I. (2023). Pollinator competition and the structure of floral resources. *Echography, 2023*, e06651.

Stevens, M. (2005). The role of eyespots as anti-predator mechanisms, principally demonstrated in the Lepidoptera. *Biological Reviews, 80*, 573–588.

Tamura, Y., Osawa, T., Tabuchi, K., Yamasaki, K., Niiyama, T., Sudo, S., Ishigooka, Y., Yoshioka, A., & Takada, M. B. (2022). Estimating plant-insect interactions under climate change with limited data. *Scientific Reports, 12*, 10554.

Tan, K., Dong, S., Li, X., Liu, X., Wang, C., Li, J., & Nieh, J. C. (2016). Honey bee inhibitory signaling is tuned to threat severity and can act as a colony alarm signal. *PLOS Biology, 14*, e1002423. https://doi.org/10.1371/journal.pbio.1002423.

Teixeira, A. C., Ribeiro, J., Morais, R., Sousa, J. J., & Cunha, A. (2023). A systematic review on automatic insect detection using deep learning. *Agriculture, 13*, 713. https://doi.org/10.3390/agriculture13030713.

Togola, A., Datinon, B., Laouali, A., Traoré, F., Agboton, C., Ongom, P. O., Ojo, J. A., Pittendrigh, B., Boukar, O., & Tamò, M. (2023). Recent advances in cowpea IPM in West Africa. *Frontiers in Agronomy, 5*, 1220387. https://doi.org/10.3389/fagro.2023.1220387.

Turlings, T. C., Alborn, H. T., Loughrin, J. H., & Tumlinson, J. H. (2000). Volicitin, an elicitor of maize volatiles in oral secretion of *Spodoptera exigua*: Isolation and bioactivity. *Journal of Chemical Ecology, 26*, 189–202.

War, A. R., Buhroo, A. A., Hussain, B., Ahmad, T., Nair, R. M., & Sharma, H. C. (2020). Plant defense and insect adaptation with reference to secondary metabolites. *Co-Evolution of Secondary Metabolites*, 795–822.

Wielkopolan, B., Jakubowska, M., & Obrępalska-Stęplowska, A. (2021). Beetles as plant pathogen vectors. *Frontiers in Plant Science, 12*, 2241.

Woodcock, B. A., Garratt, M. P. D., Powney, G. D., Shaw, R. F., Osborne, J. L., Soroka, J., Lindström, S. A. M., Stanley, D., Ouvrard, P., Edwards, M. E., Jauker, F., McCracken, M. E., Zou, Y., Potts, S. G., Rundlöf, M., Noriega, J. A., Greenop, A., Smith, H. G., Bommarco, R., & Pywell, R. F. (2019). Meta-analysis reveals that pollinator functional diversity and abundance enhance crop pollination and yield. *Nature Communications, 10*, 1481. https://doi.org/10.1038/s41467-019-09393-6.

Yadav, K., Anant, S. S., & Rai, A. (2023). *Gene Editing in Agriculture: The Transformation of Future Food Production.*

Yahner, R. H. (2012). Olfactory communication. In *Wildlife Behavior and Conservation* (pp. 113–120). Springer New York. https://doi.org/10.1007/978-1-4614-1518-3_13.

CHAPTER 2

Plant–Pathogen Interactions: When Insects Become Vectors of Destruction

MUHAMMAD ZULQAR NAIN DARA,[1,*] ARZLAN ABBAS,[1] AROGE TEMITOPE,[1] LIU PENGWEI,[2] KOMAL SHOUKAT,[3] and LEI LI[4]

[1]*College of Plant Protection, Jilin Agricultural University, Changchun, Jilin, China*

[2]*QiLu ZhongKe Academy of Modern Microbiology Technology – Crop Precision Breeding Center, Jinan, Shandong, China*

[3]*Government College University, Faisalabad, Punjab, Pakistan*

[4]*Jilin Agricultural University, Changchun, Jilin, China*

*Corresponding author

ABSTRACT

Plant-pathogen interactions are dynamic and complex, involving multiple biotic and abiotic factors. Insects often act as vectors, transmitting pathogens from one host to another and thereby contributing to crop losses and ecosystem disruption. The coevolution of plants, pathogens, and insect vectors, along with environmental and anthropogenic influences, shapes the dynamics of disease transmission. Transmission mechanisms vary depending on the pathogen, the mode of acquisition and inoculation, and the nature of the vector–host plant interaction. Many insect vectors, such as aphids, whiteflies, leafhoppers, and beetles, transmit pathogens either through contaminated mouthparts or by harboring pathogens internally. When these insects feed on plants, they introduce pathogens into plant tissues, facilitating infection. The severity of plant disease depends on host

Insect-Plant Interactions, Volume 1: Influence of Biotic and Abiotic Factors.
Jamin Ali, Adil Tonğa, Adnan Shakeel, and Nasheeman Ashraf (Eds.)

DOI: 10.1201/9781779644282-2

susceptibility, pathogen virulence, and ecological conditions that influence disease development and spread. This book chapter provides a concise overview of insect-mediated plant disease transmission, focusing on vector biology, transmission strategies, and the implications of these interactions for plant health. It also explores how understanding these mechanisms can inform more effective plant protection and integrated pest management strategies. The aim is to synthesize key concepts in plant-pathogen-insect interactions and highlight their relevance to sustainable agriculture and disease control.

2.1 INTRODUCTION

Plant pathogens, such as viruses, bacteria, fungi, and nematodes, are spread by a wide variety of insects, including whiteflies, thrips, aphids, and planthoppers. Plant pathogens, such as viruses, infect and multiply in insect cells, but each virus is particular to its insect carrier. Some of them can also be transferred mechanically, utilizing plant sap from infected plants. One of the primary means through which plant pathogens spread is via the intermediary of vectors. Insects that transmit plant pathogens across plants and become sources of disease-causing plant pathogens, known as insect vectors, have a significant indirect and direct economic influence on agricultural productivity. Vector-borne plant infections have complicated transmission methods that include many factors, including host plants, insect vectors, plant pathogens, and endosymbiotic bacteria. However, there is little information about the association of plant pathogens and their vector insects at the biochemical and molecular level.

Insect vectors play an important role in the transmission and spread of plant pathogens, which later result in losses in terms of affecting crop productivity, damaging plant health, and igniting evolution and changes in plant pathogens. For example, *Aphis fabae* reproduces more on virus-infected *Beta vulgaris*, as noted by J.S. Kennedy in 1951; this finding hints at the need for more study into the evolutionary and epidemiological consequences for both the vector and the virus (Kennedy, 1951). As a result of this discovery, a growing body of research has been motivated to investigate comparable impacts involving vector-borne plant diseases. There is evidence to suggest that plant diseases may affect their vectors to increase their rate of dissemination. Economically relevant plant illnesses are often caused by vector-dependent plant infections; therefore, these results have important implications for ecology, evolution, and practical uses of pathogens, especially in agriculture.

Both managed and natural ecosystems are plagued by insect-borne plant diseases, including viruses, bacteria like mollicutes (phytoplasmas and spiroplasmas), and fungi. There are several economically relevant pathogens that cause devastating diseases in agriculture and forestry. The epidemiology of these infections is affected by variables such as the movement of vector populations, the routes they take to spread, the hosts they prefer, and the diets they maintain. These connections provide the groundwork for dynamic interactions between the plant, vector, disease, and the environment. Rapid growth in the literature on the direct and indirect effects of plant viruses and other diseases on their vectors is being driven by the economic significance of these pathogens and the intense interest in understanding the possible evolutionary or coevolutionary processes that may underlie these effects (Belliure, 2005; Bosque-Pérez and Eigenbrode, 2011; Carmo-Sousa, 2016). This is based on a model originally created for plant viruses, with a focus on the aphid-borne viral pathosystem (Ingwell, 2012; Jiménez-Martínez, 2004; Luan, 2014).

2.2 WHAT ARE PLANT PATHOGENS?

Plants cover a significant portion of the Earth's surface, including trees, grasses, and flowers, and play vital roles in maintaining ecosystem balance and providing food for both animals and humans. Additionally, plants, whether in the wild or cultivated, act as potent biofertilizers for the soil. When they decompose, they enrich the soil with essential organic matter. Consequently, taking care of plants is a crucial and ongoing responsibility. The study of plant pathogens falls under the field of biology known as plant pathology. Plant pathology is dedicated to understanding and mitigating plant diseases arising from both biotic and/or abiotic origins. Biotic diseases result from phytopathogenic microbes that cause infections and later spread diseases in hosts, while abiotic diseases are caused by various environmental factors. In this chapter, we focus on plant pathogens, particularly phytopathogenic (a parasitic organism on a plant host) microbes such as bacteria, viruses, fungi, mollicutes, and so on.

Plant pathogens, especially their insect vectors, will be the main subject of this chapter. An insect vector is an insect that spreads illness to other living things or individuals. Plant pathology, the scientific discipline focused on studying plant diseases and their causes, is essential for the economic and hygienic well-being of humans and animals. Phytopathogens can be categorized into two types: biotic factors, which include all microbes and parasitic plants, and abiotic factors, encompassing environmental elements. Plant pathology is closely linked to other sciences, including entomology (the study of insects), bacteriology (the study of bacteria), mycology (the study of fungi), virology (the study of viruses and viroids), and weed science (the study of weeds). These interdisciplinary connections are essential due to the detrimental impacts of insects, bacteria, fungi, viruses/ viroids, and weeds on plants, respectively.

Insect vectors are responsible for the majority of plant virus transmissions. Insects that feed by sucking and those that chew or bite may both spread viruses. Both "mechanical" and "biological" modes of transmission are possible. The latter scenario describes an association between a particular insect and a viral infection. Insects that are known to spread a certain virus are referred to as vectors for that virus. Mechanical transmission occurs when an insect carries a pathogen on its body or in its digestive tract.

2.3 INSECTS AS VECTORS OF PLANT PATHOGENS

An insect vector is an organism with the ability to transfer pathogens from one host to another. In the context of plant diseases, insects that acquire disease-causing organisms while feeding on diseased plants or through contact and then transmit them to healthy plants are referred to as insect vectors. Such insects include *Aphis craccivora* Koch, *Cestius phycitis*, bees, wasps, flies, ants, and aphids, among others.

Vectors are responsible for introducing pathogens like bacteria or viruses into plants, causing infections. Insects, mites, and nematodes serve as vectors, facilitating the movement of plant pathogens among immobile plants. It is important to note that many insects and arthropods may contain plant pathogens but lack the capability to transmit them to plants, so they are not considered vectors. For certain significant plant diseases, mobile vectors are essential. Nearly all plant viruses and a group of wall-free, plant-pathogenic bacteria called mollicutes have recognized or suspected vectors.

There are certain insects known as carriers (passive) that only pick up diseases externally on their bodies, usually on their legs or mouthparts, through contact with infected or contaminated plant parts (Figure 2.1). Even though some insects are capable of carrying the disease, they are unable to actively spread it to a new host in a way that causes infection. For the pathogen to establish itself in the new host, such as flies (bacteria) or beetles (fungi), it typically requires an additional factor, such as entering a wound or being consumed. A true insect vector is one that possesses the capacity to actively spread the infection, frequently through internal harboring and growth.

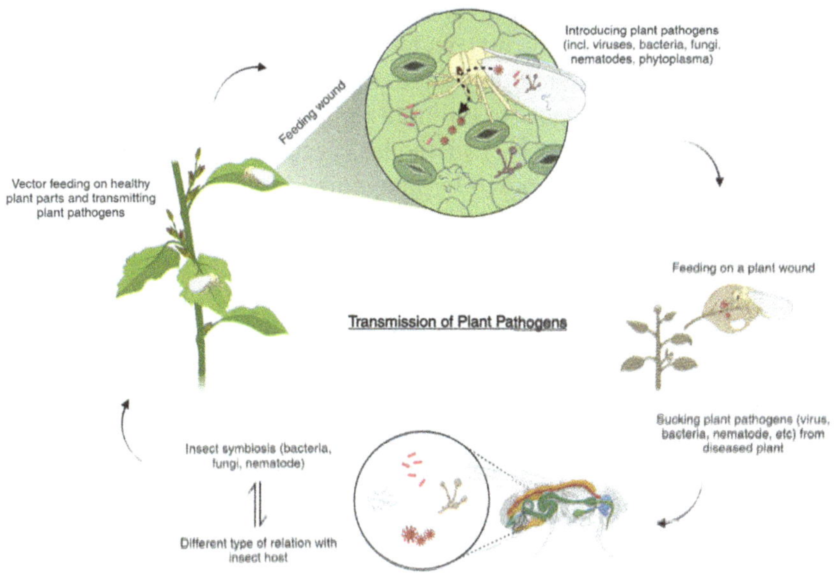

FIGURE 2.1 Flow diagram of transmission of plant pathogens carried by various insect vectors that may acquire it from affected plant part (cut or wound) and transfer to a health host plant.

2.4 INSECT VECTORS AND THEIR ROLE IN TRANSMITTING PLANT DISEASES

Plant pathogens are transmitted either by contamination through soil, through contact, or other biological agencies. Insects and other arthropods like mites are responsible for the spread of most plant diseases; mechanical techniques and soil pollution account for only a small percentage.

Biting and chewing insects are also linked to disease transmission, as are piercing-sucking insects. Aphids, leafhoppers, whiteflies, and mealybugs are all examples of Hemiptera, whereas Thysanoptera (thrips), Coleoptera (beetles), Orthoptera (grasshoppers), and Dermaptera (earwigs) are all examples of other orders of insect vectors. Around 90% of plant diseases are transmitted only by insects belonging to the order Homoptera.

2.5 HOW DO INSECTS TRANSMIT PLANT PATHOGENS?

Insects play a critical role in the transmission of plant pathogens, with significant impacts on agricultural productivity and global food security. These complex relationships involve various feeding mechanisms, biotic interactions, and environmental factors. A plant disease begins with a pathogen's invasion and multiplication inside the plant's tissue. The most common ways for pathogens to infiltrate a plant are via artificial or natural holes like stomata, lenticels, hydathodes, nectar-producing glands, and leaf scars; or by wounds caused by things like insects, nematodes, tools, machinery, and extreme weather conditions.

Aphids, mealybugs, and whiteflies have needle-like mouthparts that allow them to directly penetrate plant phloem (nutrient-rich sap). This permits infections to enter the phloem, where they can be acquired and disseminated. The chewing mouthparts of beetles, grasshoppers, and caterpillars cause superficial bruises on plant tissue. Although these wounds are mostly harmful to the plant, they can allow fungal spores or bacterial inoculums to enter. When bees and butterflies pollinate flowers, they collect nectar. Although they are not immediately dangerous, they can mistakenly transport pollen containing fungal spores or bacterial cells, allowing the infection to spread farther (Alexander et al., 2014).

Plant pathogens, such as viruses, bind to the outer surface of the insect's mouth cavity. The virus is mechanically transferred without acquisition after feeding on complete plants. Lettuce mosaic virus, for example, is spread by aphids. Plant wounds are required for pathogens such as bacteria and fungi to enter inside the plant. Through feeding activity, insects can directly transfer these pathogens. For example, the codling moth transmits *Erwinia amylovora*, which is the causal agent of apple fire (Bendix, 2018).

Viruses enter the gut of the insect during feeding, grow within its internal organs, and are then secreted with the saliva, where they might infect new plants. *Citrus tristeza* virus, for instance, is spread by citrus psyllids.

The infectivity of pathogens can be increased by insect saliva, which encourages the germination of fungal spores or the growth of bacteria. For example, cucumber beetle saliva induces the germination of *Fusarium oxysporum* spores (Bendix, 2018).

Plant cell walls are broken down by salivary enzymes, which make it easier for pathogens to enter plant tissues. Insects inject toxins into plants while they feed, weakening their defensive systems and enhancing the growth of pathogens. Insect saliva contains sugars and amino acids that plant pathogens love and use to increase their growth. The interaction between the pathogen and the vector insect determines the pathogen's capability to survive and replicate within the vector. Host-seeking and then feeding preferences of insects impact their interactions with specific plant species and associated pathogens (Wang and Blanc, 2021).

Humans transfer pathogens through cultivation, grafting, pruning, and transporting diseased plant material. Animals are other common agents that transfer pathogens to plants. However, insect vectors play a very crucial role in the transmission from an infected plant to a healthy plant and can cause fatal diseases that can severely affect plants and lead to significant yield losses. The history of this mode of transmission has often resulted in some of the worst plant disease epidemics in new areas, with the local host plant stock not being adapted to the pathogen and consequently remaining highly susceptible.

Insect vectors are responsible for the majority of plant virus transmissions. Insects that feed by sucking and those that chew or bite may both spread viruses. Both "mechanical" and "biological" modes of transmission are possible. The latter scenario describes an association between a particular insect and a viral infection. Insects that are known to spread a certain virus are referred to as vectors for that virus. Mechanical transmission occurs when an insect carries a pathogen on its body or in its digestive tract.

When insects feed on plants, they often cause damage that allows pathogens to enter the plant or root system (Figure 2.1). By providing a host for the disease to overwinter in, insects may contribute to its continued existence.

2.6 TYPES OF TRANSMISSION

There are certain types of plant pathogen transmission by vectors in plants which are given in sub sections.

2.6.1 CONTINUOUS AND DISCONTINUOUS TRANSMISSION

Disease transmission is said to be "continuous" when it persists normally throughout a region where the pathogen is established, via growth, multiplication, and spread. However, sometimes infected material is transported to a new region or far from a country by mankind for the aim of introducing new plants, crop types, and so forth, in areas where the disease has never existed before. Such transmission is considered "discontinuous" and is thus unnatural.

2.6.2 DIRECT AND INDIRECT TRANSMISSION

Disease transmission techniques and their corresponding control strategies may be conveniently divided into two categories.

2.6.2.1 DIRECT TRANSMISSION

Transmission of disease by infected seed or other planting materials such as cuttings, sets, tubers, bulbs, and so forth. Disease-causing bacteria enter plants through natural openings such as stomata, lenticels, hydathodes, nectar-producing glands, and leaf scars, or by wounds caused mostly by bad weather, human beings, machinery and agricultural equipment, insect species, and nematodes.

2.6.2.2 INDIRECT TRANSMISSION

Different types of indirect transmission include the disease spreading itself by persistent development or particular structures of the pathogen being conveyed independently by natural forces such as wind, water, animals, insects, mites, nematodes, birds, and so forth.

2.7 TYPES OF INSECT-BORNE PLANT PATHOGENS

Insect-borne plant pathogens fall into two main categories: noncirculative and circulative, based on whether they are retained in the vector's hemocoel.

Within these categories, there are further subdivisions depending on the specifics of their retention and replication within the vector.

2.7.1 NON-CIRCULATIVE PATHOGENS

These infections are connected with certain areas (external; mouthparts, legs, and body surfaces including internal, that is, digestive system, salivary glands, and reproductive system) of the insect vector's body, most often the mouthparts and foregut but also structures like the mycangia for fungal pathogens. Pathogen particles aid in this interaction by attaching to proteins on the vector's mouthparts or foregut. The length of time a host is exposed to a non-circulatory infection determines whether the pathogen is considered nonpersistent or semipersistent. Probing, test biting, or feeding are all rapid and easy ways to acquire or transmit nonpersistent diseases, but swallowing is necessary for semipersistent viruses.

2.7.2 CIRCULATIVE PATHOGENS

These infections spend considerable time in the hemolymph of the vector. The intestinal lining cells provide an entry point for circulating viruses and bacteria that cause disease. In order to infect a new host, they move to the salivary glands and ducts. Circulative pathogens can only be acquired and spread (after a latency period) by prolonged feeding. Depending on whether or not they multiply inside the vectors, circulative pathogens may be further categorized as non-propagative or propagative (some viruses and all bacterial pathogens). In most cases, a circulating infection will only infect a single species of mosquito or another closely similar vector.

2.8 VECTORS RESPONSIBLE FOR THE TRANSMISSION OF VIRUSES

Most plant viruses need the assistance of a vector in order to spread from plant to plant. The study of virus-vector interactions has shown common modes of transmission. At least one of the structural virion proteins is required for transmission in all plant viruses investigated so far. While some viruses can replicate and spread using just their capsid proteins, others need a "helper" protein to aid with retention by connecting the virus to the vector's surface. All viruses, whether they are able to replicate or not, have a similar life cycle and are acquired by vectors, starting in the insect's digestive system. Even within the same virus family, differences exist in how viruses spread from infected gut epithelial cells to other organs.

Similarities between viral infection cycles in insect and plant hosts have been highlighted by the discovery of virus movement in vectors via tubule structures made of non-structural virus proteins and the participation of the actin cytoskeleton in this process. These tubules provide a novel approach for insects to overcome midgut obstructions in certain cases. The use of the vitellogenin absorption route for viral invasion of insect eggs is a relatively new finding in the vector-borne transmission of plant viruses. Transovarial transmission of viruses is similar to that of other insect-associated microorganisms. Identification of vector molecules that interact with and react to viruses is a new subject, despite the fact that the viral components of vector interactions are well-defined in most commercially relevant systems. The adoption of cutting-edge research tools should help us figure out how these vector parts really work to spread viruses. Many different vectors may be disrupted in their ability to spread disease by focusing on the similarities between the roles played by viral structural proteins and the first locations of virus entry or retention (Whitfield, 2015b).

The majority of plant-infecting viruses rely on various insects for transmission to their host plants (Table 2.1). These viruses, and the responsible insects (that play their role as vectors to transmit them to host plants), interact in terms of duration and specificity. But some common patterns have emerged, as plant viruses typically contain structural proteins on the surface of their virions that play a crucial role in transmission. In some cases, there are additional non-structural helper proteins that serve as bridges, facilitating the attachment of the virion to the vector's binding site.

Our current knowledge of the fundamental mechanics of viral transmission has been greatly expanded by recent scientific findings, allowing for the creation of novel approaches to interrupting the cycle of infection. For example, the majority of plant-infecting viruses rely on insect vectors for transmission to their host plants. These virus-vector interactions can vary in terms of duration and specificity, but some common patterns have emerged; likewise, plant viruses typically contain structural proteins on

TABLE 2.1 Plant Viruses That are Transmitted Through Different Vectors with its Mechanism of Transmission

SL. No.	Virus	Genus	Vectors	Mechanism	References
1.	Alfalfa mosaic virus (AMV)	*Alfamovirus*	Aphid (*Aphidoidea*)	Non-persistent – stylet-borne	Garran and Gibbs (1982)
2.	Alpinia mosaic virus (AlpMV)	*Macluravirus*	Aphid (*Aphidoidea*)	Non-persistent – stylet borne	Liou (2003)
3.	Angelonia flower break mosaic virus (AnFBMV)	*Carmovirus*	Thrips (*Thysanoptera*)	Transmission	Winter (2006)
4.	Anthriscus yellows virus (AnYV)	*Waikavirus*	Aphid (*Aphidoidea*)	Non-persistent – foregut-borne	Elnagar and Murant (1976)
5.	Banana bunchy tops virus (BBTV)	*Babuvirus*	Aphid (*Aphidoidea*)	Persistent-circulative	Bressan and Watanabe (2011)
6.	Barley mild mosaic virus (BMMV)	*Bymovirus*	Fungus	Transmission	Shukla (1998)
7.	Barley yellow dwarf virus (BYDV)	*Luteovirus*	Aphid (*Aphidoidea*)	Persistent-circulative	Gray and Gildow (2003)
8.	Barley yellow striate mosaic virus (BYSMV)	*Cytorhabdovirus*	Planthopper (*Pterodictya reticularis*)	Persistent-propagative	Kluth (2002)
9.	Beet leaf curl virus (BLCV)	*Rhabdovirus*	True bug (*Hemiptera*)	Persistent circulative	Proeseler (1978)
10.	Beet necrotic yellow vein virus (BNYVV)	*Bromovirus*	Fungus	Transmission	Verchot-Lubicz (2007)
11.	Black currant reversion virus (BCRV)	*Nepovirus*	Mite	Transmission	Susi (2004)
12.	Blueberry shock virus (BlSHV)	*Ilarvirus*	Honey bee (*Apis mellifera*)	Non-persistent – stylet-borne	Boylan-Pett (1991)
13.	Broad bean wilt virus-1 (BBWV-1)	*Fabavirus*	Aphid (*Aphidoidea*)	Non-persistent – stylet-borne	Ferriol (2013)

TABLE 2.1 *(Continued)*

SL. No.	Virus	Genus	Vectors	Mechanism	References
14.	Brome mosaic virus (BMV)	*Bromovirus*	Beetle (*Coleoptera*)	Transmission	Gáborjányi and Szabolcs (1987)
15.	Carrot mottle virus (CMoV)	*Umbravirus*	Aphid (*Aphidoidea*)	Persistent-circulative	Stubbs (1952)
16.	Cauliflower mosaic virus (CaMV)	*Caulimovirus*	Aphid (*Aphidoidea*)	Non-persistent – foregut-borne	Hohn (2007)
17.	Centrosema mosaic virus (CenMV)	*Potexvirus*	True bug (*Hemiptera*)	Non-persistent – stylet-borne	Van Velsen and Crowley (1961)
18.	Cereal yellow dwarf virus (CYDV)	*Polerovirus*	Aphid (*Aphidoidea*)	Persistent – circulative	Gray and Gildow (2003)
19.	Cherry mottle leaf virus (CMOLV)	*Trichovirus*	Mite	Transmission	Ma (2014)
20.	Citrus leprosis virus-C (CILV-C)	*Cilevirus/ Rhabdovirus*	Mite	Transmission	Melzer (2012)
21.	Citrus tristeza virus (CTV)	*Closterovirus*	Aphid (*Aphidoidea*)	Non-persistent – foregut-borne	Roy and Brlansky (2009)
22.	Citrus yellow mosaic badnavirus (CYMBV)	*Badnavirus*	Aphid (Needs Confirmation)	Non-persistent – foregut-borne	Ghosh (2014)
23.	Citrus yellow mosaic virus (CYMV)	*Badnavirus*	Mealybug (*Pseudococcidae)*	Persistent – circulative	Gaddam (2011)
24.	Cocksfoot mild mottle virus (CFMMV)	*Sobemovirus*	Beetle (*Coleoptera*)	Transmission	Mohamed and Mossop (1981)
25.	Coconut foliar decay virus (CFDV)	*Nanovirus*	Planthopper (*Pterodictya reticularis*)	Persistent – circulative	Wefels (2015)
26.	Colocasia bobone disease virus (CBDV)	*Rhabdovirus*	Planthopper (*Pterodictya reticularis*)	Transmission	Gollifer, 1977, Palomar (1987)

TABLE 2.1 *(Continued)*

SL. No.	Virus	Genus	Vectors	Mechanism	References
27.	Cowpea mild mottle virus (CMMV)	*Carlavirus*	Whitefly (*Bemisia tabaci*)	Foregut-borne non-persistent – stylet-borne	Iwaki (1982); Melzer (2012)
28.	Cucumber green chlorotic mottle virus (CGCMV)	*Tobamovirus*	Beetle (*Coleoptera*)	Non-persistent – foregut-borne	Lecoq and Katis (2014)
29.	Cucumber mosaic virus (CMV)	*Cucumovirus*	Aphid (*Aphidoidea*)	Non-persistent – stylet borne	Ullman (1991)
30.	Cucumber yellows virus (CYV)	*Crinivirus*	Whitefly (*Bemisia tabaci*)	Non-persistent – foregut-borne	Huang (2010); Yamashita (1979)
31.	Cucurbit chlorotic yellows virus (CCYV)	*Crinivirus*	Whitefly (*Bemisia tabaci*)	Non-persistent – foregut-borne	Huang (2010)
32.	Cucurbit yellow stunting disorder virus (CYSDV)	*Crinivirus*	Whitefly (*Bemisia tabaci*)	Non-persistent – foregut-borne	Celix (1996)
33.	Cymbidium mosaic virus (CymMV)	*Potexvirus*	Cockroach (*Periplaneta americana*)	Non-persistent – stylet-borne	Allen (2010)
34.	Faba bean necrotic stunt virus (FBNSV)	*Nanovirus*	Aphid (*Aphidoidea*)	Persistent – circulative	Sicard (2015)
35.	Grapevine leaf roll associated virus-3 (GVLRaV-3)	*Closterovirus*	Scale insect (*Coccoidea*)	Transmission	Tsai (2010)
36.	Grapevine leaf roll associated viruses (GVLRaV)	*Ampelovirus*	Mealybug (*Pseudococcidae*)	Transmission	Tsai (2010)
37.	Grapevine virus-A (GV-A)	*Trichovirus*	Mealybug (*Pseudococcidae*)	Non-persistent – foregut-borne	Tsai (2010)

TABLE 2.1 *(Continued)*

SL. No.	Virus	Genus	Vectors	Mechanism	References
38.	Grapevine virus A and B (GV-A; GV-B)	*Vitivirus*	Mealybug (*Pseudococcidae*)	Transmission	Le Maguet (2013)
39.	Lettuce big vein virus (LBVV)	*Ophiovirus*	Fungus	Transmission	Sasaya (2008)
40.	Lettuce necrotic yellows virus (LNYV)	*Cytorhabdovirus*	Aphid (*Aphidoidea*)	Persistent-propagative	Dietzgen (2006)
41.	Lily symptomless virus (LSV)	*Carlavirus*	Aphid (*Aphidoidea*)	Non-persistent – stylet borne	Asjes (2000)
42.	Maize chlorotic mottle virus (MCMV)	*Machlomovirus*	Thrips (*Thysanoptera*) (only in Hawaii)	Non-persistent – foregut-borne	Lukanda (2014)
43.	Maize chlorotic mottle virus (MCMV)	*Machlomovirus*	Beetle; Thrip (*Thysanoptera*) (in Hawaii only)	Non-persistent – stylet-borne	Jensen (1985); Lukanda (2014)
44.	Maize mosaic virus (MMV)	*Nucleorhabdovirus*	Planthopper (*Pterodictya reticularis*)	Persistent – propagative	McEwen and Kawanishi (1967)
45.	Maize rayado fino virus (MRFV)	*Marafivirus*	Leafhopper (*Cicadellidae*)	Persistent – propagative	Edwards (2016)
46.	Maize red leaf stripe virus (MRLStV)	*Tenuivirus*	Mite	Transmission	Skare (2006)
47.	Maize streak virus (MSV)	*Mastrevirus*	Leafhopper (*Cicadellidae*)	Persistent-circulative	Shepherd (2010)
48.	Melon necrotic spot virus (MNSV)	*Carmovirus*	Fungus	Transmission	Wada (2008)
49.	Oat sterile virus (OSV)	*Fijivirus*	Planthopper (*Pterodictya reticularis*)	Persistent – propagative	Ammar and Nault (2002)
50.	Orchid fleck virus (OFV)	*Dichorhavirus*	Mite	Transmission	Kondo (2003)
51.	Parsnip yellows fleck virus (PYFV)	*Sequivirus*	Aphid (*Aphidoidea*)	Non-persistent – foregut-borne	Elnagar and Murant (1976)

TABLE 2.1 *(Continued)*

SL. No.	Virus	Genus	Vectors	Mechanism	References
52.	Pea enation mosaic virus-1 (PEMV-1)	*Enamovirus*	Aphids (*Aphidoidea*)	Persistent – circulative	Salgueiro and Hull (1999)
53.	Pepino mild mottle virus (PepMMoV)	*Potexvirus*	Fungus	Transmission	King (2012)
54.	Plum latent virus (PLV)/ Prunus necrotic ring spot virus (PNRSV)	*Ilarvirus*	Mite	Transmission	Proeseler (1968)
55.	Potato leaf roll virus (PLRV)	*Luteovirus*	Aphids (*Aphidoidea*)	Persistent – circulative	Sylvester (1956)
56.	Potato mop top virus (PMTV)	*Pomovirus*	Fungus	Transmission	Merz (2008)
57.	Potato mosaic virus-M (PMV-M)	*Carlavirus*	True bug (*Hemiptera*)	Transmission	Turka (1978)
58.	Potato virus S (PVS)	*Carlavirus*	Aphid (*Aphidoidea*)	Non-persistent – stylet borne	Burrows and Zitter (2005)
59.	Potato virus X (PVX)	*Potexvirus*	Beetle (*Coleoptera*)	Non-persistent – stylet-borne	Nagaich (1972)
60.	Potato virus Y (PVY)	*Potyvirus*	Mite	Non-persistent – stylet-borne	Van Hoof (1980)
61.	Potato virus Y (PVY)	*Potyvirus*	Aphid (*Aphidoidea*)	Non-persistent – stylet-borne	Van Hoof (1980)
62.	Raspberry latent virus (RpLV)	*Reovirus/ Rasalvirus (proposed)*	Aphid (*Aphidoidea*)	Persistent – propagative	Quito-Avila (2012)
63.	Rice black-streaked dwarf virus (RBSDV)	*Fijivirus*	Planthopper (*Pterodictya reticularis*)	Persistent – circulative	Lee and Kim (1985)
64.	Rice dwarf virus (RDV)	*Phytoreovirus*	Leafhopper (*Cicadellidae*)	Persistent – propagation	Omura (1998)
65.	Rice grassy stunt virus (RGSV)	*Tenuivirus*	Planthopper (*Pterodictya reticularis*)	Persistent – propagative	Zheng (2014)

TABLE 2.1 *(Continued)*

SL. No.	Virus	Genus	Vectors	Mechanism	References
66.	Rice ragged stunt virus (RRSV)	*Oryzavirus*	Planthopper (*Pterodictya reticularis*)	Persistent – circulative	Gray and Banerjee (1999)
67.	Rice stripe virus (RSV)	*Tenuivirus*	Planthopper (*Pterodictya reticularis*)	Persistent – circulative	Gray and Banerjee (1999); Lee and Kim (1985)
68.	Rice tungro spherical virus (RTSV)	*Waikavirus*	Leafhopper (*Cicadellidae*)	Non-persistent – foregut-borne	Hibino (1983)
69.	Rice yellow mottle virus (RYMV)	*Sobemovirus*	True Fly (*Diopsis thoracica*)	Non-persistent – stylet-borne	Koudamiloro (2015)
70.	Rice yellow mottle virus (RYMV)	*Sobemovirus*	Grasshopper (*Caelifera*)	Non-persistent – stylet-borne	Koudamiloro (2015)
71.	Rose rosette virus (RRV)	*Emaravirus*	Mite	Transmission	Laney (2011)
72.	Ryegrass mosaic virus (RGMV)	*Rymovirus*	Mite	Transmission	Gamliel-Atinsky (2009)
73.	Sowbane mosaic virus (SoMV)	*Sobemovirus*	Thrips (*Thysanoptera*)	Non-persistent – foregut-borne	Hardy and Teakle (1992)
74.	Sowbane mosaic virus (SoMV); Velvet tobacco mottle virus (TYMOV)	*Sobemovirus*	True bug (*Hemiptera*)	Persistent – circulative	Gibb and Randles (1991)
75.	Strawberry crinkle latent ring spot virus (SCLRSV)	*Nucleorhabdovirus*	Aphid (*Aphidoidea*)	Persistent – propagative	Krezal (1982)
76.	Subterranean clover stunt virus (SCSV)	*Nanovirus*	Aphid (*Aphidoidea*)	Persistent – circulative	Johnstone and McLean (1987)
77.	Sugar beet curly top virus (SCTV)	*Curtovirus*	Leafhopper (*Aphidoidea*)	Persistent – circulative	Hosseini (2010)

TABLE 2.1 *(Continued)*

SL. No.	Virus	Genus	Vectors	Mechanism	References
78	Sweet potato mild mottle virus (SPMMV)	*Ipomovirus*	Whitefly (*Bemisia tabaci*)	Persistent – circulative	Dombrovsky (2014)
79	Swollen shoot of cocoa (CSSV)	*Badnavirus*	Mealybug (*Plannococcus aceris*)	Non-persistent – foregut-borne	Quainoo (2008)
80	Tobacco streak virus TSV)	*Ilarvirus*	Thrips (*Thysanoptera*)	Transmission	Sdoodee and Teakle (1987)
81	Tomato chlorosis virus (TOCV)	*Closterovirus*	Whitefly (*Bemisia tabaci*)	Non-persistent – foregut-borne	Navas-Castillo (2000)
82	Tomato pseudo curly top virus (TPCTV)	*Topocuvirus*	Treehopper (*Ceresa taurine*)	Persistent – circulative	Ammar and Nault (2002)
83.	Tomato spotted wilt virus (TSWV)	*Tospovirus*	Thrips (*Thysanoptera*)	Persistent – propagative	Goldbach and Peters (1996)
84.	Tomato Torrado virus (ToTV)	*Torradovirus*	Whitefly (*Bemisia tabaci*)	Non-persistent – stylet-borne	Verbeek (2014)
85.	Tomato yellow leaf curl virus (TYLCV)	*Begomovirus*	Whitefly (*Bemisia tabaci*)	Persistent – propagative	Ghanim (1998)
86.	Turnip crinkle virus (TCV)	*Carmovirus*	Beetle (*Coleoptera*)	Non-persistent – stylet-borne	Brunt (1996)
87.	Turnip yellow mosaic virus (TYMV)	*Tymovirus*	Beetle (*Coleoptera*)	Non-persistent – stylet-borne	Markham and Smith (1949)
88.	Turnip yellow Mosaic virus (TYMV)	*Tymovirus*	Earwig (*Dermaptera*)	Non-persistent – stylet-borne	Raccah and Fereres (2009)
89.	Wheat streak mosaic virus (WStMV)	*Tritimovirus*	Mite	Transmission	Paliwal (1980)

the surface of their virions that play a crucial role in transmission. In some cases, there are additional non-structural helper proteins that serve as bridges, facilitating the attachment of the virion to the vector's binding site. Viruses have a specific affinity for certain sites within or on vectors, where they remain until they are eventually transmitted to their plant hosts. The factors within viruses that determine vector transmission represent promising targets for translational research aimed at disrupting transmission or reducing vector populations (Whitfield, 2015a).

A number of terminologies characterizing transmission connections based on acquisition and inoculation thresholds, as well as retention by the vectors, emerged from the first studies of the biology of insect transmission of plant viruses. There are four types of insect-vector plant virus transmission, including non-persistent, semi-persistent, circulating, and propagating (Ng and Falk, 2006). The relationship between plant viruses and the insects that transmit them has received more attention in recent times (Blanc, 2014). Some penetrate the insect stomach and circulate or proliferate within the insect vector body (circulative), whereas others attach to particular cuticular regions without accessing cells (non-circulative).

Viruses that are not actively being circulated are kept in the insect's stylet or foregut. The phloem limits the infectivity of non-propagative circulative viruses, which often enter the insect via the midgut or hindgut. Certain circulatory plant viruses gain access to the salivary glands through the hemolymph. Conversely, viruses (circulative propagative) enter the insect via its filtering organs in its anterior midgut. Some viruses (propagative) that are capable of replication utilize the hemolymph system to spread. Midgut cells and other insect tissues are used by propagating viruses for replication. Some viruses that replicate rapidly are restricted to the phloem of plants, whereas others may be found throughout the plant. In order to complete their journey, viruses must first enter the hemolymph before moving on to the salivary glands through the nerve tissue (neurotropic pathway) or the connective tissues. One kind of reovirus travels from cell to cell in the midgut by way of tubules, while another utilizes tubules to cross the basal lamina (Purcell and Almeida, 2005).

In the case of persistent transmission, an infected vector may continue to spread a pathogen for many days or weeks. For example, the beet leafhopper (*Circulifer tenellus*) is the only insect species known to transmit the beet curly top virus (BCTV), and the incubation time for the virus in beet leafhoppers is just a few hours. Their ability to spread this Gemini virus is proportional to the amount of time they spend feeding on a plant that has been infected by it. The beet curly top virus is a non-replicating virus that may be spread from one vector to another (Carter, 1973b). To complete the vector transmission process, pathogens must proliferate inside the vector until they infect enough of the vector's salivary glands, which might take a long time. Some plant diseases, such as viruses and bacteria, may cause damage to the vector because they replicate in a variety of vector organs and tissues that are probably not necessary for transmission. Reoviruses, like the rice dwarf virus, may replicate in both the insects that spread them and the plants they infect. Pathogens may be guaranteed to survive the vector's developmental molts by multiplying and circulating inside the vector's bodily cavity (hemocoel). Once the vector has shed its exoskeleton, transmission of viruses that are not persistent or just intermittent is terminated (Carter, 1973a).

2.9 VECTORS RESPONSIBLE FOR BACTERIAL PATHOGENS

Pathogenic bacteria carried by Hemipteran insects (aphids, leafhoppers, whiteflies, and psyllids) and the plants they infect have developed complex techniques for interaction. The study of these infections, which live in plant vascular tissue, offers a great chance to learn about new molecular processes governing intracellular pathogenesis and aids in the management of some of the most dangerous developing diseases worldwide (Huang, 2020).

During vector feeding, pathogenic bacteria such as *Liberibacter, Spiroplasma,* and *Phytophthora* are injected directly into the phloem or xylem, respectively. *Liberibacter* and *phytoplasma* are delivered into the phloem, where they remain confined inside sieve elements, migrate intracellularly via sieve pore channels, and accumulate in sink tissues. *Xylella fastidiosa* cells have unequal distribution and are able to travel across xylem vessels through the pit membrane.

Research on bacterial vector-borne diseases is difficult due to factors such as the difficulty in cultivating the pathogens, the need for an insect vector for transmission, and the absence of model systems (Jiang, 2019). However, novel biology that governs intracellular infections may be uncovered by studying the interaction between pathogens, vectors, and plants in complex systems, which is not possible in simpler models. One possible adaptation that permits plant infections to escape immune detection is the ability to transmit germs straight into vascular tissues like the xylem or phloem. Their pathogenicity mechanisms are likely different from those of other well-characterized Gram-negative bacteria (Kruse, 2019; Toruño, 2016), which means they may reveal novel biological insights. Diseases spread by vectors can only be controlled by learning more about the interactions between pathogens, their hosts, and the vectors that carry them. Furthermore, we focus on bacterial vector-borne plant diseases and discuss recent developments as well as open topics that need to be investigated. At least four distinct waves of pathogenicity evolution have occurred in bacteria.

Due to their small genome size, limited metabolic capacity, and life cycle that requires two hosts, bacteria that are carried by vectors play an essential role as models for research in fields including phytopathology, biochemistry, evolution, and synthetic biology. Due to the loss of metabolic genes during genome reduction, these bacteria must rely on their hosts to provide them with a wide variety of metabolites through the use of a wide variety of transporters, most notably ABC transporters (Kube, 2012; Oshima, 2004).

These bacterial vector-borne infections have had devastating effects on the citrus, grape, and olive sectors in recent decades. Candidatus *Liberibacter asiaticus* (CLas) causes Huanglongbing disease (HLB), the most damaging disease of citrus, which has cost the Florida citrus industry an estimated $7.8 billion since 2007 (Hodges, 2018), and the phytoplasma disease lime witches' broom threatens the lime industry in the Middle East (Donkersley, 2018). The local spread of coconut phytoplasmas has been responsible for the death of millions of palm trees over several decades (Geoff, 2016). Pierce's disease of grapevine (*Xylella fastidiosa*) has had a devastating impact on grape production, resulting in losses of around $100 million per year in California (Tumber, 2014). Recently, a new disease, olive rapid decline syndrome, caused by *X. fastidiosa*, has damaged olive orchards in Europe. It is predicted that the olive sector in Italy might lose up to €5.2 billion, and the olive business in Spain could lose up to €16.86 billion over the next 50 years if no disease control measures are taken (Schneider, 2020) (Table 2.2).

TABLE 2.2 Insect-Vectors That Are Responsible for Transmitting Bacterial, Fungal and Phytoplasma Pathogens and Causing Destructive Diseases in Plants

SL. No.	Insect Species	Vector (Order)	Causal Organism	Disease	Plant Specie	References
1	*Frankliniella fusca*	Thrips (*Thysanoptera*)	Bacterium (*Pantoea ananatis*)	Bacterial central rot of onion	*Allium cepa*	Dutta (2014)
2	*Apis malleifera*	Honey bees (*Hymenoptera*)	Bacterium (*Erwinia amylovora*)	Fire blight of apple/pear	*Malus domestica/ Pyrus communis*	Vanneste (1996)
3	*Acyrthosiphon pisum*	Pea aphid (*Hemiptera*)	Bacterium (*Pantoea stewartii*)	Stewart's wilt	*Zea mays*	Stavrinides (2010)
4	*Rhagoletis pomonella*	Apple maggot (*Diptera*)	Bacterium (*Pseudomonas melophthora*)	Bacterial rot of apples	*Malus domestica*	Cole (1959)
5	*Bactrocera oleae*	Olive fly (*Diptera*)	Bacterium (*Pseudomonas savastanoi*)	Olive knot disease	*Olea europaea*	Young (2004)
6	*Xylebrcis glabratus*	Red bay Ambrosia beetle (*Coleoptera*)	Bacterium (*Raffaelea lauricola*)	Laurel wilt	*Laurus nobilis*	Fraedrich (2008)
7	*Pollinators*	Ants, Bees, Wasps (*Hymenoptera*)	Fungus (*Monilina vaccinia corymbosi*)	Mummy berries of blueberry	*Vaccinium sect. Cyanococcus*	Batra and Batra (1985)
8	*Forficula auriculidae*	Earwigs (*Dermaptera*)	Fungus (*Sclerotinia fructigena*)	Brown rot of apple	*Malus domestica*	Croxall (1951)
9	*Blastophaga psenes*	Fig wasp (*Hymenoptera*)	Fungus (*Fusarium moniliforme var fici*)	Endosepsis of fig	*Ficus carica*	Michailides and Morgan (1998)
10	*Pectinophora gossypiella and Helicoverpa armigera* and *Dysdercus koenigii*	Bollworms and cotton stainers (*Lepidoptera* and *Hemiptera*)	Fungus (*Fusarium* sp. and other)	Boll rot of cotton	*Gossypium herbaceum*	Chinthagunta (2009)

TABLE 2.2 *(Continued)*

SL. No.	Insect Species	Vector (Order)	Causal Organism	Disease	Plant Specie	References
11	*Aphis fabae sp. Cirsiiacanthoidis*	Aphids (*Hemiptera*)	Fungus (*Puccinia punctiformis*)	Rust	Various plants	Kluth (2002)
12	*Drosophila subobscura*	True flies (*Diptera*)	Fungus (*Sclerotinia fructigena*)	Brown rot of plum/peach/ cherry	Various plants	Lack (1989)
13	*Monochamus alternatus* and *M Scutellus*	Beetles (*Coleoptera*)	Nematode (*Bursaphelenchus xylophilus*)	Pine wilt	*Pinaceae*	Takasu (2009)
14	*Diaphorina citri*	*Citrus psylla* (*Hemiptera*)	Phytoplasma	Citrus greening	*Citrus sps*	Manjunath (2008)
15	*Cacopsylla pyri*	*Citrus psylla* (*Hemiptera*)	Phytoplasma	Pear decline	*Pyrus communis*	Carraro (2001)
16	*Cactopsylla picta*	*Psylla* (*Hemiptera*)	Phytoplasma	Apple proliferation	*Malus domestica*	Tedeschi (2006)
17	*Haplaxius crudus*	American palm cixiid planthopper (*Hemiptera*)	Phytoplasma	Lethal yellowing of coconut	*Cocos nucifera* (L.)	Gurr (2015)
18	*Empoasca papayae and E stevensi*	Leafhopper (*Hemiptera*)	Rickettsia	Papaya bunchy tops	*Carica papaya*	Haque and Parasram (1973)
19	Not known	Leafhopper (unconfirmed report) (*Hemiptera*)	Rickettsia	Carrot proliferation	*Daucus carota*	Franova (2000)

2.10 INSECTS AS VECTORS OF FUNGAL PATHOGENS

Some insects may further assist infection by spreading diseases to host plants, since fungi and oomycetes have been shown to have a tight connection with root damage caused by herbivores. Insects are well-known to spread plant diseases as vectors. Fungal and oomycete infections are often transmitted by insects in an accidental fashion, as opposed to the very specialized interactions formed by viruses, phytoplasmas, and other pathogens with insects (Agrios, 2008). Insects may unwittingly spread disease to healthy plant tissues by carrying spores or mycelia on their bodies or eating infected ones (Agrios, 2008). Numerous instances exist of insects spreading fungal infections in above-ground ecosystems. Some of these exchanges are more generalized than others, and many of them entail unintentional transmission. The elm bark beetle (*Hylurgopinus rufipes* Eich.) is attracted to infected elm trees by the fungus responsible for Dutch elm disease because the fungus causes an overexpression of attractive socio-chemicals in damaged trees (McLeod, 2005). It is challenging to establish a strong relationship between root-feeding insects and the dispersion of fungi and oomycetes due to the restricted mobility of subterranean insects and the difficulty of recognizing interactions in the rhizosphere. As a result, scientists know very little about the vast majority of the insects that potentially serve as vectors for soil-dwelling fungi and oomycetes. When insects go from plant to plant in search of food, they also physically transmit the fungi and oomycetes that inhabit their feeding wounds. Both *Fusarium avenaceum* and *Fusarium* spp. identified from head capsules of *S. hispidulus* larvae have been connected to the spread of clover root borers (Leath and Hower, 1993). There is some evidence that root borers transfer fungi like *Kabatiella caulivora* (Kirch.) Karak and *Colletotrichum trifolii* Bain, which they may have ingested and then excreted. Given that injured roots are a source of attraction for root borers (Leath and Hower, 1993), this correlation is especially fascinating since it shows that as the root borer population rises, so does their exposure to phytopathogenic fungi and, therefore, their effectiveness as carriers would also rise.

Commercial greenhouses are closed ecosystems where fast infection and production losses may result from insect-facilitated disease transmission. Oomycete infections, which cannot spread via the air but instead depend on below-ground reproductive structures and the mechanical movement of water, soil, or diseased plant debris (Braun, 2012), may find this to be especially significant. Several common greenhouse pests, such as fungus gnats (*Bradysia impatiens*), shore flies (*Scatella stagnalis* Fallén; Diptera: Ephydridae), and moth flies (*Psychoda* spp.; Diptera: Psychodidae), are known to carry oomycetes and fungi in their larval stages (El-Hamalawi, 2008).

Even after being digested by a larva, oospores, fungal spores, and other propagules may still infect and spread in healthy plants (Hyder, 2009). Some pathogens seem to gain by transiting the gastrointestinal tract; for example, oospores from one strain of *Pythium aphanidermatum* (Edson) Fitzp. and chlamydospores generated by the fungus *Thielaviopsis basicola* (Berk. and Broome) are more infectious after being excreted (Stanghellini, 1999). In a greenhouse, transmission is more likely to occur in seedling flats, hydroponic troughs, and other situations where the disease may spread on its own due to the restricted mobility of larvae (Braun, 2012). The potential for disease spread would be significantly amplified if live propagules could be retained from the larval to highly mobile adult stage. Shore flies may transmit *Pythium* spp., *T. basicola, Verticillium dahliae* Kleb, and *Fusarium* spp., although fungus and moth flies cannot (El-Hamalawi, 2008; Goldberg and Stanghellini, 1990). Therefore, it seems doubtful that *B. impatiens* and *Psychoda* spp. play a significant role in the transmission of soil-borne diseases in a greenhouse environment. However, further investigation is required to determine whether or not they can reliably preserve oospores and fungal propagules over time.

Insects spread fungal pathogens unintentionally. Insects may pick up the fungus or its spores from infected plants they visit, either via direct contact (when walking on the plants) or indirect contact (when feeding on the plants). Insects may spread disease to plants by carrying spores and mycelia on their bodies or by eating infected plants (Willsey, 2017). Beetles, in particular, have specialized organs called mycangia for transporting fungus (Amman, 1983).

2.11 VECTORS RESPONSIBLE FOR PHYTOPLASMAS

Hundreds of commercial and native plant species are susceptible to phytoplasma-caused or -associated plant diseases, which may cause anything from minimal to serious damage. The ability of managers to make educated choices that protect crops and endangered endemic plants is hindered by the fact that insect vectors, notably leafhoppers, planthoppers, and psyllids, have only been found for a small fraction of phytoplasma diseases. Our understanding of insect vector-phytoplasma interactions has greatly expanded over the last two decades, enabling more precise forecasting of the distribution and prevalence of phytoplasma-related illnesses (Weintraub and Beanland, 2006).

Bacteria belonging to the phylum Tenericutes, class Mollicutes, and only found in the internal organs of phloem-feeding Hemipteran insects and the phloem sieve components of vascular plants are called phytoplasmas. Phytoplasmas, along with three vector-borne Mollicutes species in the genus Spiroplasma (Kwon, 1999), and the Alphaproteobacteria "*Candidatus Liberibacter*" taxa, have a unique life cycle among known plant pathogenic bacteria and are obligate residents of hosts representing two distantly related kingdoms, plants (Plantae) and insects (Animalia). A Hemipteran insect picks up the bacteria from an infected plant, carries them around in its body, allows them to replicate over the course of several days, and then inoculates other plants (i.e., a persistent-propagative transmission mode) so that they can spread throughout the environment (Hogenhout, 2008; Huang, 2020).

The ability of a phytoplasma vector to effectively spread the pathogen relies on numerous characteristics:

1. Both nymphs and adult vector insects exhibit similar feeding behaviors and are often found in the same physical location, making it possible for both nymphs and adults to transmit phytoplasmas.
2. These vectors have a specific diet, feeding exclusively on phloem sap, which is where phytoplasmas are primarily located.
3. There is a propagative and persistent relationship between these vectors and phytoplasmas. Multiple generations of phytoplasmas may be found inside an infected insect, and the vector insect species is chronically infected.
4. Transovarial transmission of obligatory symbiotic prokaryotes occurs in insect vectors and is passed on to the next generation. Transovarial transmission of phytoplasmas may be possible through similar methods.

Since phytoplasmas are confined to the phloem, only phloem-feeding insects have the potential to acquire and transmit them. However, among phloem-feeding insects, only a relatively small number of species have been confirmed as vectors of phytoplasmas. These vectors encompass various leafhoppers, four families of planthoppers, and two genera of psyllids.

Phytoplasmas are acquired by phloem-feeding insects during their feeding on phytoplasma-infected plants, but a significant amount of time is required for the insect to acquire a sufficient titer of phytoplasma to establish an infection within its own body. This period is termed the "acquisition access period" (AAP) and can vary from minutes to hours.

The latent period (LP), also known as the incubation period, is the time it takes for a feeding insect to get infected with phytoplasma and transfer it to a plant. When an insect vector is exposed to phytoplasma during the LP, the insect's digestive system becomes infected with the pathogen. Once within the insect, these cells enter the hemolymph and begin circulating, ultimately infecting and multiplying in different insect organs. Phytoplasmas infect particular cells in the salivary glands, multiply, and are then transferred to plants by injections of saliva into the phloem tissue of the plant during feeding.

Researchers are exploring a promising approach for controlling phytoplasma transmission by insect vectors. This approach involves manipulating the symbiotic bacteria present in these insects. Many arthropods, including the insect vectors in question, rely on symbiotic microorganisms for essential life functions. It is hypothesized that the symbiotic bacteria on the insect's body may be manipulated genetically to stop the spread of infections like phytoplasmas. Another hypothesis is that the carbohydrate-binding capabilities of substances like plant lectins may be utilized to disrupt the insect's diet and limit the vector's efficacy, since phytoplasma vectors receive nutrition by feeding on phloem tissue.

2.12 EFFECTS OF INSECT-TRANSMITTED PLANT PATHOGENS

2.12.1 ECONOMIC LOSSES

Worldwide, pests are responsible for a yearly loss of 20%–40% of agricultural yield. According to the FAO, plant diseases annually cost the world economy around $220 billion, and invasive insects about $70 billion. In addition to pests and diseases, weeds pose a serious threat to crop production across the world (USDA, 2023). Plant pathogens and pests pose a significant threat to global crop yields, with wheat, maize, rice, potato, and soybean collectively experiencing yield losses of up to 40% worldwide (Savary, 2019). The economic impact of plant diseases causes approximately USD 220 billion annually (Savary, 2019). Among these pathogens, viruses are a particularly costly problem, accounting for nearly half of all plant disease-related expenses, with an annual global cost exceeding USD 30 billion (Nicaise, 2014). Rice cultivation, which supports nearly half of the world's population across 100 countries, faces substantial risks from various vector-transmitted viruses, resulting in annual losses of USD 1.5 billion.

Research has mostly focused on viruses affecting cash crops under the 26 families of plant viruses that were defined by the International Committee on Taxonomy of Viruses in 2019 (Sastry, 2019). One economically critical example is Cassava mosaic begomovirus, a DNA virus belonging to the Geminiviridae family, which causes a loss of 25 million tons annually. Famine is inevitable as this virus spreads since half a billion people depend on cassava for their daily calories. In West Africa, where 70% of the world's cacao is grown, an additional virus, Cacao swollen shoot badnavirus (CSSV), a member of the Caulimoviridae family, has become endemic in the countries of Togo, Nigeria, and Ghana. If cacao farms disappeared, it would have a devastating effect on local economies and cause a worldwide scarcity of cacao. Costly eradication measures have been established to protect the cacao sector in light of the loss of 200 million trees to CSSV infection (Nicaise, 2014).

In addition, SARS-CoV-2 is only one of several positive-sense single-stranded RNA viruses that have shared characteristics with economically destructive plant virus families, including Bromoviridae, Closteroviridae, Luteoviridae, and Potyviridae. These viruses can cause epidemics in fruits, vegetables, and animal feed. *Citrus tristeza* closterovirus, a member of the Closteroviridae family, is responsible for the death of 100 million citrus trees throughout the globe and represents a serious danger to the lucrative international citrus sector (Nicaise, 2014).

Damage to crops by plant diseases is seldom an issue in the industrialized world, and as a result, there is less of a sense of urgency among policymakers and the general public to fund research into innovative means of preventing and treating crop infection. However, in the coming decades, we will need to boost agricultural production to meet the needs of a growing population, even as we fight the effects of climate change on agriculture, such as drought, floods, and the increasing incidence of plant diseases.

Current efforts to produce resistant crops cannot keep up with the number of diseases endangering the global food supply unless they receive more attention and financing (He and Krainer, 2020).

2.12.2 ENVIRONMENTAL CHALLENGES

Research conducted by the Secretariat (2021) concludes that pathogen transmission through vectors in agricultural and forestry ecosystems will be more likely as a result of climate change, particularly in the colder Arctic, boreal, temperate, and subtropical areas. For instance, a single abnormally warm winter might facilitate the spread of exotic pathogens and pests.

Insect vectors have already expanded due to the warmer environment, such as the fall armyworm (which feeds on a rising variety of crops, such as maize, sorghum, and millet) and the Tephritid fruit fly (which damages fruit and other crops and carries pathogens). It is anticipated that desert locusts, the world's most devastating migratory pest, may shift their range and migration patterns as a result of climate change. In the previous decade, the number of international travel and trade increased, making it the single most significant factor in the spread of new plant diseases. Once an insect population has established itself in a new area, it is almost impossible to completely wipe it out, and doing so may be both time-consuming and costly.

The proliferation and intensification of insect vectors due to climate change pose a danger to global food supplies. Small farmers, those whose income is directly tied to plant health, and those who reside in nations plagued by food shortages are particularly at risk. Adjustments to plant protection techniques and better controls to prevent the spread of plant pathogens through insect vectors during travel are also crucial. Additional investment in upgrading national phytosanitary systems and structures is also highlighted, as is the need for further study into the effect of climate change on insects that facilitate the spread of plant diseases.

2.12.3 HUMAN HEALTH RISKS

Plants play a crucial role in providing sustenance to both humans and livestock, accounting for more than 80% of the food we consume. However, the well-being and availability of plants for consumption are constantly under threat from diseases and pests. These threats result in substantial losses in crop yields, sometimes as high as 30%, leading to staggering economic losses amounting to billions of dollars globally. Beyond economic concerns, plant diseases and insect vectors have profound implications for food security, affecting both the quantity and safety of our food supply (Boa, 2015).

The impact of plant pathogens extends beyond agriculture; it reaches into human and animal health. Measures taken to combat these issues, such as the use of pesticides, can have adverse effects on the well-being of agricultural workers and consumers (Savary, 2017). Moreover, these efforts can contribute to the development of resistance (antimicrobial and antifungal) in the pathogens. Food plants have the ability to carry serious microbiological toxins and infections that can harm people. Foodborne infections, for instance, are known to impact 600 million individuals worldwide or 33 million Disability Adjusted Life Years (DALYs) annually, posing a severe threat to human health (Fisher, 2018; Ramakrishnan, 2019). Additionally, food plants themselves can become carriers of harmful human pathogens and microbial-based toxins.

Emergence of new pathogen variants and the expansion of known pest species into new geographic areas can disrupt global food production and place a significant burden on the world economy. In light of these challenges, it is crucial to adopt effective pest and disease management strategies, including sustainable pesticide management practices. Furthermore, there is a pressing need to quantify the extent of crop losses and their broader impacts on humans, plants, animals, and land use to inform and guide necessary actions (Wilkinson, 2011).

KEYWORDS

- **economic impact of plant diseases**
- **host vulnerability**
- **phytopathogen-insect interactions**
- **phytoplasma vectors**
- **transmission mechanism**
- **vector-borne diseases**

REFERENCES

Agrios, G. N. (2008). Transmission of plant diseases by insects. *Encyclopedia of Entomology*, 3853–3885.

Alexander, H. M., Mauck, K. E., Whitfield, A. E., Garrett, K. A., & Malmstrom, C. M. J. (2014). Plant-virus interactions and the agro-ecological interface. *European Journal of Plant Pathology*, *138*, 529–547. https://doi.org/10.1007/s10658-013-0317-1.

Allen, C. (2010). Virus transmission in orchids through the feeding damage of Australian cockroach, *Periplaneta australasiae*. *International Orchid Symposium*, *878*, 375–379.

Amman, G. D. (1983). *Mountain Pine Beetle Dynamics in Lodgepole Pine Forests: Part II. Population Dynamics*. Intermountain Forest and Range Experiment Station, U.S. Department of Agriculture.

Ammar, E. D., & Nault, L. R. (2002). Virus transmission by leafhoppers, planthoppers and treehoppers (*Auchenorrhyncha*, *Homoptera*). *Advances in Botanical Research*, *36*, 141–167. https://doi.org/10.1016/S0065-2296(02)36062-2.

Asjes, C. J. (2000). Control of aphid-borne *Lily symptomless virus* and *Lily mottle virus* in *Lilium* in the Netherlands. *Virus Research*, *71*, 23–32. https://doi.org/10.1016/S0168-1702(00)00185-4.

Batra, L. R., & Batra, S. W. (1985). Floral mimicry induced by mummy-berry fungus exploits host's pollinators as vectors. *Science*, *228*, 1011–1013. https://doi.org/10.1126/science.228.4702.1011.

Belliure, B., Janssen, A., Maris, P. C., Peters, D., & Sabelis, M. W. (2005). Herbivore arthropods benefit from vectoring plant viruses. *Ecology Letters*, *8*, 70–79. https://doi.org/10.1111/j.1461-0248.2004.00699.x.

Bendix, C., & Lewis, J. D. (2018). The enemy within: Phloem-limited pathogens. *Molecular Plant Pathology*, *19*(1), 238–254. https://doi.org/10.1111/mpp.12526.

Blanc, S., Drucker, M., & Uzest, M. (2014). Localizing viruses in their insect vectors. *Annual Review of Phytopathology*, *52*, 403–425. https://doi.org/10.1146/annurev-phyto-102313-045920.

Boa, E., Danielsen, S., & Haesen, S. (2015). Better together: Identifying the benefits of a closer integration between plant health, agriculture and One Health. *One Health: The Theory and Practice of Integrated Health Approaches*, 258–271.

Bosque-Pérez, N. A., & Eigenbrode, S. D. (2011). The influence of virus-induced changes in plants on aphid vectors: Insights from luteovirus pathosystems. *Virus Research*, *159*, 201–205. https://doi.org/10.1016/j.virusres.2011.04.020.

Boylan-Pett, W., Ramsdell, D., Hoopingarner, R., & Hancock, J. (1991). Honey bee foraging behavior and the transmission of the pollen-borne *Blueberry leaf mottle virus* in highbush blueberry. *XV International Symposium on Small Fruit Virus Diseases*, *308*, 99–108.

Braun, S., Sanderson, J., & Wraight, S. (2012). Larval *Bradysia impatiens* (Diptera: Sciaridae) potential for vectoring *Pythium* root rot pathogens. *Phytopathology*, *102*, 283–289. https://doi.org/10.1094/PHYTO-09-11-0262.

Bressan, A., & Watanabe, S. (2011). Immunofluorescence localization of *Banana bunchy top virus* (family *Nanoviridae*) within the aphid vector, *Pentalonia nigronervosa*, suggests a virus tropism distinct from aphid-transmitted lute viruses. *Virus Research*, *155*, 520–525. https://doi.org/10.1016/j.virusres.2010.12.005.

Brunt, A., Crabtree, K., Dallwitz, M., Gibbs, A., Watson, L., & Zurcher, E. (1996). *Viruses of Plants: Descriptions and Lists from the VIDE Database*. CAB International.

Burrows, M. E., & Zitter, T. A. (2005). *Virus Problems of Potatoes*. Department of Plant Pathology, USDA-ARS, Cornell University.

Carmo-Sousa, M., Moreno, A., Plaza, M., Garzo, E., & Fereres, A. (2016). *Cucurbit aphid-borne yellows virus* (*CABYV*) modifies the alighting, settling, and probing behavior of its vector *Aphis gossypii*, favoring its own spread. *Annals of Applied Biology*, *169*, 284–297. https://doi.org/10.1111/aab.12300.

Carraro, L., Loi, N., & Ermacora, P. (2001). The life cycle of pear declines phytoplasma in the vector *Cacopsylla pyre*. *Journal of Plant Pathology*, *83*(2), 87–90. https://www.jstor.org/stable/41998044.

Carter, J. J. (1973a). The mode of transmission of *Tipula iridescent virus*: I. Source of infection. *Journal of Invertebrate Pathology*, *21*, 123–130.

Carter, J. J. (1973b). The mode of transmission of *Tipula iridescent virus*: II. Route of infection. *Journal of Invertebrate Pathology*, *21*, 136–143.

Celix, A., Lopez-Sese, A., Almarza, N., Gomez-Guillamon, M., & Rodriguez-Cerezo, E. (1996). Characterization of *Cucurbit yellow stunting disorder virus*, a *Bemisia tabaci*-transmitted *Closterovirus*. *Phytopathology*, *86*, 1370–1376.

Chinthagunta, L., Verma, O., Piyush, D., Naik, S., & Ashok, K. (2009). Incidence of boll rot, boll and locule damage in different Bt cotton crops. *International Journal of Plant Protection*, *2*, 171–175.

Cole, M. (1959). Bacterial rotting of apple fruit. *Annals of Applied Biology*, *47*, 601–611. https://doi.org/10.1111/j.1744-7348.1959.tb07292.x.

Croxall, H., Collingwood, C., & Jenkins, J. (1951). Observations on brown rot (*Sclerotinia fructigena*) of apples in relation to injury caused by earwigs (*Forficula auricularia*). *Annals of Applied Biology*, *38*, 833–843. https://doi.org/10.1111/j.1744-7348.1951.tb07853.x.

Dietzgen, R. G., Callaghan, B., Wetzel, T., & Dale, J. L. (2006). Completion of the genome sequence of *Lettuce necrotic yellows virus*, type species of the genus *Cytorhabdovirus*. *Virus Research*, *118*, 16–22. https://doi.org/10.1016/j.virusres.2005.10.024.

Dombrovsky, A., Reingold, V., & Antignus, Y. (2014). *Ipomovirus*–an atypical genus in the family *Potyviridae* transmitted by whiteflies. *Pest Management Science*, *70*, 1553–1567. https://doi.org/10.1002/ps.3735.

Donkersley, P., Blanford, J. M., Queiroz, R. B., Silva, F. W., Carvalho, C. M., Al-Sadi, A. M., & Elliot, S. L. (2018). Witch's broom disease of lime (*Candidatus Phytoplasma aurantifolia*): Identifying high-risk areas by climatic mapping. *Journal of Economic Entomology*, *111*, 2553–2561. https://doi.org/10.1093/jee/toy248.

Dutta, B., Barman, A., Srinivasan, R., Avci, U., Ullman, D., Langston, D., & Gitaitis, R. (2014). Transmission of *Pantoea ananatis* and *P. agglomerans*, causal agents of center rot of onion (*Allium cepa*), by onion thrips (*Thrips tabaci*) through feces. *Phytopathology*, *104*, 812–819. https://doi.org/10.1094/PHYTO-07-13-0199-R.

Edwards, M. C., Weiland, J. J., Todd, J., Stewart, L. R., & Lu, S. (2016). ORF43 of *Maize rayado fino virus* is dispensable for systemic infection of maize and transmission by leafhoppers. *Virus Genes*, *52*, 303–307. https://doi.org/10.1007/s11262-016-1287-0.

El-Hamalawi, Z. (2008). Acquisition, retention, and dispersal of soilborne plant pathogenic fungi by fungus gnats and moth flies. *Annals of Applied Biology*, *153*, 195–203. https://doi.org/10.1111/j.1744-7348.2008.00247.x.

Elnagar, S., & Murant, A. (1976). Relations of the semi-persistent viruses, *Parsnip yellow fleck* and *Anthriscus yellows*, with their vector, *Cavarietta aegopodii*. *Annals of Applied Biology*, *84*, 153–167. https://doi.org/10.1111/j.1744-7348.1976.tb01745.x.

Ferriol, I., Rubio, L., Pérez-Panadés, J., Carbonell, E. A., Davino, S., & Belliure, B. (2013). Transmissibility of *Broad bean wilt virus 1* by aphids: Influence of virus accumulation in plants, virus genotype, and aphid species. *Annals of Applied Biology*, *162*, 71–79. https://doi.org/10.1111/j.1744-7348.2012.00579.x.

Fisher, M. C., Hawkins, N. J., Sanglard, D., & Gurr, S. J. (2018). Worldwide emergence of resistance to antifungal drugs challenges human health and food security. *Science*, *360*, 739–742. https://doi.org/10.1126/science.aap7999.

Fraedrich, S., Harrington, T., Rabaglia, R., Ulyshen, M., Mayfield III, A., Hanula, J., Eickwort, J., & Miller, D. (2008). A fungal symbiont of the redbay ambrosia beetle causes a lethal wilt in redbay and other Lauraceae in the southeastern United States. *Plant Disease*, *92*, 215–224. https://doi.org/10.1094/PDIS-92-2-0215.

Franova, J., Karešová, R., Navrátil, M., Šimková, M., Valova, P., & Nebesářová, J. (2000). A carrot proliferation disease associated with rickettsia-like organisms in the Czech Republic. *Journal of Phytopathology*, *148*, 53–55. https://doi.org/10.1111/j.1439-0434.2000.tb04624.x.

Gáborjányi, R., & Szabolcs, J. (1987). *Brome mosaic virus* transmission by cereal leaf beetle (*Oulema melanopus*, Coleoptera, Chrysomelidae). *Cereal Research Communication*, *15*, 259–264.

Gaddam, S. A., Kotakadi, V. S., & Saigopal, D. (2011). Antigenic relationships of *Citrus yellow mosaic virus* by immunological methods. *Asian Journal of Plant Science Research*.

Gamliel-Atinsky, E., Freeman, S., Sztejnberg, A., Maymon, M., Ochoa, R., Belausov, E., & Palevsky, E. (2009). Interaction of the mite *Aceria mangiferae* with *Fusarium mangiferae*, the causal agent of mango malformation disease. *Phytopathology*, *99*, 152–159. https://doi.org/10.1094/PHYTO-99-2-0152.

Garran, J., & Gibbs, A. (1982). Studies on *Alfalfa mosaic virus* and alfalfa aphids. *Australian Journal of Agricultural Research*, *33*, 657–664.

Geoff, G. M., Johnson, A. C., Ash, G. J., Wilson, B., Ero, M., Pilotti, C., Dewhurst, C., & You, M. (2016). Coconut lethal yellowing diseases: A phytoplasma threat to palms of global economic and social significance. *Frontiers in Plant Science*, *7*, 1521. https://doi.org/10.3389/fpls.2016.01521.

Ghanim, M., Morin, S., Zeidan, M., & Czosnek, H. (1998). Evidence for transovarial transmission of *Tomato yellow leaf curl virus* by its vector, the whitefly, *Bemisia tabaci*. *Virology*, *240*, 295–303. https://doi.org/10.1006/viro.1997.8937.

Ghosh, D. K., Bhose, S., Mukherjee, K., Aglave, B., Warghane, A., Motghare, M., Baranwal, V., & Dhar, A. K. (2014). Molecular characterization of *Citrus yellow mosaic badnavirus* (CMBV) isolates revealed the presence of two distinct strains infecting citrus in India. *Phytoparasitica*, *42*, 681–689. https://doi.org/10.1007/s12600-014-0409-2.

Gibb, K. S., & Randles, J. W. (1991). Transmission of *Velvet tobacco mottle virus* and related viruses by the mirid *Cyrtopeltis nicotianae*. *Advances in Disease Vector Research*, *7*, 255–269.

Goldbach, R., & Peters, D. (1996). Molecular and biological aspects of tospoviruses. In *The Bunyaviridae* (pp. 129–157). Springer.

Goldberg, N., & Stanghellini, M. (1990). Ingestion-egestion and aerial transmission of *Pythium aphanidermatum* by shore flies (*Ephydrinae: Scatella stagnalis*). *Phytopathology*, *80*, 1244–1246. https://doi.org/10.1094/Phyto-80-1244.

Gollifer, D., Jackson, G., Dabek, A., Plumb, R., & May, Y. (1977). The occurrence and transmission of viruses of edible aroids in the Solomon Islands and the Southwest Pacific. *PANS*, *23*, 171–177. https://doi.org/10.1080/09670877709412425.

Gray, S. M., & Banerjee, N. (1999). Mechanisms of arthropod transmission of plant and animal viruses. *Microbiology and Molecular Biology Reviews*, *63*, 128–148. https://doi.org/10.1128/mmbr.63.1.128-148.1999.

Gray, S., & Gildow, F. E. (2003). Luteovirus-aphid interactions. *Annual Review of Phytopathology*, *41*, 539–566. https://doi.org/10.1146/annurev.phyto.41.012203.105815.

Gurr, G. M., Bertaccini, A., Gopurenko, D., Krueger, R. R., Alhudaib, K. A., Liu, J., & Fletcher, M. J. (2015). Phytoplasmas and their insect vectors: Implications for date palm. In *Sustainable Pest Management in Date Palm: Current Status and Emerging Challenges* (pp. 287–314). https://doi.org/10.1007/978-3-319-24397-9_10.

Haque, S., & Parasram, S. (1973). *Empoasca stevensi*, a new vector of bunchy top disease of papaya. *Plant Disease Reporter*, *57*, 412–413.

Hardy, V., & Teakle, D. (1992). Transmission of *Sowbane mosaic virus* by *Thrips tabaci* in the presence and absence of virus-carrying pollen. *Annals of Applied Biology*, *121*, 315–320. https://doi.org/10.1111/j.1744-7348.1992.tb03444.x.

He, S., & Krainer, K. M. C. (2020). Pandemics of people and plants: Which is the greater threat to food security? *Molecular Plant*, *13*, 933–934. https://doi.org/10.1016/j.molp.2020.06.007.

Hibino, H. (1983). Transmission of two rice tungro-associated viruses and rice waika virus from doubly or singly infected source plants by leafhopper vectors. *Plant Disease*, *67*, 774–777.

Hodges, A. W., Rahmani, M., & Spreen, T. (2018). Economic contributions of the Florida citrus industry in 2015–16: *FE1021, 7/2017. EDIS, 2018*.

Hogenhout, S. A., Oshima, K., Ammar, E. D., Kakizawa, S., Kingdom, H. N., & Namba, S. (2008). Phytoplasmas: Bacteria that manipulate plants and insects. *Molecular Plant Pathology*, *9*, 403–423. https://doi.org/10.1111/j.1364-3703.2008.00472.x.

Hohn, T. (2007). Plant virus transmission from the insect point of view. *Proceedings of the National Academy of Sciences*, *104*, 17905–17906. https://doi.org/10.1073/pnas.0709178104.

Hosseini, S. (2010). Interaction between the chromatin of *Beet curly top virus* and TFL2 protein.

Huang, L. H., Tseng, H. H., Li, J. T., & Chen, T. C. (2010). First report of *Cucurbit chlorotic yellows virus* infecting cucurbits in Taiwan. *Plant Disease, 94*, 1168. https://doi.org/10.1094/PDIS-94-9-1168B.

Huang, W., Reyes-Caldas, P., Mann, M., Seifbarghi, S., Kahn, A., Almeida, R. P., Béven, L., Heck, M., Hogenhout, S. A., & Coaker, G. (2020). Bacterial vector-borne plant diseases: Unanswered questions and future directions. *Molecular Plant, 13*, 1379–1393. https://doi.org/10.1016/j.molp.2020.08.010.

Hyder, N., Coffey, M. D., & Stanghellini, M. E. (2009). Viability of oomycete propagules following ingestion and excretion by fungus gnats, shore flies, and snails. *Plant Disease, 93*, 720–726. https://doi.org/10.1094/PDIS-93-7-0720.

Ingwell, L. L., Eigenbrode, S. D., & Bosque-Pérez, N. A. (2012). Plant viruses alter insect behavior to enhance their spread. *Scientific Reports, 2*, 578. https://doi.org/10.1038/srep00578.

Iwaki, M., Thongmeearkom, P., Prommin, M., Honda, Y., & Hibi, T. (1982). Whitefly transmission and some properties of *Cowpea mild mottle virus* on soybean in Thailand. *Plant Disease, 66*, 365–368. https://doi.org/10.1094/PD-66-365.

Jensen, S. (1985). Laboratory transmission of *Maize chlorotic mottle virus* by three species of corn rootworms. *Plant Disease, 69*, 864–868. https://doi.org/10.1094/PD-69-864.

Jiang, Y., Zhang, C. X., Chen, R., & He, S. Y. (2019). Challenging battles of plants with phloem-feeding insects and prokaryotic pathogens. *Proceedings of the National Academy of Sciences, 116*, 23390–23397. https://doi.org/10.1073/pnas.1915396116.

Jiménez-Martínez, E., Bosque-Pérez, N., Berger, P., & Zemetra, R. S. (2004). Life history of the bird cherry-oat aphid, *Rhopalosiphum padi* (Homoptera: Aphididae), on transgenic and untransformed wheat challenged with *Barley yellow dwarf virus*. *Journal of Economic Entomology, 97*, 203–212. https://doi.org/10.1093/jee/97.2.203.

Johnstone, G., & McLean, G. (1987). Virus diseases of subterranean clover. *Annals of Applied Biology, 110*, 421–440. https://doi.org/10.1111/j.1744-7348.1987.tb03274.x.

Kennedy, J. (1951). Benefits to aphids from feeding on galled and virus-infected leaves. *Nature, 168*, 825–826. https://doi.org/10.1038/168825a0.

King, A. M., Adams, M. J., Carstens, E. B., & Lefkowitz, E. J. (2012). *Virus Taxonomy* (9th ed.). International Committee on Taxonomy of Viruses.

Kluth, S., Kruess, A., & Tscharntke, T. (2002). Insects as vectors of plant pathogens: Mutualistic and antagonistic interactions. *Oecologia, 133*, 193–199. https://doi.org/10.1007/s00442-002-1016-3.

Kondo, H., Maeda, T., & Tamada, T. (2003). *Orchid Fleck Virus*: *Brevipalpus californicus* mite transmission, biological properties, and genome structure. *Experimental and Applied Acarology, 30*, 215–223. https://doi.org/10.1023/B:APPA.0000006550.88615.10.

Koudamiloro, A., Nwilene, F. E., Togola, A., & Akogbeto, M. (2015). Insect vectors of *Rice yellow mottle virus*. *Journal of Insects, 2015*. http://dx.doi.org/10.1155/2015/721751.

Krezal, H. (1982). Investigation on the biology of *Chaetosiphon fragaefolii*, the most important vector of strawberry viruses in West Germany. *Acta Horticulturae, 129*, 63–68.

Kruse, A., Fleites, L. A., & Heck, M. (2019). Lessons from one fastidious bacterium to another: What can we learn about *Liberibacter* species from *Xylella fastidiosa*? *Insects, 10*, 300. https://doi.org/10.3390/insects10090300.

Kube, M., Mitrovic, J., Duduk, B., Rabus, R., & Seemüller, E. (2012). Current view on phytoplasma genomes and encoded metabolism. *The Scientific World Journal, 2012*. https://doi.org/10.1100/2012/185942.

Kwon, M. O., Wayadande, A. C., & Fletcher, J. (1999). *Spiroplasma citri* movement into the intestines and salivary glands of its leafhopper vector, *Circulifer tenellus*. *Phytopathology, 89*, 1144–1151. https://doi.org/10.1094/PHYTO.1999.89.12.1144.

Lack, K. (1989). The spread of apple brown rot (*Monilinia fructigena*) by insects. *Annals of Applied Biology, 115*, 221–227.

Laney, A. G., Keller, K. E., Martin, R. R., & Tzanetakis, I. E. (2011). A discovery 70 years in the making: Characterization of the *Rose rosette virus*. *Microbiology Society, 92*(7), 1727–1732.

Le Maguet, J., Fuchs, J. J., Chadœuf, J., Beuve, M., Herrbach, E., & Lemaire, O. (2013). The role of the mealybug *Phenacoccus aceris* in the spread of *Grapevine leafroll-associated virus-1* in two French vineyards. *European Journal of Plant Pathology, 135*, 415–427. https://doi.org/10.1007/s10658-012-0099-x.

Leath, K., & Hower, A. (1993). Interaction of *Fusarium oxysporum* f. sp. *medicaginis* with feeding activity of clover root curculio larvae in alfalfa. *Plant Disease, 77*, 799–802. https://doi.org/10.1094/PD-77-0799.

Lecoq, H., & Katis, N. (2014). Control of cucurbit viruses. In *Advances in Virus Research*. Elsevier.

Lee, K., & Kim, S. (1985). The effects of *Rice black-streaked dwarf virus* on vector (*Laodelphax striatellus*) and host plants. *Korean Journal of Plant Pathology, 1*, 190–194.

Liou, R., Yan, H., & Hong, J. (2003). Molecular evidence that aphid-transmitted *Alpinia mosaic virus* is a tentative member of the genus *Macluravirus*. *Archives of Virology, 148*, 1211–1218. https://doi.org/10.1007/s00705-003-0009-9.

Luan, J. B., Wang, X. W., Colvin, J., & Liu, S. S. (2014). Plant-mediated whitefly–begomovirus interactions: Research progress and future prospects. *Bulletin of Entomological Research, 104*, 267–276. https://doi.org/10.1017/S000748531400011X.

Lukanda, M., Owati, A., Ogunsanya, P., Valimunzigha, K., Katsongo, K., Ndemere, H., & Kumar, P. L. (2014). First report of *Maize chlorotic mottle virus* infecting maize in the Democratic Republic of the Congo. *Plant Disease, 98*, 1448. https://doi.org/10.1094/PDIS-05-14-0484-PDN.

Ma, Y., Li, J., Li, G., & Zhu, S. (2014). First report of *Cherry mottle leaf virus* infecting cherry in China. *Plant Disease, 98*, 1161. https://doi.org/10.1094/PDIS-01-14-0058-PDN.

Manjunath, K. Á., Halbert, S., Ramadugu, C., Webb, S., & Lee, R. (2008). Detection of *'Candidatus Liberibacter asiaticus'* in *Diaphorina citri* and its importance in the management of citrus huanglongbing in Florida. *Phytopathology, 98*, 387–396. https://doi.org/10.1094/PHYTO-98-4-0387.

Markham, R., & Smith, K. M. (1949). Studies on the virus of turnip yellow mosaic. *Parasitology, 39*, 330–342. https://doi.org/10.1017/S0031182000083918.

McEwen, F. L., & Kawanishi, C. Y. (1967). Insect transmission of corn mosaic: Laboratory studies in Hawaii. *Journal of Economic Entomology, 60*, 1413–1417. https://doi.org/10.1093/jee/60.5.1413.

McLeod, G., Gries, R., Von Reuss, S. H., Rahe, J. E., McIntosh, R., König, W. A., & Gries, G. (2005). The pathogen causing Dutch elm disease makes host trees attract insect vectors. *Proceedings of the Royal Society B: Biological Sciences, 272*, 2499–2503. https://doi.org/10.1098/rspb.2005.3202.

Melzer, M. J., Sether, D. M., Borth, W. B., & Hu, J. S. (2012). Characterization of a virus infecting *Citrus volkameriana* with citrus leprosis-like symptoms. *Phytopathology, 102*, 122–127. https://doi.org/10.1094/PHYTO-01-11-0013.

Merz, U. (2008). Powdery scab of potato—Occurrence, life cycle and epidemiology. *American Journal of Potato Research, 85*, 241–246. https://doi.org/10.1007/s12230-008-9019-1.

Michailides, T. J., & Morgan, D. P. (1998). Spread of endosepsis in Calimyrna fig orchards. *Phytopathology, 88*, 637–647. https://doi.org/10.1094/PHYTO.1998.88.7.637.

Mohamed, N., & Mossop, D. (1981). *Cynosurus* and *Cocksfoot mottle viruses*: A comparison. *Journal of General Virology, 55*, 63–74. https://doi.org/10.1099/0022-1317-55-1-63.

Navas-Castillo, J., Camero, R., Bueno, M., & Moriones, E. (2000). Severe yellowing outbreaks in tomato in Spain associated with infections of *Tomato chlorosis virus*. *Plant Disease, 84*, 835–837. https://doi.org/10.1094/PDIS.2000.84.8.835.

Ng, J. C., & Falk, B. W. (2006). Virus-vector interactions mediating nonpersistent and semipersistent transmission of plant viruses. *Annual Review of Phytopathology, 44*, 183–212. https://doi.org/10.1146/annurev.phyto.44.070505.143325.

Nicaise, V. (2014). Crop immunity against viruses: Outcomes and future challenges. *Frontiers in Plant Science, 5*, 660. https://doi.org/10.3389/fpls.2014.00660.

Omura, T., Yan, J., Zhong, B., Wada, M., Zhu, Y., Tomaru, M., Maruyama, W., Kikuchi, A., Watanabe, Y., & Kimura, I. (1998). The P2 protein of *Rice dwarf phytoreovirus* is required for adsorption of the virus to cells of the insect vector. *Journal of Virology, 72*, 9370–9373.

Oshima, K., Kakizawa, S., Nishigawa, H., Jung, H. Y., Wei, W., Suzuki, S., Arashida, R., Nakata, D., Miyata, S. I., & Ugaki, M. (2004). Reductive evolution suggested from the complete genome sequence of a plant-pathogenic phytoplasma. *Nature Genetics, 36*, 27–29. https://doi.org/10.1038/ng1277.

Paliwal, Y. (1980). Relationship of *Wheat streak mosaic virus* and *Barley stripe mosaic virus* to vector and nonvector eriophyid mites. *Archives of Virology, 63*, 123–132. https://doi.org/10.1007/BF01320769.

Proeseler, G. (1968). Übertragungsversuche mit dem latenten *Prunus-Virus* und der Gallmilbe *Vasates fockeui* Nal. *Journal of Phytopathology, 63*, 1–9. https://doi.org/10.1111/j.1439-0434.1968.tb02364.x.

Proeseler, G. (1978). Die Akquisitions-und Zirkulationszeit des *Rübenkräusel-Virus*. *Archives of Phytopathology and Plant Protection, 14*, 95–98. https://doi.org/10.1080/03235407809436307.

Purcell, A. H., & Almeida, R. P. (2005). Insects as vectors of disease agents. *Encyclopedia of Plant Crop Science, 10*, 1–5.

Quainoo, A., Wetten, A., & Allainguillaume, J. (2008). Transmission of *Cocoa swollen shoot virus* by seeds. *Journal of Virological Methods, 150*, 45–49. https://doi.org/10.1016/j.jviromet.2008.03.009.

Quito-Avila, D. F., Lightle, D., Lee, J., & Martin, R. R. (2012). Transmission biology of *Raspberry latent virus*, the first aphid-borne reovirus. *Phytopathology, 102*, 547–553. https://doi.org/10.1094/PHYTO-12-11-0331.

Raccah, A., & Fereres, A. (2009). Plant virus transmission by insects.

Ramakrishnan, B., Venkateswarlu, K., Sethunathan, N., & Megharaj, M. (2019). Local applications but global implications: Can pesticides drive microorganisms to develop antimicrobial resistance? *Science of the Total Environment, 654*, 177–189. https://doi.org/10.1016/j.scitotenv.2018.11.0.

Roy, A., & Brlansky, R. (2009). Population dynamics of a Florida Citrus tristeza virus isolate and aphid-transmitted subisolates: Identification of three genotypic groups and recombinants after aphid transmission. *Phytopathology, 99*, 1297–1306.

Salgueiro, S., & Hull, R. (1999). Comparison of *Pea enation mosaic virus* (PEMV) isolates: Sequence of coat protein and aphid transmission protein genes of a European isolate. *Journal of Phytopathology, 147*, 391–396.

Sasaya, T., Fujii, H., Ishikawa, K., & Koganezawa, H. (2008). Further evidence of *Mirafiori lettuce big-vein virus* but not of *Lettuce big-vein associated virus* with big-vein disease in lettuce. *Phytopathology, 98*, 464–468.

Sastry, K. S., Mandal, B., Hammond, J., Scott, S. W., & Briddon, R. W. (2019). *Encyclopedia of Plant Viruses and Viroids.* Springer.

Savary, S., Bregaglio, S., Willocquet, L., Gustafson, D., Mason D'Croz, D., Sparks, A., Castilla, N., Djurle, A., Allinne, C., & Sharma, M. (2017). Crop health and its global impacts on the components of food security. *Food Security, 9*, 311–327.

Savary, S., Willocquet, L., Pethybridge, S. J., Esker, P., McRoberts, N., & Nelson, A. (2019). The global burden of pathogens and pests on major food crops. *Nature Ecology & Evolution, 3*, 430–439.

Schneider, K., Van Der Werf, W., Cendoya, M., Mourits, M., Navas-Cortés, J. A., Vicent, A., & Oude Lansink, A. (2020). Impact of *Xylella fastidiosa* subspecies *pauca* in European olives. *Proceedings of the National Academy of Sciences, 117*, 9250–9259.

Sdoodee, R., & Teakle, D. (1987). Transmission of *tobacco streak virus* by *Thrips tabaci*: A new method of plant virus transmission. *Plant Pathology, 36*, 377–380.

Secretariat, I., Gullino, M., Albajes, R., Al-Jboory, I., Angelotti, F., Chakraborty, S., Garrett, K., Hurley, B., Juroszek, P., & Makkouk, K. (2021). *Scientific Review of the Impact of Climate Change on Plant Pests.* FAO on behalf of the IPPC Secretariat.

Shepherd, D. N., Martin, D. P., Van Der Walt, E., Dent, K., Varsani, A., & Rybicki, E. P. (2010). *Maize streak virus*: An old and complex "emerging" pathogen. *Molecular Plant Pathology, 11*, 1–12.

Shukla, D., Ward, C., Brunt, A., & Berger, P. (1998). *Potyviridae* family. *AAB Descriptions of Plant Viruses, 366.*

Sicard, A., Zeddam, J. L., Yvon, M., Michalakis, Y., Gutiérrez, S., & Blanc, S. (2015). Circulative nonpropagative aphid transmission of nano viruses: An oversimplified view. *Journal of Virology, 89*, 9719–9726.

Skare, J. M., Wijkamp, I., Denham, I., Rezende, J. A., Kitajima, E. W., Park, J. W., Desvoyes, B., Rush, C. M., Michels, G., & Scholthof, K. B. G. (2006). A new eriophyid mite-borne membrane-enveloped virus-like complex isolated from plants. *Virology, 347*, 343–353.

Stanghellini, M., Rasmussen, S., & Kim, D. (1999). Aerial transmission of *Thielaviopsis basicola*, a pathogen of corn-salad, by adult shore flies. *Phytopathology, 89*, 476–479.

Stavrinides, J., No, A., & Ochman, H. (2010). A single genetic locus in the phytopathogen *Pantoea stewartii* enables gut colonization and pathogenicity in an insect host. *Environmental Microbiology, 12*, 147–155.

Stubbs, L. (1952). Further host range and transmission studies with a virus disease of carrot endemic in Australia. *Australian Journal of Biological Sciences, 5*, 399–408.

Susi, P. (2004). *Black Currant Reversion Virus*, a mite-transmitted nepovirus. *Molecular Plant Pathology, 5*, 167–173.

Sylvester, E. S. (1956). Beet mosaic and beet yellows virus transmission by the green peach aphid. *Journal of the American Society of Sugar Beet Technologists, 9*, 56–61.

Takasu, F. (2009). Individual-based modeling of the spread of pine wilt disease: Vector beetle dispersal and the Allee effect. *Population Ecology, 51*, 399–409.

Tedeschi, R., Ferrato, V., Rossi, J., & Alma, A. (2006). Possible phytoplasma transovarial transmission in the psyllids *Cacopsylla melanoneura* and *Cacopsylla pruni*. *Plant Pathology, 55*, 18–24.

Toruño, T. Y., Stergiopoulos, I., & Coaker, G. (2016). Plant-pathogen effectors: Cellular probes interfering with plant defenses in spatial and temporal manners. *Annual Review of Phytopathology, 54*, 419–441.

Tsai, C. W., Rowhani, A., Golino, D. A., Daane, K. M., & Almeida, R. P. (2010). Mealybug transmission of grapevine leafroll viruses: An analysis of virus–vector specificity. *Phytopathology, 100*, 830–834.

Tumber, K., Alston, J., & Fuller, K. (2014). Pierce's disease costs California $104 million per year. *California Agriculture, 68*, 20–29.

Turka, I. (1978). *Lygus rugulipennis* Popp. (Heteroptera, Miridae), vector of potato viruses. *Trudy.*

Ullman, D., Cho, J., & German, L. (1991). Occurrence and distribution of cucurbit viruses in the Hawaiian Islands. *Plant Disease, 75*, 367–370.

Van Hoof, H. (1980). Aphid vectors of potato virus YN. *Netherlands Journal of Plant Pathology, 86*, 159–162. https://doi.org/10.1007/BF019897108.

Van Velsen, R., & Crowley, N. (1961). Centrosema mosaic: A plant virus disease transmitted by both aphids and plant bugs. *Nature, 189.*

Vanneste, J. (1996). Honey bees and epiphytic bacteria to control fire blight, a bacterial disease of apple and pear. *Biocontrol News and Information, 17*, 67N–78N.

Verbeek, M., Van Bekkum, P. J., Dullemans, A. M., & Van Der Vlugt, R. A. (2014). Torradoviruses are transmitted in a semi-persistent and stylet-borne manner by three whitefly vectors. *Virus Research, 186*, 55–60. https://doi.org/10.1016/j.virusres.2013.12.003.

Verchot-Lubicz, J., Ye, C. M., & Bamunusinghe, D. (2007). Molecular biology of potexviruses: Recent advances. *Journal of General Virology, 88*, 1643–1655. https://doi.org/10.1099/vir.0.82667-0.

Wada, Y., Tanaka, H., Yamashita, E., Kubo, C., Ichiki-Uehara, T., Nakazono-Nagaoka, E., Omura, T., & Tsukihara, T. (2008). The structure of melon necrotic spot virus determined at 2.8 Å resolution. *Acta Crystallographica Section F: Structural Biology and Crystallization Communications, 64*, 8–13. https://doi.org/10.1107/S1744309107066481.

Wang, X. W., & Blanc, S. (2021). Insect transmission of plant single-stranded DNA viruses. *Annual Review of Entomology, 66*, 389–405. https://doi.org/10.1146/annurev-ento-060920-094531.

Wefels, E., Morin, J., & Randles, J. (2015). Molecular evidence for a persistent-circulative association between coconut foliar decay virus and its vector *Myndus taffini*. *Australasian Plant Pathology, 44*, 283–288. https://doi.org/10.1007/s13313-014-0343-2.

Weintraub, P. G., & Beanland, L. (2006). Insect vectors of phytoplasmas. *Annual Review of Entomology, 51*, 91–111. https://doi.org/10.1146/annurev.ento.51.110104.151039.

Whitfield, A. E., Falk, B. W., & Rotenberg, D. (2015). Insect vector-mediated transmission of plant viruses. *Virology, 479*, 278–289. https://doi.org/10.1016/j.virol.2015.03.026.

Wilkinson, K., Grant, W. P., Green, L. E., Hunter, S., Jeger, M. J., Lowe, P., Medley, G. F., Mills, P., Phillipson, J., & Poppy, G. M. (2011). Infectious diseases of animals and plants: An interdisciplinary approach. *Philosophical Transactions of the Royal Society B: Biological Sciences, 366*(1573), 1933–1942.

Willsey, T., Chatterton, S., & Cárcamo, H. (2017). Interactions of root-feeding insects with fungal and oomycete plant pathogens. *Frontiers in Plant Science, 8*, 1764. https://doi.org/10.3389/fpls.2017.01764.

Winter, S., Hamacher, A., Engelmann, J., & Lesemann, D. E. (2006). Angelonia flower mottle, a new disease of *Angelonia angustifolia* caused by a hitherto unknown carmovirus. *Plant Pathology, 55*, 820.

Yamashita, S., Doi, Y., Yora, K., & Yoshino, M. (1979). Cucumber yellows virus: Its transmission by the greenhouse whitefly, *Trialeurodes vaporariorum* (Westwood), and the yellowing disease of cucumber and muskmelon caused by the virus. *Japanese Journal of Phytopathology, 45*, 484–496.

Young, J. (2004). Olive knot and its pathogens. *Australasian Plant Pathology, 33*, 33–39. https://doi.org/10.1071/AP03074.

Zheng, L., Mao, Q., Xie, L., & Wei, T. (2014). Infection route of rice grassy stunt virus, a tenuivirus, in the body of its brown planthopper vector, *Nilaparvata lugens* (Hemiptera: Delphacidae) after ingestion of virus. *Virus Research, 188*, 170–173. https://doi.org/10.1016/j.virusres.2014.04.008.

CHAPTER 3

Plant-Mediated Interactions Between Sucking Pests, Herbivores, and Natural Enemies

AMANDEEP KAUR* and VIJAY KUMAR

Department of Entomology, Punjab Agricultural University, Ludhiana, Punjab, India

*Corresponding author

ABSTRACT

Plants play a crucial role in various ecological interactions, whether with the environment, insects, or both. Tritrophic interactions among plants, herbivores, and natural enemies involve complex chemical and visual cues. Herbivores like whiteflies, aphids, and thrips cause direct damage to plants by sucking sap, weakening the plant, and facilitating the transmission of plant viruses. In response, plants employ both direct and indirect defense mechanisms, such as antibiosis, antixenosis, and tolerance. These mechanisms include physiological responses like the production of volatile compounds (e.g., salicylic acid, jasmonic acid), which activate defense systems to reduce insect damage. Additionally, plants regulate nutrient content to strengthen resistance against pests and pathogens. Indirect defenses involve the release of volatile organic compounds (VOCs) that attract natural enemies, including parasitoids and predators, to target herbivores. Plant-mediated interactions are regulated by enzymes and gene expression, helping plants counter herbivore attacks. These defense strategies offer insights into how plants reduce pest impacts, providing valuable information for developing control strategies against sucking pests.

Insect-Plant Interactions, Volume 1: Influence of Biotic and Abiotic Factors.
Jamin Ali, Adil Tonğa, Adnan Shakeel, and Nasheeman Ashraf (Eds.)

DOI: 10.1201/9781779644282-3

This chapter explores these complex interactions and suggests avenues for further research to enhance pest management approaches.

3.1 INTRODUCTION

Action taken by three entities at three levels is called tritrophic interaction. As the name indicates, tritrophic interaction is the interaction between host, pest, and natural enemies. About half a million insect species inhabit and interact with plants, and a variety of chemicals are involved in this tritrophic interaction (Wu and Baldwin, 2010). Plant-mediated interaction occurs through herbivore-induced changes in the plant defense system, which can be defined as the effects of plant traits on herbivore preference or performance (Karban and Myers, 1989). A single host shares a number of entries to feed, reproduce, and interact with each other. However, we lack sufficient data on the role of interaction among the herbivores feeding on the same host and interacting with different trophic levels. Two herbivores elicit changes in plant quality and can be interspecific to another herbivore or intraspecific, which may be positive, negative, or zero. Other resistance factors also affect their presence or density of herbivores (McNutt and Underwood, 2016).

Tritrophic interaction is an important part of the plant ecosystem, as plants must bear the load of both biotic and abiotic factors. The third trophic level (natural enemies), which acts on the second trophic level (herbivores), induces defense in the first trophic level (plants). To deal with these interactions, plants create different traits that play a resistant role against herbivores and lure natural enemies that reduce the herbivore load on the plant. Changes in the plant's external characteristics repel or deter the herbivore from settling or completing its biology, while on the other hand, the internal defense system gets induced by releasing different Herbivore-Induced Plant Volatiles (HIPVs) in response to herbivore attack. Understanding this interaction will enhance the eco-friendly management of these herbivores by modulating the plant system.

Plant-mediated interactions vary from plant to plant and area to area. The presence of morphological characters like trichomes, thickness, and waxy coatings repels herbivores. On the other hand, plants induce secondary metabolites such as terpenes, phenolics, and nitrogen- and sulfur-containing compounds like jasmonic acid (JA) and salicylic acid

(SA) to protect themselves from a diversity of insect pests (herbivores and pathogens) (Belete, 2018). Insects have long been known to exhibit a variety of feeding behaviors, including polyphagy, oligophagy, and monophagy. Whiteflies, aphids, and thrips are polyphagous pests that can be found on most field crops, causing direct damage to the plants by feeding on plant cell sap and indirectly by transmitting viruses from infected to healthy plants, acting as mediators called vectors. This alters photosynthesis, retards chloroplast development, and reduces chlorophyll content, as viruses interfere with the carbohydrate metabolism of the host (Ertunc, 2020) and cause up- and down-regulation of enzymes (Bahar et al., 2021). In the present scenario, these polyphagous pests have become a significant challenge for humankind, as it has become difficult to manage them. These pests have gained resistance against many insecticides by changing their behavior, penetration, metabolism, or altering the target site to defend against different modes of action of these insecticides. To combat the hurdles created by herbivores, integrated pest management (IPM) is very important, as plant-mediated interactions among the plants, herbivores, natural enemies, and the environment will enhance new technologies to tackle pests in their own ways in the field. The use of attractants, repellents, or foliar applications will change the environment between the plants and herbivores. The use of biostimulants, such as plant growth-promoting rhizobacteria (PGPR), enhances the plant defense system as well (Pereira et al., 2021). These strategies are used by the plants to reduce pest attacks. This interaction will help to explore deeper studies to open new windows to find the changes in both plants and insects and guide the control strategies against these sucking pests at various levels. Therefore, in order to improve the management techniques against insect attacks, various mechanisms are involved in the plant system that reduce infestation and help the plant to survive.

3.2 PLANT MEDIATE INTERACTION OF SUCKING PESTS (WHITEFLY, APHID, AND THRIPS)

Insects that attack plants are classified as having chewing or sucking types of mouthparts. Among the chewing types are the insect orders Lepidoptera, Coleoptera, Orthoptera, and others, where immature stages feed on leaves, flowers, and fruits, causing a variety of symptoms. Sucking insects include the insect orders Hemiptera, Diptera, Thysanoptera, and others. It has been reported that sap-sucking pests have been flourishing throughout the globe, affecting a vast array of plants while simultaneously developing a wide range of pesticide resistance (CABI, 2021). The mouthparts are modified into long stylets that must penetrate superficially into the plant tissue to suck the sap or deeply into the plant vascular bundles via intra- or intercellular paths (Kaloshian and Walling, 2005). Sucking pests like whiteflies, aphids, mealybugs, soft scales, and psyllids pierce the epidermis of the plant's upper layer. The three major phases play a role in feeding strategies: labial exploration, stylet penetration, and sucking the sap (Cheng et al., 2013; Gullan and Cranston, 2010; Hao et al., 2008; Spiller, 1990). Sap suckers cause less mechanical damage than chewing insects or leaf miners (Schoonhoven et al., 2005). Some sap suckers transmit viruses that cause deformation and stunted growth of the plant (Figure 3.1). Studying their biology and interaction with plant systems will enhance knowledge to come up with new ideas to deal with these notorious herbivores.

3.2.1 WHITEFLY

Whitefly is a small flyer, polyphagous and sap-sucking pest that has become a major pest for most crops. It belongs to the order Hemiptera and family Aleyrodidae, colonizing more than 1,000 host plants and becoming successful vectors of over 300 plant viruses. *B. tabaci* is a cryptic species complex made up of ecologically and genetically distinct groups, distinguished by biochemical or molecular polymorphism markers and characteristics. The life cycle of a whitefly includes four stages: egg, nymph, pseudo pupa, and winged adult, lasting for 15–20 days. It reproduces and spreads quickly; within one year, they can produce 15 generations (Brown, 1994). It has the potential to affect the biochemistry, physiology, and development of infested plants. Whitefly attacks crops and causes direct and indirect losses. Nymphs and adults suck the cell sap directly from the phloem by inserting their proboscis deep into the leaf tissue and produce honeydew as excreta that deposits on the upper surface of the leaf, leading to the development of black sooty mold that covers the leaf surface, making the plant unable to perform photosynthesis and reducing the vitality of the crop. Whitefly also acts as a vector for various plant viral diseases. Whiteflies carrying virus particles indirectly transmit the

FIGURE 3.1 General damaging symptoms caused by sucking insect pests.

plant virus from infected to healthy plants. A deep phloem feeder causes less injury to the plant externally and induces changes in the physiology of the plant by triggering HIPVs. Whitefly populations vary from species to species; among the collected 266 samples of *B. tabaci* from 31 regions across India, they were grouped into Asia I, Asia II-5, Asia II-7, and Asia II-8, clustered into the MEAM 1 group (Chowda-reddy et al., 2021). Biotype B was observed to be the more efficient vector of phytovirus and further observed that MEAM-1 has a wide host range and excretes a higher quantity of honeydew compared to other types. Morphological traits of plants with high trichome density influence *B. tabaci* density and were recorded in Asia I, Asia II-1, and Asia II-7 on several cotton cultivars (Thomas et al., 2014). Thus, being small in size, having a wide host range, being an efficient vector, and being resistant to a number of insecticides pose a great challenge for humans to manage this pest.

3.2.2 APHID

Aphids, small sap-sucking plant lice, belong to the superfamily Aphidoidea. There are approximately 5000 known aphid species that infest hundreds of host plant families but colonize only a small group of closely related plant species (Ogawa and Miura, 2014). The topography, climate, and profuse vegetation in an area have come together to form an ideal ecological complex, allowing the assemblages to develop into an incredibly diverse and rich aphid fauna (Blackman and Eastop, 2008). They have a broad host range and significantly reduce the vitality of plants via nutrient depletion and feeding damage to the plant tissue (Smith and Chuang, 2014). Their life cycle is complicated as they reproduce and survive on two hosts with polymorphic characteristics. On the primary host, the gravid wingless female lays eggs during late summer, which endure through the winter season. After hatching from overwintering eggs, they find a secondary host in the spring, and females reproduce without fertilization during warm conditions as they produce young ones (viviparity). A number of offspring can be seen at one spot feeding gregariously. The young ones develop into wingless adults and reproduce throughout the season on the secondary host, while a few winged adults fly to nearby plants, new areas, or back to the primary host for overwintering. Aphids exhibit host alteration characteristics in response to the nourishment provided by their host plants (Sahoo et al., 2022). Aphids also adapt to seasonal variations in the nutritional flow content of their host plants by changing their host (Dixon, 1973). Nutritional flow is greater in woody plants during summer than in winter, but nitrogen content in phloem sap is low. As a result, the aphid population starts shifting towards herbaceous plants to fulfill their needs (Blackman and Eastop, 2008). Aphids that alternate between hosts during favorable conditions must return to woody plants before the start of winter (Debnath and Chakrabarti, 2020).

3.2.2.1 DIRECT DAMAGE

Searching for a new host is highly hazardous, as 99% of winged aphids fail to find a suitable host (Ward et al., 1998). Aphids, while settling on a host, may use both visual and volatile cues to detect the host plant (Döring, 2014). Without piercing the cuticle, an aphid assesses the suitability of a plant based on surface molecules that include lipids and secondary metabolites (Smith and Severson, 1992). In plants that lack such defensive responses, aphids secrete saliva and imbibe their stylets again to evaluate the molecules with the chemoreceptor (Miles, 1972). If a suitable host is found, the aphid sucks the cell sap by inserting its proboscis into the deep phloem. Aphid feeding may be recognized by the plant through mechanical damage to cells and shifts in turgor (Will et al., 2013). However, if many aphids feed in a highly stealthy manner, this suppresses recognition (Tjallingii, 2006; Walling, 2008; Will et al., 2013). After acceptance of the host, the aphid pierces the cuticle and penetrates its stylet across the extracellular space towards the sieve tubes of the phloem, secreting distinct types of saliva at various stages of feeding (Zust and Agrawal, 2022). While feeding, aphids intake cell sap that has excessive sugar and other plant nutrients. To remove the excess sugar, they excrete honeydew containing high sugar content that drops onto the upper surface of the lower canopy leaves, inviting the growth of black sooty mold. Other damage symptoms include leaf curling, chlorosis, and ultimately stunting of the plant. Direct feeding on plant tissue causes a minor infestation. On the other hand, it causes indirect loss by transmitting plant viruses, leading to significant losses. Phytophagous pests also impact the photosynthetic activity of the host plant, which largely depends on the species specificity of the host and insect. The intensity of photosynthesis can be enhanced by abiotic and biotic stress. The activation of compensatory mechanisms in the plant depends on the number of aphids and their feeding duration. Longer exposure to aphids reduces the photosynthetic activity of the host (Nietupski et al., 2022).

3.2.2.2 INDIRECT DAMAGE

Aphids spread more than half of the known phytovirus species, and transmission depends upon vector activity (Fingu-Mabola, 2021). Primary host migrants acquire phytoviruses in early spring and transmit them to secondary host or crop plants, while summer migrants transmit the same virus within the field from one plant to another, more or less distant, over several generations (Watson, 1967). Aphid performance can be categorized in two forms: (i) presence of nutrients in the host; (ii) presence or absence of toxins and deterrents in the plant. For example, the presence of free essential (lysine, arginine, and threonine) and non-essential (homocysteine and glycine) amino acids, as well as soluble carbohydrates, enhances the performance of *Aphis gossypii* (Gadhave et al., 2019). Aphid saliva contains a number of proteins that facilitate infestation. Aphid products, like saliva and honeydew, also contain factors that suppress host defenses. Pea aphid-infested faba beans excrete honeydew that suppresses the activation of the jasmonate response (Schwartzberg and Tumlinson, 2014).

3.2.3 THRIPS

Thrips are small sap-sucking insects that feed on the underside of the leaves and are difficult to see. Among plant herbivores, approximately 7700 species of thrips have been recorded, and only less than 1% are reported to be pests of crops (Kawale and Mahajan, 2021). Hot climatic conditions are favorable for thrips to survive and multiply. Thrips belong to the order Thysanoptera in the family Thripidae and thrive in hot climates because they can multiply quickly. Being a polyphagous pest, they are difficult to manage under field and greenhouse conditions; a number of sprays were applied that cause a heavy load on products. They feed on a wide variety of crop plants, including ornamental vegetables, and are known to include Western flower thrips (*F. occidentalis*), onion thrips (*Thrips tabaci*), and melon thrips (*Thrips palmi*) (Mouden and Leiss, 2021). Thrips can be present on flowers, fruits, leaves, and new shoots. They lay eggs in the leaf tissue, which is difficult to see with the naked eye. The first and second instars feed on the leaf tissue, while the third instar, also called a pseudo pupa, falls to the soil and pupates inside it before emerging as an adult to cause a new infestation and give birth to a new generation. The mouth of thrips is asymmetrical, as the right mandible is reduced or vestigial in nature. The mode of feeding differs from that of other sucking insects. With its stylets, thrips probe plant tissue, which comprises three behavioral phases: (i) piercing; (ii) salivation; and (iii) ingestion of cell sap (Kindt et al., 2006). There are two types of feeding: shallow type feeding, where thrips are restricted to epidermal cells and a few layers of mesophyll cells, and penetrating type feeding, which is restricted to deep mesophyll tissues. As a result, the attack area fills up with air, and in severe damage, the plant remains stunted. Thrips also transmit viral, fungal, and bacterial diseases. They attack the new shoots/terminals that are succulent, and the crop's top canopy has the highest population. Thrips cause distortion of leaves, malformation of seedlings, and reduce the growth and vitality of plants. They transmit four different groups of plant diseases. The toxic saliva is injected into the tissue while feeding, and as a result, the bronzing silvery surface of leaves becomes conspicuous. Additionally, the transmission of bacterial and fungal diseases is of the mechanical type but nonspecific. The losses are not only from direct feeding but also from transmitting phytoviruses. Deadly tospoviruses of the genus *Tospovirus*, family Bunyaviridae, are transmitted by thrips. Some thrips are predaceous in nature, feeding on soft-bodied insects, including phytophagous thrips.

3.3 INTRODUCTION TO HERBIVORE-INDUCED CHANGES

Plant defenses can be categorized into direct and indirect mechanisms, which can be displayed constitutively regardless of the presence of herbivores. However, these defenses are more often triggered upon herbivore attack due to the trade-off between plant growth and defense (Zust and Agrawal, 2017). Such herbivore-induced plant responses may influence the initial attacker's behavior and growth, as well as host-plant suitability for other herbivores, even when these are temporally or spatially separated, thereby mediating interspecific and intraspecific competition between plants and herbivores or herbivores to herbivores, which may be positive or negative depending on the characteristics of the responding herbivore species. The herbivore-induced resistance system reacts differently to inhibit and lower the second trophic level to static.

3.3.1 MECHANISMS OF HERBIVORE-INDUCED PLANT RESPONSE

3.3.1.1 IMPACTS OF PLANT TRAITS

Plant-mediated interactions between insect herbivores are diverse in nature and include various mechanisms such as plant allelochemistry, nutrition, morphology, and altered natural-enemy interactions. Plant-mediated interactions occur through three pathways that induce plants to resist herbivore attacks. The first response is between herbivores and plants, resulting from damage caused to the plant and the subsequent influence of the plant's response to other species, which may be positive or negative (Karban and Myers, 1989). There is a variety of interactions in the first and second trophic levels that include competition, amensalism, commensalism, and facilitation. The second type of interaction involves when plants induce their defense system intra-specifically after herbivores cause damage, positively or negatively affecting conspecifics (McNutts and Underwood, 2016). The third type of plant-mediated interaction includes plant constitutive defenses, pre-damage responses, or levels of resistance (Denno et al., 1995). Thus, the three components of plant qualities involved in plant-herbivore interactions are interspecific, intraspecific, and constitutive resistance. Direct defense mechanisms in plants, which initially come into contact with herbivores, are the visual traits of the first trophic level. Plant structures serve as the first line of defense against any attack. Morphological characteristics of the plant (trichomes, waxy layers, etc.), production of defensive proteins (chitinase, proteinase inhibitors, etc.), plant toxins (phytoalexins, alkaloids, glucosinolates, etc.), or the effects of elicitors (help the emitter) or effectors (help the receiver) play key roles in plant-insect interactions (Dicke et al., 2009; Kaplan et al., 2008). Mechanisms of resistance in plants include antixenosis, antibiosis, and tolerance (Emden, 2002). Numerous examples have been cited to show the role of leaf traits in attracting or repelling herbivores. Glossy leaves, plumule, and leaf sheath pigmentation depict resistance against sorghum shootfly (*Altherigona soccata*). The presence of trichomes also affects various biological parameters (Chamarathi et al., 2011). Ovipositional behavior, feeding, and the formation of the next generation were highly affected by the modification of trichomes. Dense trichomes affect insect movement and activity on the leaf surface, thereby reducing their access to the leaf epidermis. Trichomes may be straight, hooked, branched, or unbranched, glandular, or non-glandular (Belete, 2018). The trichome density increases upon insect attack, varying between 25% and 100%, and in special cases, it can increase to 500%–1000% (War et al., 2012). In studies of antixenosis in soybean, trichome density, size, and angle determine resistance against whiteflies (Valle et al., 2012). The presence of a greater number of leaf trichomes and spines on stems and leaves reduces aphid populations on brinjal. Additionally, leaf pH and ash content negatively affect the abundance of aphids on the tested germplasms (Yasmin et al., 2022). Physical defense by trichomes was more pronounced in young leaves, where aphids performed poorly, while chemical defense by benzoxazinoids showed a complex response at the leaf and tissue levels that altered aphid feeding behavior and identified a resistant wild emmer wheat accession that helps improve aphid resistance in cultivated wheat (Singh et al., 2021). The attack of insect pests or any pathogens depends on the shape of the leaf and the surface of the plant. Okra-shaped leaves repel the silver whitefly, and glabrous leaf traits reduce oviposition preference. Less hairy and very hairy leaves were less preferred by whiteflies for laying eggs. Moderately hairy leaf characteristics were preferred by adults because they provide a micro-congenial environment by trapping a layer of more humid air that adheres to the leaf surface, making it easier for structures to stay attached to the leaf surface (Miyazaki et al., 2013).

3.3.1.2 IMPACT OF PLANT BIOCHEMISTRY

Plants contain a number of phytochemicals that deter herbivores from completing their life journey, known as antibiosis. Exudates from trichomes alter the abundance, distribution, and infestation of associated insects (Amin et al., 2020). The plant component phenylalanine ammonia-lyase contributes to the mechanical strength of the plant cell wall by synthesizing lignin and suberin, forming a physical barrier that blocks the further spread of pathogens or infections (Nagadze et al., 2012; Singh et al., 2014). A high content of total phenols in the resistant cultivar Ravi of *Gossypium hirsutum* against CLCuBuV has been reported. The attack of viruses, bacteria, and fungi triggers the production of phenols and phenolic compounds, which are considered defense-related compounds in plants (Siddique et al., 2014). Total soluble protein, phenylalanine ammonia-lyase activity, peroxidase (POD) activity, protease activity, catalase activity, and polyphenol oxidase (PPO) activity enhance the plant defense system. Increased content of tannins and flavonoids has a negative correlation with the bio-protection of host plants against whiteflies (Taggar et al., 2014). The presence of phytochemicals such as terpenes, zingiberene, curcumene, p-cymene, α-terpinene, and α-phellandrene from wild tomato plants repels *B. tabaci* (Bleeker et al., 2009). Some plants without HIPVs can be suppressed by fall armyworms (*Spodoptera frugiperda*) (De Lange et al., 2020).

3.3.1.3 PHYSIOLOGICAL CONSEQUENCES OF PLANT

In the case of antixenosis and antibiosis, the plant interferes with the biology of the insect, whereas post-damage, also called tolerance, refers to a situation where no harm to the herbivore from the plant side is observed. In general, five primary physiological mechanisms have been categorized by which plants may tolerate insect attack: (i) increased net photosynthesis rate after injury; (ii) high growth rate; (iii) increased branching or tillering after the release of apical dominance; and (iv) pre-existing high levels of carbon storage in roots and the ability to reallocate carbon after injury from roots to shoots (Strauss and Agrawal, 1999). Tolerant soybean genotypes suffered less damage compared to the susceptible ones by producing more seeds (Sulistyo and Inayati, 2016). High levels of phosphorus, potash, and polyphenol content were less preferred by *P. maidis* (Mote and Shahane, 1994). A significant increase in superoxide anion radicals (O_2) was observed in maize crop seedlings infested by *Rhophalosiphum padi* (Bird cherry oat aphid) (Sytykiewiez et al., 2014). Accordingly, aphid infestation also resulted in a significant increase in transcript abundances of genes encoding Glutathione-S-Transferase (GSTs) in the resistant maize plants, relative to the susceptible variety, suggesting a potential role of GST in limiting the adverse effects of oxidative stress within the resistant maize. As GST is central to redox balance in the plant system, it implicates resistance to external bodies (Perez and Brown, 2014). Upon the attack of a herbivore, the plant defense can be expected to impart growth, either by depressing the growth of existing meristems or by reducing the formation of new growth (Koch et al., 2016).

3.3.1.4 HERBIVORE-INDUCED PLANT VOLATILES (HIPVS)

Plants free from any damage maintain a baseline level of VOCs that are released into the atmosphere from the surface of leaves and/or from storage sites in the leaves, which often include monoterpenes, sesquiterpenes, and aromatics, accumulating in high quantities in specialized glands or trichomes (Pare and Tumlinson, 1997a). When leaves are damaged mechanically, the plant releases a blend of saturated and unsaturated six-carbon alcohols, aldehydes, and esters produced by the autolytic oxidative breakdown of membrane lipids. The VOCs released by the plant mediate interactions that differ from those of undamaged and mechanically damaged plant parts (Pare and Tumlinson, 1999). Searching for a host by the natural enemy is a significant task without any cues. To overcome this, it develops the ability to learn the chemical cues associated with its host (Lewis and Tumlinson, 1988). The settling and feeding of herbivores activate the plant defense mechanism to release phytochemicals, rendering the entire plant attractive to natural enemies of the herbivore (Dicke et al., 1993). The interaction with the host releases VOCs that are recognized by the parasitic wasp, allowing her to familiarize herself with particular host location cues. For example, the aphid parasitoid *Aphidius ervi* experiences oviposition on the plant host complex, significantly enhancing the flight and landing response of parasitic females compared to those that are not allowed to sting but are exposed to undamaged or host-damaged plants (Du et al., 1997). In addition, the wasp learns the volatile odors associated with food sources and uses them to locate necessary food (Lewis and Takasu, 1990).

Herbivores search the host and insert their mouthparts into the plant tissue, which triggers the plant defense system by emitting VOCs. HIPVs, also called VOCs, play a key role in these interactions as they can attract natural enemies (predators or parasitoids) to herbivore-attacked plants (Turling and Erb, 2018). Moreover, increasing HIPVs in crop fields promotes biological control by recruiting different natural enemies in the crop field (Croijmans et al., 2022). As an above-ground reaction of plant defense, a similar response occurs below ground, with insects damaging roots producing volatiles that attract entomopathogenic nematodes (Rasman et al., 2004). Insect elicitors also trigger the plant-mediated interaction by inviting third trophic level organisms to act upon them, as they are biologically active compounds present in insect saliva that are regurgitated while feeding and are recognized by the plant, subsequently inducing the plant defense (Chen and Mao, 2020; Snoeck et al., 2022). These elicitors also lure natural enemies that make direct exposure to herbivores. Plant-mediated interactions of co-existing arthropods, like whiteflies or aphids, with omnivorous thrips form food webs in the ecosystem, creating a third tritrophic level by influencing the preferences and performance of natural enemies (Vaello et al., 2019). The production of terpenes has been associated with the enhanced attraction of predators of herbivorous pests (Dicke et al., 1990; Francis et al., 2004; Kollner et al., 2008). One example to explain the attraction of natural enemies is the volatile chemicals emitted from cucumber plants infested by *F. occidentalis*, which attracted the predatory mite *Neoseiulus cucumeris* and the general predatory bug *Orius laevigatus* (Venzon et al., 1999), as one of the most efficient predators of thrips in agricultural ecosystems (Weintraub et al., 2011). *Solanum melongena* emitted VOCs upon infestation by *T. palmi*, luring the predatory bug *Orius sauteri* (Mochizuki and Yano, 2007). These compounds also act as attractants for natural enemies of aphids, such as predators and parasitoids (Du et al., 1998). Many plants release (E)-β-farnesene when attacked by aphids, a volatile compound that is generally used by aphids as an alarm pheromone that elicits dropping off plants or moving away from the feeding site (Petrescu et al., 2001). In an example of citrus, when infested by California red scale (*Aonidiella aurantii*), VOCs were identified that attract the predator *Rhyzobius lophthae*, and 22 VOCs were found to be involved, with some of them present in high concentrations in lemon, orange, and tangerine. Among these VOCs, *R. lophthae* were attracted to methyl salicylate and D-limonene at dosages of 1 and 10 μl/ml using an olfactometer (Alsabte et al., 2022).

3.3.1.5 ADAPTIVE SIGNIFICANCE AND EVOLUTIONARY ASPECTS

Inbuilt resistance is one aspect of the plant defense system that occurs without any interaction with herbivores, but herbivore-induced changes in plants are of greater significance. External and internal changes occur in the plant when a herbivore infests it. Moreover, ups and downs in resistance levels decrease the potential for herbivores to adapt to specific plant resistant traits and necessitate a focus on changing plant quality (Poelman et al., 2023). Induced resistance, along with plant defense, provides information to other community members to act upon. Neighboring plants also perceive the next incoming attack by receiving information through plant-induced volatiles from the damaged plant (Moreira and Abdala-Roberts, 2019) or cues while depositing eggs on or in the plant tissue (Griese et al., 2021). With evolution, natural selection prevails when the plant correctly evaluates the herbivore attack, accurately adapts its induced response, and reduces herbivore fitness to counter its attack and associated fitness costs (Agrawal, 2011; Karban and Orrock, 2018). Within species, biotypes from different geographical areas also show variation in feeding and inducing the plant defense system. A few examples cite the plant-mediated interactions caused by sucking pests:

1. It has been observed that when whitefly infests the plant, it activates the plant's defense system, resulting in low fecundity on newly formed leaves in the top canopy. *T. vaporariorum* feeding on husk tomato leaves activates the defense mechanism, resulting in impaired nymphal development in primed newly formed leaves that remain undamaged, but not in the local leaves (Canales et al., 2022). In cassava, a whitefly-resistant genotype was characterized by the accumulation of higher levels of long-chain alcohol wax components, phenylpropanoids, flavonoids, and lignin during early and late stages of infestation (Perez-Fons et al., 2019). As plants modify, the whitefly also forms new types. Thc B biotype feeds better on cabbage and cucumber, whereas the Q biotype infests more readily on tomato, cotton, and poinsettia than the B biotype. Thus, behavior and performance play a major role in distinguishing species with respect to host range. The intraspecific competition within species alters their competitive interactions (Liu et al., 2012).
2. Plants of upland switchgrass (*Panicum virgatum*) serve as suitable hosts for greenbug aphid (Schizaphis, GB) and yellow sugarcane aphid (*Sipha flava*, YSA), while lowland cultivars with improved agronomic practices exhibit multiple species resistance that reduces aphid populations. It has been found that GB-infested plants at 15 days after infestation show high levels of ABA,

whereas YSA infestation shows elevated SA, and JA was observed, inducing significant enrichment of flavonoids associated with an upregulation of many genes at 155 days of infestation. Together, these clues provide data and show the interplay of metabolism and remodeling accompanying the defense response to aphids (Koch et al., 2020).

3. Crop sweet pepper plants were infested by major greenhouse pests *F. occidentalis*, *B. tabaci*, and the mired pirate *Orius laevigatus*, which induce defense in plant-mediated interactions and enhance the emission of HIPVs that attract zoo phytophagous mirids *Nesidiocoris tenuis* (Reuter) and *Macrolophus pygmaeus* (Rambur). These VOCs also enhance the activity of *Encarsia formosa* (Gahan) against *B. tabaci* (Bouagga et al., 2018) and activate the defensive transduction pathways by regulating SA and JA pathways (Bouagga et al., 2017). External application of HIPVs [(Z)-3-hexenol, (Z)-3-hexenyl acetate, (Z)-3-hexenyl propanoate, (Z)-3-hexenyl butanoate, methyl salicylate, and methyl jasmonate] activates the immune system of sweet pepper plants against *F. occidentalis*, luring *Orius laevigatus*, the main natural enemy, using the Y-tube method (Riahi et al., 2022).

Plant-mediated interaction considers mechanisms promoting positive interactions (facilitation via induced susceptibility) and negative interspecific interactions (competition via induced resistance) such as:

1. **Positive Interspecific Interaction:** The attack of one herbivore enhances the activity of another herbivore.

   a. In Sudan Gezira, the attack of whitefly has little effect at the time of boll opening due to leaf age structure and reduces photosynthesis, but fruit damage by Heliothis larvae enhances cotton vegetative growth and alters leaf age composition that benefits B. tabaci development and density (Baumgartner et al., 1986).

   b. In Arabidopsis, infestation with the whitefly Bemisia tabaci suppresses the upregulation of many genes induced by P. xylostella caterpillars (Zhang et al., 2013b).

   c. The double-infested plant with Striped Stem Borer and BPH releases volatiles that attract their females to lay eggs and repel the respective parasitoids (Liu et al., 2023).

2. **Negative Interspecific Interactions:** These can be defined as competition via induced resistance.
   a. Aphids on chili plants may induce the production of secondary metabolites such as monoterpenes (cymene; 1,8-cineole), sesquiterpenes (β-cadinene, α-copaene), and methyl salicylate (MeSA), which deter *B. tabaci* from settling on their host plants (Saad et al., 2015).
   b. *Bemisia argentifolii* feeding induces increases in defensive proteins in cabbage (*Brassica oleracea*) with negative consequences for the growth and survival of the cabbage looper *Trichoplusia ni* and in *L. esculentum* with negative effects on the oviposition, density, and survival of the leaf miner *Liriomyza trifolii* (Mayer et al., 2002).
   c. Spider-mite infestation of cotton plants increased resistance against conspecific mites and whiteflies but also enhanced susceptibility to aphids (Agrawal et al., 2000).

3.4 INDUCED DEFENSE IN PLANT IN RESPONSE TO SUCKING INSECT HERBIVORE

3.4.1 PHYSICAL DEFENSE

The direct response of plants to herbivores depends on plant traits (War et al., 2012), while the indirect response depends on herbivores' natural enemies (Dicke, 2015). Insect attacks trigger plant defenses by secreting plant volatiles that lure natural enemies (Figure 3.2). HIPVs are emitted from attacked plant tissue through three different and non-exclusive mechanisms. In the first mechanism, plants leave behind wounds that create direct contact between plant cell contents and the atmosphere. Secondly, they may diffuse through the cell membrane to the cuticle and thereby reach the atmosphere. Thirdly, HIPVs may translocate or diffuse from the interior of the cell into the apoplastic space and then be released through stomata into the atmosphere (Turlings and Erb, 2018). In general, JA-mediated signaling pathways are implicated in the regulation of anti-herbivore defense, while the SA pathway is associated with defense responses against pathogens (Smith et al., 2009). The phytohormone JA is considered a key player in the defense regulatory network that is effective against herbivores. The interaction of JA with other hormones, such as ET, ABA, and SA, allows for fine-tuning of the spatiotemporal dynamics of the defense response (Schuman and Baldwin, 2016). External application of nutrients like calcium slightly induces the activity of plant defense-related enzymes, whereas a more dominant defense response is triggered by herbivore attacks, a phenomenon generally referred to as priming. Thrips attacks on chrysanthemums experience a stronger induction of JA signaling (Escobar et al., 2019), while JA or its volatile derivative methyl jasmonate, commonly tested synthetic elicitors, do not show much progress in the development of products based on this mechanism (Chen et al., 2020). Terpenes also play a role in plant defense against herbivores, as they are important constituents of HIPVs. Plant-induced chemicals called VOCs are airborne signals that can induce defense responses in systemic undamaged parts of the plant and also key defense responses in neighboring plants (Sharma et al., 2017). Elicitors from insects induce plant defense and complex physiological and biochemical processes such as activation of the SA and JA pathways, Ca^{2+} flux, reactive oxygen species (ROS) burst, mitogen-activated protein kinase (MAPK) activation, and other chemical responses (Wang et al., 2023). Insect herbivores secrete saliva containing a number of active molecules into plant cells after landing to weaken their defense; these chemicals are known as effectors (beneficial to the receiver) (Chen and Mao, 2020).

3.4.2 SALICYLIC ACID PATHWAY

It is a plant hormone that stimulates plant defense against several biotic and abiotic stresses (Prodhan et al., 2018). Upon attack by herbivores, the plant defense is triggered and releases HIPVs involving ET, SA, and JA, which are activated by insect oral secretions. The application of SA raises the amounts of proline, soluble carbohydrates, and soluble proteins, as well as SOD, POD, CAT, and ascorbate peroxidase (APX) (Liu et al., 2016) and also regulates basal defense in plant tissue by triggering the Systemic Acquired Resistance (SAR) through genes (Fu-ZQ, 2013; Gao et al., 2015). For the activation of SAR, SA is a prerequisite (Ellili et al., 2017). It also synthesizes, activates, and signals that result in the accumulation of ROS, callose deposition, activation of pathogenesis-related

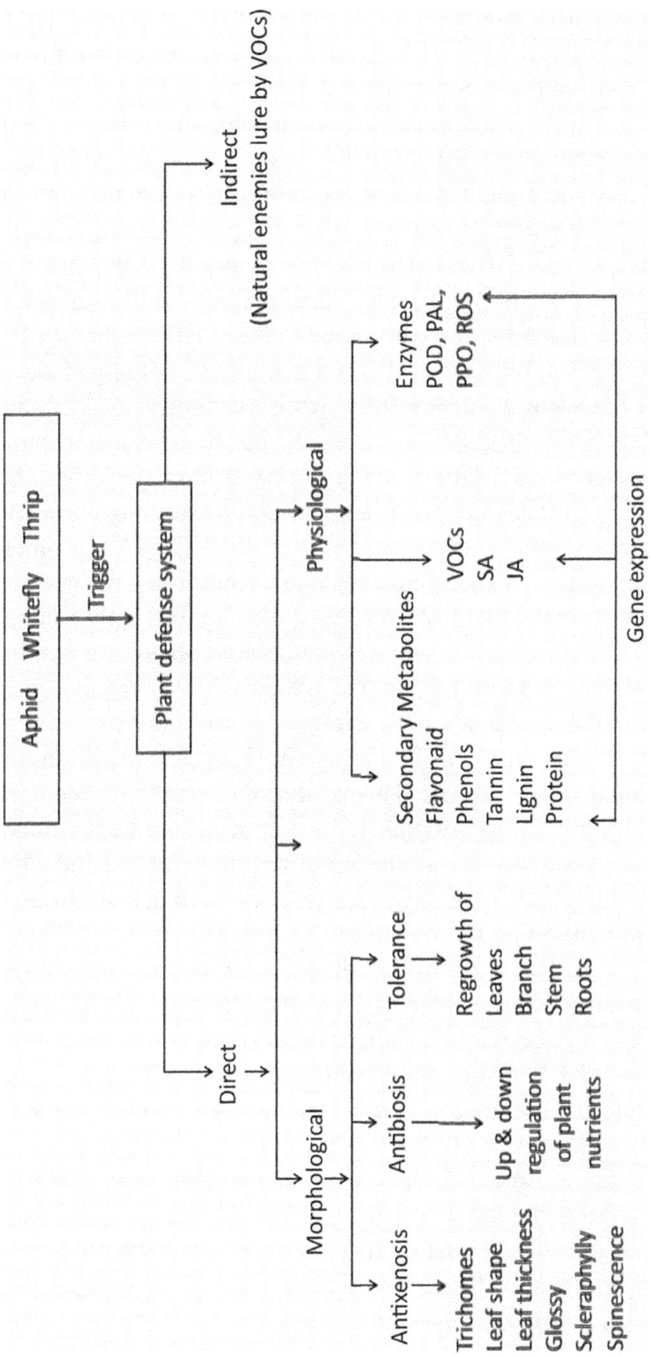

FIGURE 3.2 Plant defense system triggered by sucking pests.

(PR) proteins, and induction of hypersensitive response (HR) (Luna et al., 2011; Voigt et al., 2014; Yi et al., 2014). An antagonistic effect has been observed for SA-mediated plant defense-related pathways, especially in the case of insects, necrotrophs, and viruses (Guerreiro et al., 2016; Yuan et al., 2017). It has been noted that whitefly nymph infestation activates plant defense by suppressing JA through the activation of SA signaling to enhance whitefly performance (Zarate et al., 2007; Zhang et al., 2013a). Herbivores manipulate the plant defense in one way, but on the other side, specialized whitefly parasitoids are adapted to this change and are attracted to these HIPVs that are typically released after SA upregulation (Birkett et al., 2003; Zhang et al., 2013b). The use of MeSA was successfully achieved, attracting lacewings and other predators (Rodriguez-Saona et al., 2011). The foliar application of SA increased the activity of many antioxidant enzymes in wheat and mitigated the adverse impacts of several abiotic stressors and regulated the plant photosynthesis process (Zafar et al., 2021). As in other insect attacks, the plant shows a defense response similar to a thrips attack. For example, the tomato plant shows resistance against aphid attacks due to the presence of tomatine and non-elucidated phenolic compounds as putative defense metabolites against western flower thrips. In addition to this, glandular trichomes with sticky resinous substances (acylsugar) also play a role as antixenosis against WFT, which lowers oviposition (Bac-Molenaar et al., 2019). Thrips attack triggers the plant immune system by JA production, which is later transported to the undamaged areas of the plant. JA acts as the main regulator of the induced thrips defense system (Sarde et al., 2019). Some research has observed that the artificial application of JA overamplifies defense rather than mimicking the natural response, and unlike adults, WFT larvae were observed to tolerate naturally induced levels of JA (Ataide et al., 2019). Maharijaya et al. (2019) categorized two capsianosides and flavonoids as resistant traits against WFT using metabolite quantitative trait loci mapping, whereas sucrose and malonylated flavones glycosides are related to susceptibility (Macel et al., 2019). In addition, olfactory cues (volatile emissions) play a direct role in plant defense (Ren et al., 2020). Thrips attack also affects the biochemistry through the upregulation and downregulation of nutritional host quality (Qian et al., 2018; Yu et al., 2018) by inducing defense. Thrips also act as vectors by transmitting plant pathogens, viz., viruses, fungi, and bacteria.

3.4.3 JASMONIC ACID PATHWAY

Herbivore feeding triggered the plant general mode to defense by activation of JA, which led to the production of more compounds that can serve as direct and indirect defense systems (Howe and Jander, 2008; Okada et al., 2015). The interaction of JA and other phytochemicals plays a key role in fine-tuning the magnitude and nature of the final downstream defense response (Pieterse et al., 2012; Schuman and Baldwin, 2016). The activation of JA pathways triggered plant resistance against thrips attacks, as the exogenous application of JA reduces plant susceptibility to herbivores (Abe et al., 2009; El-Wakeil et al., 2010), while plants insensitive to JA or lacking accumulation of JA are more prone to thrips attacks (Abe et al., 2009). Understanding the plant defense mechanism in plant systems, new strategies from different angles have emerged to bypass the dilemma of plants to grow or defend. Wang et al. (2019) approached identifying components of the JA pathway that mainly involve the regulation of defense but not in growth, such as JAV1, JAZ2, and NaJAZi. Silencing NAJAZi results in increased resistance to herbivores due to elevated levels of defense compounds such as (E)-α-bergamotene, trypsin PIs, and defensins (Li et al., 2017a). Several phytohormones play a crucial role in plant growth, development, and the implementation of plant defense strategies against pests. Other plant hormones, such as auxins, ABA, cytokinins, gibberellins, and brassinosteroids, which have been thoroughly described to regulate plant development and growth, have recently emerged as key regulators of plant immunity (Denance et al., 2013).

3.5 ROLE OF HONEYDEW IN SUPPRESSION OF PLANT DEFENSE

Sucking insects, also called phloem feeders, interact with the plant system. Upon feeding on cell sap derived from phloem, insects secrete honeydew with a high content of sucrose called trehalose as excreta, which, when it falls on the dorsal surface of the leaf, facilitates the deposition and growth of black sooty mold, resulting in a reduction in plant growth by interfering with the photosynthesis process. Proteins from insects, microbes, and plants might be elicitors, effectors, or self-recognized chemicals for plant defense, which provide further insights into their roles in the multitrophic interactions of plants, insects, and microbes. It has also been observed that insect honeydew harbors many bacterial proteins from herbivores or plant symbionts or from the plant surface, which could be microbe-associated molecular patterns (MAMPs) or herbivore-associated molecular patterns (HAMPs) that alter plant defense against herbivores (Zhu et al., 2020). Insect honeydew has a high sugar content that facilitates secondary infection with black mold but also provides an opportunity for recognition by natural enemies. Other researchers reported that honeydew contains proteins like chaperonin, GroEL, and DnaK chaperones, elongation factor TU, and flagellin that might be involved in insect-plant interactions (Sabri et al., 2013). Whitefly honeydew was observed to have most of the SA converted to SAG (Salicylic Acid glycoside), which modulates the SA level in the honeydew and adds complexity, triggering plant defense against herbivores (Vandoorn et al., 2015). *A. pisum* honeydew contains plant-derived SA that suppresses herbivore-induced defenses in plants (Schwartzberg and Tumlinson, 2014). Whitefly feeding on tomato plants contains SA as 80% of SA was converted to its glycoside, salicylic acid glucoside (SAG), inside the whitefly and was reported in honeydew. When whiteflies were allowed to feed from an artificial diet spiked with labelled SA, labelled SAG also was produced.

3.6 INTERACTION BETWEEN VIRUS, HOST, AND INSECT

Vectors play a major role in the transmission of plant viruses from infected to healthy plants. As a virus cannot move on its own, there is a required source that helps it reach its destination. Insects carrying virus titers are called vectors. When an insect feeds deep into the plant tissue, making its stylet penetrate the phloem or xylem while crossing many barriers, it allows the plant virus to reach its infection point (Irwin et al., 2007). Plant vessels, xylem, carry water and low nutrient levels compared to phloem, which consists of companion cells that provide metabolic and regulatory components to cell sap and sieve elements, forming long hollow tube-like structures throughout the plant that carry cell sap. The phloem tissue consists of companion cells, providing metabolic and regulatory components to the phloem sap, and sieve elements, forming a long-distance transport system throughout the plant (Lucas, 2006; Will et al., 2013). Because this specialized transport system offers access to a rich source of carbohydrates, proteins, and amino acids, numerous viral and bacterial microbes specifically colonize the phloem (Bove and Garnier, 2003; Lough and Lucas, 2006). Old references confirm that viral infection causes many physiological and biochemical changes in the plant, and many viruses induce metabolic changes within the plant. Experiments through PCR have confirmed that begomoviruses cause changes in phytochemical concentration, resulting in the activation of the plant defense system. However, transmission efficiency depends on various plant characteristics, such as plant genotype, insect or vector species, presence of virus titers, or geographical origin. Vectors insert their proboscis and transmit the virus into healthy plants. Upon infection, the virus causes various phenotypic, physical, and physiological changes, including alterations in color, size, texture, amino acids, and phytohormone levels, distortion of cell structure, decreased photosynthesis in lower leaves, and reduced nutrient uptake in the plant, resulting in delayed plant growth and development (Mauck et al., 2014; Zao et al., 2016). Orders belonging to Hemiptera, which include mostly sucking pests such as aphids, whiteflies, and hoppers, involve most of the phytopathogenic virus vectors (Richard and Davis, 1977) belonging to the families Begomovirus and Mastreviruses (Bahar et al., 2020).

Whitefly is a phloem feeder and can transmit more than 300 plant pathogenic viruses, including cotton leaf curl virus, yellow vein mosaic virus, tomato yellow leaf curl virus, and so forth. Several aspects that deal with the transmission of viruses from infected hosts to healthy plants, such as the effects of pre-acquisition/post-acquisition, the relative efficiency of males and females, and the number of vectors in transmission, have been reported by Indian researchers (Chandsekharan and Shashank, 2017). For example, *B. tabaci* transmits Pumpkin yellow vein mosaic viral disease (PYVMVD) in pumpkin, and 100% transmission of the disease was achieved by increasing the number of whiteflies up to 10; a single whitefly is sufficient to transmit Squash leaf curl China virus (SLCCNV) affecting *Cucurbita moschata* Duchesne. The acquisition and inoculation access periods (AAP and IAP) of the disease were 30 minutes each, with a significant positive relationship between AAP and IAP and a retention period of 6 and 7 days during the spring and rainy seasons, respectively, after which all of the whiteflies had died. Female whiteflies showed higher efficiency in transmitting the PYVMVD (Singh et al., 2023). On the basis of biotype Asia 1, *B. tabaci* has a significantly higher transmission rate for tomato leaf curl New Delhi virus (ToLCNDV) compared to Asia II-1 (Anokhe et al., 2015). Vanitha and Umesha (2008) proved that the early and changing levels of expression of different enzymes are an important characteristic of plant resistance to any pathogen. For example, tomato yellow leaf curl China virus in plants interferes with JA, resulting in increased performance of *B. tabaci*. Zhang et al. (2012) reported that mutualism between the vector and virus is a consequence of changes in defense against begomovirus. Upon the attack of a plant virus on a healthy plant, a number of secondary compounds are involved as defense mechanisms against the intruder virus, such as PAL, protein, protease, SOD, PPO, and MDA, which play an active role in disease resistance, as reported in ClCuBuV, and can be considered biochemical markers to study the interaction between plant virus compatibility (Siddique et al., 2014). The virus uses the plant system by utilizing various strategies that are more conducive to its multiplication and spread inside the plant system. A number of changes occur in the interaction of the plant, virus, and vector, which mainly include the up- and down-regulation of photosynthetic pigments, an increase in the concentration of phenolic compounds, elevation of starch content in the leaf, and up- and down-regulation of antioxidative enzymes, including GPX, PPO, APX, SOD, and CAT. The secretion of these phytochemicals leads to the initiation of a HR, characterized by thickening of the leaf, lignification under the leaf surface, blocking of stomatal openings, systematic cell death, generation of ROS, and activation of SA and JA resistant pathways in the plant (Bahar et al., 2020). Tomato leaf curl virus (TLCV) infected plants show a decrease in insoluble and soluble sugars in the stem and leaf (Yadav and Yadav, 2019). The POX enzyme triggers other enzymes like chitinase and POD that help in structural defense by controlling the replication of the virus; furthermore, polymerization, suberification, cell wall elongation, resistance, and wound healing are all processes that provide structural defense. Sometimes, viral infection in healthy plants increases the activity of SOD and PPO, resulting in the enhancement of ROS that ultimately triggers the plant defense mechanism (Bahar et al., 2020). On a molecular basis, TMV-Cg (a strain of crucifer) shows that the virus coat protein (CgCP) may suppress signaling of SA by stabilizing DELLA proteins without altering the SA or JA levels in the plant (Rodriguez et al., 2014b) and also reduces plant growth and delays floral transition timing (Rodriguez et al., 2014b). Similarly, for GA signaling, DELLA proteins regulate negatively and suppress the SA-triggered defense response, thus displaying antagonism between SA and JA (Naseem et al., 2015). SA inhibits Tomato bushy stunt virus (TBSV) replication in one system by competitive binding of cytosolic glyceraldehyde 3-phosphate dehydrogenase (GAPDH) (Tian et al., 2015). A number of examples have been cited by various scientists that show resistance in different cultivars of crops against viral pathogens, producing a HR upon infection; the SA levels became several times higher in infected plants compared to non-infected plants (Islam et al., 2018), causing disturbances in primary metabolites and antioxidant enzyme activity within the plant system (Bahar et al., 2020).

1. Aphid-transmitted viruses belong to 19 of the 70 known virus genera and approximately 275 viruses (Nault, 1997). Approximately 75% of viruses with aphids as vectors are transmitted in a non-persistent manner, which causes a number of important diseases. Host selection and feeding behavior are important factors in virus epidemiology, and the suitable host range and life cycle characteristics of aphid species are also key in determining the rate of spread of the virus. During feeding, aphids penetrate their stylet deep into the epidermal layer where the virus titer is high. As the feeding period increases beyond the epidermal layer, the subsequent transmission rate decreases rapidly. The effect of pre-acquisition fasting of aphids for 15 minutes–4 hours can result in increased efficiency (Hull, 2002). Aphids lose their infectivity after molting since the stylet and cuticular lining of the foregut are shed as the viruses are stylet-born. Aphids are viruliferous only for a few minutes to hours (Nikos et al., 2007). Viruses transmitted by aphids activate secondary metabolite plant defenses against them. Aphid survival on CMV-infected tobacco plants (carrying the 2b protein) increases, while the survival of aphids decreases when they feed upon plants lacking 2b (Tungadi et al., 2017). In another example, TuMVNIA-Pro (Nuclear inclusion protein domain) changes the response of ET, suppressing aphid-induced callose defense. Among potyviruses, Nia-Pro is highly conserved, and scientists and coworkers proposed a conserved mechanism of this interaction for the transmission of viruses through aphids as vectors (Islam et al., 2019). Upon the attack of a vector on a healthy plant, the virus causes changes in the composition and the quantity of VOCs released by the plant that attract aphids. The aphid parasitoid *Aphidius colemani* (Hymenoptera: Aphelinidae) has shown higher parasitism and survivability of aphids that fed on non-infected plants. The parasitoid showed a preference for

CMV and PVY-infected plants compared to non-infected plants (Milonas et al., 2023).

2. Tospovirus is not seed-transmitted as it needs a vector (thrips) to transmit, and more than 20 tospoviruses are reported throughout the globe (Kawale and Mahajan, 2021). Thrips transmit viruses in a persistent and propagative manner, where they multiply inside the body of the vector and later transmit to the healthy host. Thrips' 1st and 2nd instars can acquire the virus from virus-infected plants and can transmit it. However, adults cannot transmit as tospoviruses need a latent period to mature enough to cause disease in healthy plants. Vector behavioral responses can be mediated by VOCs that aim to attract (Mwando et al., 2018) or repel (Wu et al., 2019). Thrips carry the tomato spotted wilt virus (TSWV) in their bodies when injected unintentionally into the plant system, which induces defense in the plant by triggering the SA pathway, indirectly enhancing the fitness and performance of the vector by exploiting SA-JA antagonism (Nachappa et al., 2020). In addition to immunological plant modulation, viruses modify vector behavioral responses to facilitate virus transmission (Harding and Nault et al., 2019; Wan et al., 2020). Pathogens are also interacting with insects, like the growth of the fungal endophyte *Hypocrea lixii*, which not only enhances resistance against *T. tabaci* but also reduces the incidence of transmission of the Iris yellow spot virus (IYSV) (Muvea et al., 2018). It has been observed that there is more probing and host plant selection behavior of thrips (Ogada et al., 2013; Shalileh et al., 2016). For example, TSWV-infected *F. occidentalis* viruliferous thrips showed more probing on infected plants compared to non-viruliferous thrips on jimsonweed (*Datura stramonium*). This may be due to the increased availability of free nutrients (Shrestha et al., 2012). Similarly, viruliferous *F. occidentalis* is more attracted to TSWV-infected pepper (*Capsicum annuum*) plants compared to healthy plants (Ogada et al., 2013; Shalileh et al., 2016) and suggested that viruses can also interfere with plant defense, possibly through the interaction of SA with JA signaling. Thaler et al. (2002) reported that JA-induced resistance in wild tomato (*S. lycopersicum* var. cerasiforme) reduces thrips feeding by 75%, but no reduction in the transmission of TSWV was observed.

3.7 HARNESSING PLANT INTERACTIONS FOR EFFECTIVE PEST MANAGEMENT

Integrated pest management involves all aspects of management, but in the present scenario, the last option of IPM, that is, chemical control, has become the first option to manage pest attacks. We are unaware of the interactions through which we can utilize other techniques to manage pest attacks. Due to this management practice, a number of problems emerge, such as pest resistance, resurgence, secondary pest outbreaks, and overall environmental pollution, while the usage of other natural compounds is declining day by day (Due et al., 2020). For the reduction of pesticides, it is essential to recognize both the challenges and the opportunities in the area of signaling, plant defense responses to herbivores, and their strategies for challenging those defense responses (Jones et al., 2022). Moreover, in recent years, studies have been conducted to identify potential sources of resistance against multiple plant antagonists, which could include the use of wild (ancestral) genotypes that are often more resistant than their domesticated counterparts in breeding programs, as well as tactics to enhance crop resistance using biostimulants or elicitors of plant defenses (Rodriguez-Saona et al., 2022). So, it's very important to study the insect-plant interaction from an IPM point of view, as this will help future generations reduce environmental pollution and adopt new generation techniques to reduce pest loads in agroecosystems.

3.7.1 USE OF FOLIAR APPLICATIONS OF HIPVS, ATTRACTANTS, OR REPELLENTS IN THE FIELD

External application of HIPVs, attractants, or repellents triggers plant defense and attracts natural enemies that help the plant system to survive and fight in different ways. A number of examples have been cited by various researchers to illustrate the use of external applications of these chemicals. The direct effects of silica NPs on a variety of pest insects have been examined in the lab (e.g., Mousa et al., 2014). It impairs the digestive tract in insect herbivores that feed on silica-treated plants (Smith et al., 1969) without posing any hazardous effects from traditional chemicals on natural enemies (Abbasi et al., 2020). The insecticidal properties of silica NPs are thought to be due to direct abrasion of the insect cuticle or sorption through the cuticular layers. It also reduces the population of insect pests and associated predators, such as Rove beetles, as the concentration of silica NPs increases. Simultaneous application of GLVs, (Z)-3-hexenyl acetate, and aromatic volatile indole has additive and prime specific responses and elicits an advantageous JA signaling pathway in comparison to individual applications (Hu et al., 2019). In another example, olfactometer-based choice tests found that Rove beetles were more likely to be attracted to herbivore-infested plants treated with silica than to water-treated plants, suggesting that silica enhances the attraction of natural enemies via HIPVs. For the cowpea aphid *A. craccivora,* a complete (100%) reduction was observed 5–15 days after applying silica NPs at concentrations of 225–425 mg/L and after 7–15 days at 75–150 mg/L (Thabet et al., 2021). Si application affects host preference by PBH, significantly reducing honeydew secretion and inhibiting oviposition and hatch rates. Foliar application of external JA impaired the feeding of WFT (Chen et al., 2018; 2020) and increased the duration of developmental stages (Zheng et al., 2019). Exposure to volatile (Z)-3-HP and MeSA on sweet pepper was found to be repellent to *F. occidentalis* but attractive to the predator *O. laevigatus* (Riah et al., 2022). However, the predator was found to be attracted to the plant exposed to (Z)-3-H, (Z)-3-HP, (Z)-3-HB, MeSA, and MeJA. Application of these HIPVs ((Z)-3-HP and MeSA) on tomato affected the attractiveness to tomato pests *Tuta absoluta, B. tabaci,* and *F. occidentalis* (Perez-Hedo et al., 2021a) and lured the nymphal parasitoid (*E. formosa*) of *B. tabaci* (Yang et al., 2020). External applications of ABA enhance resistance against plant viruses such as TMV and tobacco necrosis virus (TNV) by hindering virus movement within plant tissue (Qin et al., 2013). SA application results in the deposition of callose within PD (Yadav et al., 2014). This also inhibits virus movement and its replication via another process that requires the involvement of the mitochondrial alternative oxidase (AOX) pathway (Faoro and Gozzo, 2015). Furthermore, CMV-2b silencing suppressor interrupts JA signaling to enhance vector (*Myzus persicae*)-mediated transmission. Moreover, CMV-infected plants or other transgenic plants expressing 2b exhibited the least levels of JA (Islam et al., 2019).

3.7.2 USE OF ELICITOR

Volatile chemicals present in the saliva produced by insects while feeding are perceived by plants, which triggers the plant defense system.

In the case of aphids, Mp10 and Mp42 were two elicitors that showed decreased fecundity of aphids when feeding on plants. Cysteine protease Cathepsin B3 was also recognized as a potential elicitor protein, which suppresses aphid feeding by enhancing ROS (Guo et al., 2020).

3.7.3 USE OF BIO-STIMULANTS

The use of bio-stimulants has gained the interest of farmers and industrialists throughout the world (Yakhin et al., 2016). External application of PGPR can enhance resistance to insect pests that feed on shoots or roots (Rasmann et al., 2017). They cause repellence and weight reduction effects on larvae (Disi et al., 2019), increase nymph mortality (Dorta et al., 2020), and reduce the feeding and reproductive rates of some insects (Rashid et al., 2017). With multiple and synergistic effects of bio-stimulants, PGPR and HA not only increase the production of nutritive plants but also boost the plant defense, which requires less use of pesticides (Periera et al., 2021). Humic acid-treated plants show high egg parasitization and development. Similarly, the application of the external non-protein amino acid DL-β-aminobutyric acid (BABA) increases pea aphid (*A. pisum*) mortality on ontic bean (*Vicia faba* var. minor), pea (*P. sativum*), broad bean (*V. faba* var. major), and alfalfa (*M. sativa*) (Hodge et al., 2005).

3.7.4 BREEDING AND LARGE-SCALE SCREENING OF GERMPLASM

Securitizing the plant character among the various varieties as a genetic source of host plant resistance is an important step to initiate a breeding program against insect pests in agriculture. In some cultivars, the inbuilt resistance characters are present, but breeding (or engineering) crop plants in such a way that activates plant defense upon attack has a fitness benefit, as resources are reallocated to defense only in times of stress (Pappas et al., 2017). Mutti et al. (2006) used RNAi techniques to silence the expression of salivary elicitors, which resulted in *A. pisum* mortality. This technique was also developed in thrips (Badillo-Vargas et al., 2015). The use of UV-B light has been reported to increase constitutive and inducible plant defenses against a wide range of pests (Zavala et al., 2015).

3.8 CONCLUSION

Plant-mediated interactions between plants, herbivores, and natural enemies play a key role in balancing the ecosystem. Plant responses to herbivores exhibit a considerable degree of specificity. Herbivores trigger the plant defense system. The first herbivore-induced change in plant phenotype affects the interactions with subsequently arriving herbivores, which then further affect the phenotype in an interactive way. Based on resistance, morphological and physiological defense mechanisms repel or attract insects directly or indirectly. The presence of different leaf structures like trichomes, waxy coatings, and thickness repels herbivores, while leaf shape acts as antixenosis plant resistance. Physiological changes caused by herbivore infestation trigger the plant defense system. Upon attack by herbivores, SA and JA-mediated plant VOCs, different phytochemicals, attractants, or repellents are released that help the plant defend against herbivore attacks and enhance the activities of natural enemies that reduce the pest load. Herbivores also act as vectors that activate plant defense, as viruses present in the herbivore saliva or hemolymph, when reaching the plant system, trigger the internal defense role of the plant. Continuous application of insecticides on these insect species leads to resistance, and this chapter will help enhance the knowledge to understand more about plant defense, which will play a key role in future technology to increase immunity in plants against herbivores. By using various strategies against herbivores, external applications like biostimulants and other VOCs not only increase nutrition to plants but also boost the plant defensive mechanism. Therefore, identifying different HIPVs will enhance the IPM strategy to tackle pests in a more eco-friendly manner, and this interaction will likely help to better understand the ecology of the more complex relationships among plants, pest organisms, and natural enemies. However, many questions arise, such as: Can these VOCs be synthesized by humans? How do susceptible plants show reactions when plant-mediated interactions take place? Do all hosts show the same induced plant defense system, or does it vary from plant to plant?

KEYWORDS

- **bio-stimulants**
- **herbivores**
- **insect resistance**
- **plant defense**
- **plant pests**
- **vectors**

REFERENCES

Abbasi, A., Sufyan, M., Arif, M. J., & Sahi, S. T. (2020). Effect of silicon on tritrophic interaction of cotton, *Gossypium hirsutum* (Linnaeus), *Bemisia tabaci* (Gennadius) (Homoptera: Aleyrodidae) and the predator, *Chrysoperla carnea* (Stephens) (Neuroptera: Chrysopidae). *Arthropod-Plant Interactions, 14*, 717–725.

Abe, H., Shimoda, T., Ohnishi, J., Kugimiya, S., Narusaka, M., Seo, S., Narusaka, Y., Tsuda, S., & Kobayashi, M. (2009). Jasmonate-dependent plant defense restricts thrips performance and preference. *BMC Plant Biology, 9*, 97.

Acevedo, F. E., Rivera-Vega, L. J., Chung, S. H., Ray, S., & Felton, G. W. (2015). Cues from chewing insects–The intersection of DAMPs, HAMPs, MAMPs and effectors. *Current Opinion in Plant Biology, 26*, 80–86. https://doi.org/10.1016/j.pbi.2015.05.029.

Agrawal, A. A. (2011). Current trends in the evolutionary ecology of plant defence. *Functional Ecology, 25*, 420–432. https://doi.org/10.1111/j.1365-2435.2010.01796.x.

Agrawal, K., Karban, R., & Colfer, R. G. (2002). How leaf domatia and induced plant resistance affect herbivores, natural enemies and plant performance. *Oikos, 89*, 70–80.

Alsabte, A., Ahmed, Q. H., Kayahan, A., & Karaca, I. (2022). Effects of volatile organic compounds (VOCs) emitted by citrus infested with *Aonidiella aurantii* on the predator *Rhyzobius lophanthae* attraction. *Phytoparasitica, 50*, 645–653.

Amin, M. R., Sharmin, M. A., Miah, M. R. U., Akanda, A. M., Suh, S. J., Kwon, Y. K., & Kwon, O. (2020). Aphid population abundance and pestiferous effect on various bean plant species. *Entomological Research, 50*, 257–266.

Anokhe, A., Mandal, B., & Subramanian, S. (2015). Characterization of ToLCNDV transmission in Asia-I and Asia-II-1 genotypes of *Bemisia tabaci*. *Virocon-15, Shillong, Meghalaya.*

Arimura, G. I. (2020). Making sense of the way plants sense herbivores. *Trends in Plant Science, 26*(3), 288–298.

Ataide, L. M., Dias, C. R., Schimmel, B. C., van Erp, T., Pallini, A., & Kant, M. R. (2019). Food decisions of an omnivorous thrips are independent from the indirect effects of jasmonate-inducible plant defences on prey quality. *Scientific Reports, 9*, 1–12.

Bac-Molenaar, J. A., Mol, S., Verlaan, M. G., Van Elven, J., Kim, H. K., Klinkhamer, P. G., Leiss, K. A., & Vrieling, K. (2019). Trichome-independent resistance against western flower thrips in tomato. *Plant Cell Physiology, 60*, 1011–1024.

Badillo-Vargas, I. E., Rotenberg, D., Schneweis, B. A., & Whitfield, A. E. (2015). RNA interference tools for the western flower thrips, *Frankliniella occidentalis*. *Journal of Insect Physiology, 76*, 36–46.

Bahar, T., Qureshi, A., Qurashi, F., Abid, M., Zahra, M., & Saleemhaider, M. (2021). Changes in phyto-chemical status upon viral infections in plant: A critical review. *Phyton-International Journal of Experimental Botany, 90*(1), 75–85.

Baumgärtner, J., Delucchi, V., von Arx, R., & Rubli, D. (1986). Whitefly (*Bemisia tabaci* Genn., Stern.: Aleyrodidae) infestation patterns as influenced by cotton, weather, and *Heliothis*: Hypothesis testing by using simulation models. *Agriculture, Ecosystems & Environment, 17*(1, 2), 49–59.

Belete, T. (2018). Defense mechanisms of plants to insect pests: From morphological to biochemical approach. *Trends in Technical & Scientific Research, 2*(2), 30–38.

Birkett, M. A., Chamberlain, K., Guerrieri, E., Pickett, J. A., Wadhams, L. J., & Yasuda, T. (2003). Volatiles from whitefly-infested plants elicit a host-locating response in the parasitoid, *Encarsia formosa*. *Journal of Chemical Ecology, 29*, 1589–1600.

Blackman, R. L., & Eastop, V. F. (2008). *Aphids on the World's Herbaceous Plants and Shrubs*. Wiley Publications, Natural History Museum.

Bleeker, P. M., Diergaarde, P. J., Ament, K., Guerra, J., Weidner, M., Schütz, S., de Both, M. T. J., Haring, M. A., & Schuurink, R. C. (2009). The role of specific tomato volatiles in tomato-whitefly interaction. *Plant Physiology, 151*, 925–935.

Bouagga, S., Urbaneja, A., Rambla, J. L., Flors, V., Granell, A., Jaques, J. A., et al. (2018). Zoophytophagous mirids provide pest control by inducing direct defences, antixenosis, and attraction to parasitoids in sweet pepper plants. *Pest Management Science, 74*, 1286–1296. https://doi.org/10.1002/ps.4838.

Bouagga, S., Urbaneja, A., Rambla, J. L., Granell, A., & Pérez-Hedo, M. (2017). *Orius laevigatus* strengthens its role as a biological control agent by inducing plant defenses. *Journal of Pest Science, 91*, 55–64. https://doi.org/10.1007/s10340-017-0886-4.

Bové, J. M., & Garnier, M. (2003). Phloem- and xylem-restricted plant pathogenic bacteria. *Plant Science, 164*, 423–438. https://doi.org/10.1016/S0168-9452(03)00032-3.

Brown, J. K. (1994). Current status of *Bemisia tabaci* as a plant pest and virus vector in agroecosystems worldwide. *FAO Plant Protection Bulletin, 42*, 3–32.

CABI. (2017). *State of the World's Plants: Plant Health*. Retrieved from: https://stateoftheworldsplants.org/2017/plant-health.html (accessed on 3 April 2025).

Canales, I. V. D. M., Pahua, T. V., Ponce, O. V., Romero, R. R., Vargas, J. M. M., Ortiz, J. J. O., Winkler, R., Frierc, J. P. P. D., & Hernándeza, C. V. V. S. (2022). Systemic whitefly-induced metabolic responses in newly developed distal leaves of husk tomato plants (*Physalis philadelphica*) impairs whitefly development. *Pest Management Science, 79*(1), 368–380. https://doi.org/10.1002/ps.7206.

Chamarathi, S. K., Sharma, H. C., Vijay, P. M., & Natasu, M. L. (2011). Leaf surface chemistry of sorghum seedling influencing expression of resistance to sorghum shoot fly, *Atherigona soccata*. *Journal of Plant Biochemistry and Biotechnology, 20*, 211–216.

Chandrashekar, K., & Shashank, P. R. (2017). Indian contribution to whitefly (*Bemisia tabaci*) research. In B. Mandal, G. Rao, V. Baranwal, & R. Jain (Eds.),

A Century of Plant Virology in India (pp. 1–12). Springer, Singapore. https://doi.org/10.1007/978-981-10-5672-7_25.

Chen, C. Y., & Mao, Y. B. (2020). Research advances in plant-insect molecular interaction. *F1000 Research.* https://doi.org/10.12688/f1000research.21502.1.

Chen, G., Kim, H. K., Klinkhamer, P. G., & Escobar-Bravo, R. (2020). Site-dependent induction of jasmonic acid-associated chemical defenses against western flower thrips in *Chrysanthemum*. *Planta, 251*, 8.

Chen, G., Klinkhamer, P. G., Escobar-Bravo, R., & Leiss, K. A. (2018). Type VI glandular trichome density and their derived volatiles are differently induced by jasmonic acid in developing and fully developed tomato leaves: Implications for thrips resistance. *Plant Science, 276*, 87–98.

Chowda-Reddy, R. V., Kiran Kumar, M., Seal, S. E., Muniyappa, V., Valand, G. B., Govindappa, M. R., & Colvin, J. (2012). *Bemisia tabaci* phylogenetic groups in India and the relative transmission efficacy of tomato leaf curl Bangalore virus by an indigenous and an exotic population. *Journal of Integrative Agriculture, 11*, 235–248.

Croijmans, L., Valstar, R. T., Schuur, L., Jacobs, I., van Apeldoorn, D. F., & Poelman, E. H. (2022). Intraspecific plant variation and nonhost herbivores affect parasitoid host location behaviour. *Animal Behavior, 194*, 169–184. https://doi.org/10.1016/j.anbehav.2022.09.021.

De Lange, E. S., Laplanche, D., Guo, H., Xu, W., Vlimant, M., Erb, M., Ton, J., & Turlings, T. C. J. (2020). *Spodoptera frugiperda* caterpillars suppress herbivore-induced volatile emissions in maize. *Journal of Chemical Ecology, 46*, 344–360.

Debnath, M., & Chakrabarti, S. (2020). Diversity, bioecology and biosystematics of aphids (*Hemiptera: Aphididae*) in Darjeeling Himalaya. *International Journal of Experimental Research and Review, 21*, 10–18.

Denancé, N., Sánchez-Vallet, A., Goffner, D., & Molina, A. (2013). Disease resistance or growth: The role of plant hormones in balancing immune responses and fitness costs. *Frontiers in Plant Science, 4*, 155. https://doi.org/10.3389/fpls.2013.00155.

Denno, R. F., McClure, M. S., & Ott, J. R. (1995). Interspecific interactions in phytophagous insects: Competition reexamined and resurrected. *Annual Review of Entomology, 40*, 297–331.

Dicke, M. (2015). Herbivore-induced plant volatiles as a rich source of information for arthropod predators: Fundamental and applied aspects. *Journal of the Indian Institute of Science, 95*, 35–42.

Dicke, M., Baarlen, P. V., Wessels, R., & Dijkman, H. (1993). Herbivory induces systemic production of plant volatiles that attract predators of the herbivore: Extraction of endogenous elicitor. *Journal of Chemical Ecology, 19*, 581–599.

Dicke, M., Van Beek, T. A., Posthumus, M. A., Ben Dom, N., Van Bokhoven, H., & De Groot, A. (1990). Isolation and identification of volatile kairomone that affects acarine predator-prey interactions: Involvement of host plant in its production. *Journal of Chemical Ecology, 16*, 381–396.

Dicke, M., van Loon, J. J. A., & Soler, R. (2009). Chemical complexity of volatiles from plants induced by multiple attack. *Nature Chemical Biology, 5*, 317–324.

Dillon, F. M., Chludil, H. D., Reichelt, M., Mithöfer, A., & Zavala, J. A. (2018). Field-grown soybean induces jasmonates and defensive compounds in response to thrips feeding and solar UV-B radiation. *Environmental and Experimental Botany, 156*, 1–7.

Döring, T. F. (2014). How aphids find their host plants, and how they don't. *Annals of Applied Biology, 165*, 3–26.

Dorta, S. D. O., Balbinotte, J., Monnerat, R., Lopes, J. R. S., da Cunha, T., & Zanardi, O. Z. (2020). Selection of *Bacillus thuringiensis* strains in citrus and their pathogenicity to *Diaphorina citri* (*Hemiptera: Liviidae*) nymphs. *Insect Science, 27*, 519–530. https://doi.org/10.1111/1744-7917.12654.

Du, Y. J., Poppy, G. M., Powell, W., & Wadhams, L. J. (1997). Chemically mediated associative learning in the host foraging behavior of the aphid parasitoid *Aphidius ervi* (*Hymenoptera: Braconidae*). *Journal of Insect Behavior, 10*, 509–522.

Ellili, A., Rabier, J., Prudent, P., Salducci, M. D., Heckenroth, A., Lachaâl, M., & Laffont-Schwob, I. (2017). Decision-making criteria for plant-species selection for Phyto stabilization: Issues of biodiversity and functionality. *Journal of Environmental Management, 201*, 215–226. https://doi.org/10.1016/j.jenvman.2017.06.041.

El-Wakeil, N. E., Volkmar, C., & Sallam, A. A. (2010). Jasmonic acid induces resistance to economically important insect pests in winter wheat. *Pest Management Science, 66*, 549–554.

Emden, H. (2002). Mechanisms of resistance: Antibiosis, antixenosis, tolerance, nutrition. In D. Pimentel (Ed.), *Encyclopedia of Pest Management*. Marcel Dekker, Inc.

Ertunç, F. (2020). Physiology of virus-infected plants. In *Applied Plant Virology: Advances, Detection, and Antiviral Strategies* (pp. 199–205).

Escobar-Bravo, R., Nederpel, C., Naranjo, S., Kim, H. K., Rodríguez-López, M. J., Chen, G., Glauser, G., Leiss, K. A., & Klinkhamer, P. G. (2019). Ultraviolet radiation modulates both constitutive and inducible plant defenses against thrips but is dose and plant genotype dependent. *Journal of Pest Science*, 1–13.

Faoro, F., & Gozzo, F. (2015). Is modulating virus virulence by induced systemic resistance realistic? *Plant Science, 234*, 1–13. https://doi.org/10.1016/j.plantsci.2015.01.011.

Fingu-Mabola, J. C., & Francis, F. (2021). Aphid–plant–phytovirus pathosystems: Influencing factors from vector behavior to virus spread. *Agriculture, 11*, 502.

Francis, F., Lognay, G., & Haubruge, E. (2004). Olfactory responses to aphid and host plant volatile releases: (E)-β-farnesene, an effective kairomone for the predator *Adalia bipunctata*. *Journal of Chemical Ecology, 30*, 741–755.

Fu, Z. Q., & Dong, X. (2013). Systemic acquired resistance: Turning local infection into global defense. *Annual Review of Plant Biology, 64*, 839–863. https://doi.org/10.1146/annurev-arplant-042811-105606.

Gadhave, K. R., Dutta, B., Coolong, T., & Srinivasan, R. (2019). A non-persistent aphid-transmitted potyvirus differentially alters the vector and non-vector biology through host plant quality manipulation. *Scientific Reports, 9*, 2503.

Gao, Q. M., Zhu, S., Kachroo, P., & Kachroo, A. (2015). Signal regulators of systemic acquired resistance. *Frontiers in Plant Science, 6*. http://journal.frontiersin.org/article/10.3389/fpls.2015.00228/abstract (accessed on 3 April 2025).

Griese, E., Caarls, L., Bassetti, N., Mohammadin, S., Verbaarschot, P., Bukovinszkine'Kiss, G., Poelman, E. H., Gols, R., Schranz, M. E., & Fatouros, N. E. (2021). Insect egg-killing: A new front on the evolutionary arms race between brassicaceous plants and pierid butterflies. *New Phytologist, 230*, 341–353. https://doi.org/10.1111/nph.17145.

Guerreiro, A., Figueiredo, J., Sousa Silva, M., & Figueiredo, A. (2016). Linking jasmonic acid to grapevine resistance against the biotrophic oomycete *Plasmopara viticola. Frontiers in Plant Science, 7.* http://journal.frontiersin.org/article/10.3389/fpls.2016.00565/abstract (accessed on 3 April 2025).

Gullan, P. J., & Cranston, P. S. (2010). *The Insects: An Outline of Entomology* (p. 528). Wiley.

Hao, P., Liu, C., Wang, Y., Chen, R., Tang, M., Du, B., Zhu, L., & He, G. (2008). Herbivore-induced callose deposition on the sieve plates of rice: An important mechanism for host resistance. *Plant Physiology, 146*, 1810–1820.

Howe, G. A., & Jander, G. (2008). Plant immunity to insect herbivores. *Annual Review of Plant Biology, 59*, 41–66.

Hu, X., Su, S., Liu, Q., Jiao, Y., Peng, Y., Li, Y., & Turlings, T. C. J. (2020). Caterpillar-induced rice volatiles provide enemy-free space for the offspring of the brown planthopper. *eLife, 9*, e55421.

Irwin, M. E., Kampmeier, G. E., & Weisser, W. W. (2007). Aphid movement: Process and consequences. In H. F. van Emden & R. Harrington (Eds.), *Aphids as Crop Pests* (pp. 153–186). CAB International.

Islam, W., Qasim, M., Noman, A., Adnan, M., Tayyab, M., Farooq, T. H., Wei, H., & Wang, L. (2018). Plant microRNAs: Front line players against invading pathogens. *Microbial Pathogenesis, 118*, 9–17. https://doi.org/10.1016/j.micpath.2018.03.008.

Jones, A. C., Gary, W., & Tumlinson, F. J. H. (2022). The dual function of elicitors and effectors from insects: Reviewing the 'arms race' against plant defenses. *Plant Molecular Biology, 109*, 427–445. https://doi.org/10.1007/s11103-021-01203-2.

Kaloshian, I., & Walling, L. L. (2005). Hemipterans as plant pathogens. *Annual Review of Phytopathology, 43*, 491–521.

Kaplan, I., Halitschke, R., Kessler, A., Sardanelli, S., & Denno, R. F. (2008). Constitutive and induced defenses to herbivory in above- and belowground plant tissues. *Ecology, 89*, 392–406.

Karban, R., & Myers, J. H. (1989). Induced plant responses to herbivory. *Annual Review of Ecology and Systematics, 20*, 331–348.

Karban, R., & Orrock, J. L. (2018). A judgment and decision-making model for plant behavior. *Ecology, 99*, 1909–1919.

Kawale, S. S., & Mahajan, U. V. (2021). Thrips as a vector of plant diseases. *Just Agriculture, 1*(10).

Kindt, F., Joosten, N. N., & Tjallingii, W. F. (2006). Electrical penetration graphs of thrips revised: Combining DC- and AC-EPG signals. *Journal of Insect Physiology, 52*, 1–10.

Koch, K. G., Chapman, K., Louis, J., Heng-Moss, T., & Sarath, G. (2016). Plant tolerance: A unique approach to control hemipteran pests. *Frontiers in Plant Science, 7*, 1363. https://doi.org/10.3389/fpls.2016.01363.

Koch, K. G., Palmer, N. A., Donze-Reiner, T., Scully, E. D., Seravalli, J., Amundsen, K., Twigg, P., Louis, J., Bradshaw, J. D., Heng-Moss, T. M., & Sarath, G. (2020). Aphid-responsive defense networks in hybrid switchgrass. *Frontiers in Plant Science, 11*, 1145. https://doi.org/10.3389/fpls.2020.01145.

Köllner, T. G., Held, M., Lenk, C., Hiltpold, I., Turlings, T. C., Gershenzon, J., & Degenhardt, J. (2008). A maize (*E*)-β-caryophyllene synthase implicated in indirect defense responses against herbivores is not expressed in most American maize varieties. *The Plant Cell, 20*, 482–494.

Kumar, V., Kaur, A., & Jindal, V. (2023). Management of major insect pests of Bt cotton: Advances in agricultural entomology. In K. Ghoneim (Ed.), *Advances in Entomology* (Vol. 24, pp. 1–35). AkiNik Publications.

Leach, A., Fuchs, M., Harding, R., & Nault, B. A. (2019). *Iris Yellow Spot Virus* prolongs the adult lifespan of its primary vector, onion thrips (*Thrips tabaci*) (Thysanoptera: Thripidae). *Journal of Insect Science, 19*(8).

Lewis, W. J., & Takasu, K. (1990). Use of learned odors by a parasitic wasp in accordance with host and food needs. *Nature, 348*, 635–636.

Lewis, W. J., & Tumlinson, J. H. (1988). Host detection by chemically mediated associative learning in a parasitic wasp. *Nature, 331*, 257–259.

Li, R., Wang, M., Wang, Y., Schuman, M. C., Weinhold, A., Schäfer, M., Jiménez-Alemán, G. H., Barthel, A., & Baldwin, I. T. (2017a). Flower-specific jasmonate signaling regulates constitutive floral defenses in wild tobacco. *Proceedings of the National Academy of Sciences, 114*, E7205–E7214.

Li, Z., Han, X., Song, X., Zhang, Y., Jiang, J., Han, Q., Liu, M., Qiao, G., & Zhuo, R. (2017b). Overexpressing the *Sedum alfredii* Cu/Zn superoxide dismutase increased resistance to oxidative stress in transgenic *Arabidopsis*. *Frontiers in Plant Science, 8*, 1010.

Liu, P., Sun, F., Gao, R., & Dong, H. (2012). RAP 2.6L overexpression delays waterlogging-induced premature senescence by increasing stomatal closure more than antioxidant enzyme activity. *Plant Molecular Biology, 79*, 609–622. https://doi.org/10.1007/s11103-012-9936-8.

Liu, Q., Ted, C. J., Turlings, & Li, Y. (2023). Can herbivores sharing the same host plant be mutualists? *Trends in Ecology and Evolution, 38*(6), 509–511.

Liu, S. S., Colvin, J., & De Barro, P. J. (2012). Species concepts as applied to the whitefly *Bemisia tabaci* systematics: How many species are there? *Journal of Integrative Agriculture, 11*(2), 176–186.

Liu, X., Zhou, H., Zhao, J., Hua, H., & He, Y. (2016). Identification of the secreted watery saliva proteins of the rice brown planthopper, *Nilaparvata lugens* (Stål) by transcriptome and shotgun LC-MS/MS approach. *Journal of Insect Physiology, 89*, 60–69. https://doi.org/10.1016/j.jinsphys.2016.04.002.

Lough, T. J., & Lucas, W. J. (2006). Integrative plant biology: Role of phloem long-distance macromolecular trafficking. *Annual Review of Plant Biology, 57*, 203–232. https://doi.org/10.1146/annurev.arplant.56.032604.144145.

Luan, J. B., Yao, D. M., Zhang, T., Walling, L. L., & Yang, M. (2012). Suppression of terpenoid synthesis in plants by a virus promotes its mutualism with vectors. *Ecology Letters, 16*(2), 390–398. https://doi.org/10.1111/ele.12055.

Lucas, W. J. (2006). Plant viral movement proteins: Agents for cell-to-cell trafficking of viral genomes. *Virology, 344*, 169–184. https://doi.org/10.1016/j.virol.2005.09.026.

Luna, E., Pastor, V., Robert, J., Flors, V., Mauch-Mani, B., & Ton, J. (2011). Callose deposition: A multifaceted plant defense response. *Molecular Plant-Microbe Interactions, 24*, 183–193. https://doi.org/10.1094/MPMI-07-10-0149.

Macel, M., Visschers, I. G., Peters, J. L., Kappers, I. F., de Vos, R. C., & van Dam, N. M. (2019). Metabolomics of thrips resistance in pepper (*Capsicum* spp.) reveals monomer and dimer acyclic diterpene glycosides as potential chemical defenses. *Journal of Chemical Ecology, 45*, 490–501.

Maharijaya, A., Vosman, B., Pelgrom, K., Wahyuni, Y., de Vos, R. C., & Voorrips, R. E. (2019). Genetic variation in phytochemicals in leaves of pepper (*Capsicum*) in relation to thrips resistance. *Arthropod-Plant Interactions, 13*, 1–9.

Mauck, K. E., De Moraes, C. M., & Mescher, M. C. (2014). Biochemical and physiological mechanisms underlying effects of *Cucumber mosaic virus* on host-plant traits that mediate transmission by aphid vectors. *Plant, Cell & Environment, 37*, 1427–1439.

Mayer, R. T., Inbar, M., McKenzie, C. L., Shatters, R., & Borowicz, V. (2002). Multitrophic interactions of the silverleaf whitefly, host plants, competing herbivores, and phytopathogens. *Archives of Insect Biochemistry and Physiology, 51*, 151–169.

McNutt, D. W., & Underwood, N. (2016). Variation in plant-mediated intra- and interspecific interactions among insect herbivores: Effects of host genotype. *Ecosphere, 7*(10), e01520. https://doi.org/10.1002/ecs2.1520.

Miles, P. W. (1972). The saliva of Hemiptera. *Advances in Insect Physiology, 9*, 183–255.

Milonas, P. G., Eirini, A., Aikaterini, P., Georgios, P., Fragkopoulos, G. N., Kektsidou, O., Vassilakos, N., & Kapranas, A. (2023). Plant viruses induce plant volatiles that are detected by aphid parasitoids. *Scientific Reports, 3*, 8721.

Miyazaki, J., Stiller, W. N., & Wilson, L. J. (2013). Identification of host plant resistance to silverleaf whitefly in cotton: Implications for breeding. *Field Crops Research, 154*, 145–155.

Mochizuki, M., & Yano, E. (2007). Olfactory response of the anthocorid predatory bug *Orius sauteri* to thrips-infested eggplants. *Entomologia Experimentalis et Applicata, 123*(1), 57–62.

Moreira, X., & Abdala-Roberts, L. (2019). Specificity and context-dependency of plant–plant communication in response to insect herbivory. *Current Opinion in Insect Science, 32*, 15–21.

Mote, U. N., & Shahane, A. K. (1994). Biophysical and biochemical characters of sorghum varieties contributing resistance to delphacid, aphid, and leaf sugary exudations. *Indian Journal of Entomology, 56*, 113–122.

Mouden, S., & Leiss, K. A. (2021). Host plant resistance to thrips (Thysanoptera: Thripidae)–Current state of art and future research avenues. *Current Opinion in Plant Biology, 45*, 28–34.

Mousa, K. M., Elsharkawy, M. M., Khodeir, I. A., El-Dakhakhni, T. N., & Youssef, A. E. (2014). Growth perturbation, abnormalities, and mortality of oriental armyworm *Mythimna separata* (Walker) (Lepidoptera: Noctuidae) caused by silica nanoparticles and *Bacillus thuringiensis* toxin. *Egyptian Journal of Biological Pest Control, 24*, 283–287.

Mutti, N. S., Park, Y., Reese, J. C., & Reeck, G. R. (2006). RNAi knockdown of a salivary transcript leading to lethality in the pea aphid, *Acyrthosiphon pisum*. *Journal of Insect Science, 6*, 1–7.

Muvea, A. M., Subramanian, S., Maniania, N. K., Poehling, H. M., Ekesi, S., & Meyhöfer, R. (2018). Endophytic colonization of onions induces resistance against viruliferous thrips and virus replication. *Frontiers in Plant Science, 9*, 1785.

Mwando, N. L., Tamiru, A., Nyasani, J. O., Obonyo, M. A., Caulfield, J. C., Bruce, T. J., & Subramanian, S. (2018). *Maize chlorotic mottle virus* induces changes in host plant volatiles that attract vector thrips species. *Journal of Chemical Ecology, 44*, 681–689.

Nachappa, P., Challacombe, J. F., Margolies, D. C., Nechols, J. R., Whitfield, A. E., & Rotenberg, D. (2020). *Tomato spotted wilt virus* benefits its thrips vector by modulating metabolic and plant defense pathways in tomato. *Frontiers in Plant Science, 11*, 564–575.

Naseem, M., Kaltdorf, M., & Dandekar, T. (2015). The nexus between growth and defense signaling: Auxin and cytokinin modulate plant immune response pathways. *Journal of Experimental Botany, 66*, 4885–4896. https://doi.org/10.1093/jxb/erv297.

Ngadze, E., Icishahayo, D., Coutinho, T. A., & Van der Waals, J. E. (2012). Role of polyphenol oxidase, peroxidase, phenylalanine ammonia-lyase, chlorogenic acid, and total soluble phenols in resistance of potatoes to soft rot. *Plant Disease, 96*(2), 186–192. https://doi.org/10.1094/PDIS-02-11-0149.

Nietupski, M., Ludwiczak, E., Olszewski, J., Gabryś, B., & Kordan, B. (2022). Effect of aphid foraging on the intensity of photosynthesis and transpiration of selected crop plants in its early stages of growing. *Agronomy, 12*, 2370.

Ning, L. V., Yang, Q. Y., Li, C. C., Zhang, T. W., Ali, S., Liu, C. Z., Abid, I., & Gawwad, M. R. A. (2023). Effects of different host plants on population fitness of pea aphid (*Acyrthosiphon pisum*). *Journal of King Saud University–Science, 35*, 102764.

Ogada, P. A., Maiss, E., & Poehling, H. M. (2013). Influence of *Tomato spotted wilt virus* on performance and behavior of western flower thrips (*Frankliniella occidentalis*). *Journal of Applied Entomology, 137*, 488–498.

Ogawa, K., & Miura, T. (2014). Aphid polytheisms: Transgenerational developmental regulation through viviparity. *Frontiers in Physiology, 5*, 1.

Okada, K., Abe, H., & Arimura, G. (2015). Jasmonates induce both defense responses and communication in monocotyledonous and dicotyledonous plants. *Plant & Cell Physiology, 56*, 16–27.

Pappas, M. L., Broekgaarden, C., Broufas, G. D., Kant, M. R., Messelink, G. J., Steppuhn, A., Wäckers, F., & van Dam, N. M. (2017). Induced plant defences in biological control of arthropod pests: A double-edged sword. *Pest Management Science, 73*, 1780–1788.

Paré, P. W., & Tumlinson, J. H. (1999). Plant volatiles as a defense against insect herbivores. *Plant Physiology, 121*, 325–331.

Pereira, R. V., Filgueiras, C. C., Dória, J., Peñaflor, M. F. G. V., & Willett, D. S. (2021). The effects of biostimulants on induced plant defense. *Frontiers in Agronomy, 3*, 596–630. https://doi.org/10.3389/fagro.2021.630596.

Perez, I. B., & Brown, P. J. (2014). The role of ROS signaling in cross-tolerance: From model to crop. *Frontiers in Plant Science, 5*, 754. https://doi.org/10.3389/fpls.2014.00754.

Perez-Fons, L., Bohorquez-Chaux, A., Irigoyen, M. L., Garceau, D. C., & Morreel, K. (2019). A metabolomics characterization of natural variation in the resistance of cassava to whitefly. *BMC Plant Biology, 19*, 518.

Pérez-Hedo, M., Alonso-Valiente, M., Vacas, S., Gallego, C., Rambla, J. L., & Navarro Llopis, V. (2021). Eliciting tomato plant defenses by exposure to herbivore-induced plant volatiles. *Entomologia Generalis, 41*, 209–218. https://doi.org/10.1111/pce.14209.

Petrescu, A. S., Mondor, E. B., & Roitberg, B. D. (2001). Subversion of alarm communication: Do plants habituate aphids to their own alarm signals? *Canadian Journal of Zoology, 79*, 737–740.

Pieterse, C. M. J., Van der D., Zamioudis, C., Leon Reyes, A., & Van Wees, S. C. M. (2012). Hormonal modulation of plant immunity. *Annual Review of Cell and Developmental Biology, 28*, 489–521.

Poelman, E. H., Bourne, M. E., Croijmans, L., Cuny, M. A. C., Delamore, Z., Joachim, G., Kalisvaart, S. N., Kamps, B. J. J., Longuemare, M., Suijkerbuijk, H. A. C., & Zhang, N. X. (2023). Bringing fundamental insights of induced resistance to agricultural management of herbivore pests. *Journal of Chemical Ecology, 49*, 218–229.

Prodhan, M. Y., Munemasa, S., Nahar, M. N. E. N., Nakamura, Y., & Murata, Y. (2018). Guard cell salicylic acid signaling is integrated into abscisic acid signaling via the Ca^{2+}/CPK-dependent pathway. *Plant Physiology, 178*, 441–450.

Qian, L., He, S., Liu, X., Huang, Z., Chen, F., & Gui, F. (2018). Effect of elevated CO_2 on the interaction between invasive thrips, *Frankliniella occidentalis*, and its host kidney bean, *Phaseolus vulgaris*. *Pest Management Science, 74*, 2773–2782.

Qin, L., Mo, N., Zhang, Y., Muhammad, T., Zhao, G., Zhang, Y., & Liang, Y. (2017). *CaRDR1*, an RNA-dependent RNA polymerase, plays a positive role in pepper resistance against TMV. *Frontiers in Plant Science, 8*. http://journal.frontiersin.org/article/10.3389/fpls.2017.01068/full (accessed on 3 April 2025).

Rashid, M., Khan, A., Hossain, M. T., & Chung, Y. R. (2017). Induction of systemic resistance against aphids by endophytic *Bacillus velezensis* YC7010 via expressing phytoalexin deficient4 in *Arabidopsis*. *Frontiers in Plant Science, 8*, 211. https://doi.org/10.3389/fpls.2017.00211.

Rasmann, S., Bennett, A., Biere, A., Karley, A., & Guerrieri, E. (2017). Root symbionts: Powerful drivers of plant above-and belowground indirect defenses. *Insect Science, 24*, 947–960. https://doi.org/10.1111/1744-7917.12464.

Rasmann, S., Kollner, T. G., Degenhardt, J., Hiltpold, I., & Toepfer, S. (2005). Recruitment of entomopathogenic nematodes by insect-damaged maize roots. *Nature, 434*, 732–737.

Rdanuy, A., Albajes, R., & Turlings, T. C. (2016). Innate and learned prey searching behavior in a generalist predator. *Journal of Chemical Ecology, 42*, 497–507.

Ren, X., Wu, S., Xing, Z., Gao, Y., Cai, W., & Lei, Z. (2020). Abundances of thrips on plants in vegetative and flowering stages are related to plant volatiles. *Journal of Applied Entomology, 114*, 732–742.

Riahi, C., González-Rodríguez, J., Alonso-Valiente, M., Urbaneja, A., & Pérez-Hedo, M. (2022). Eliciting plant defenses through herbivore-induced plant volatiles' exposure in sweet peppers. *Frontiers in Ecology and Evolution, 9*, 776827. https://doi.org/10.3389/fevo.2021.776827.

Richards, O. W., & Davies, R. G. (1977). *Imms' General Textbook of Entomology* (Vols. 1, 2, 10th ed.). New York: Chapman & Hall.

Rodriguez, A., Angel, C. A., Lutz, L., Leisner, S. M., Nelson, R. S., & Schoelz, J. E. (2014a). Association of the P6 protein of *cauliflower mosaic virus* with plasmodesmata and plasmodesmal proteins. *Plant Physiology, 166*, 1345–1358. https://doi.org/10.1104/pp.114.249250.

Rodriguez, M. C., Conti, G., Zavallo, D., Manacorda, C. A., & Asurmendi, S. (2014b). *TMV-Cg* coat protein stabilizes DELLA proteins and in turn negatively modulates salicylic acid-mediated defense pathway during *Arabidopsis thaliana* viral infection. *BMC Plant Biology, 14*. https://doi.org/10.1186/s12870-014-0210-x.

Rodriguez-Saona, C., & de Lange, E. (2022). Advances in crop resistance for insect pest control. *Frontiers in Agronomy, 4*, 845961. https://doi.org/10.3389/fagro.2022.845961.

Rodriguez-Saona, C., Kaplan, I., Braasch, J., Chinnasamy, D., & Williams, L. (2011). Field responses of predaceous arthropods to methyl salicylate: A meta-analysis and case study in cranberries. *Biological Control, 59*, 294–303.

Saad, K. A., Mohammad Roff, M. N., Hallett, R. H., & Idris, A. B. (2015). Aphid-induced defences in chilli affect preferences of the whitefly, *Bemisia tabaci* (Hemiptera: Aleyrodidae). *Scientific Reports, 5*, 13697–13705.

Sabri, A., Vandermoten, S., Leroy, P. D., Haubruge, E., Hance, T., Thonart, P., De Pauw, E., & Francis, F. (2013). Proteomic investigation of aphid honeydew reveals an unexpected diversity of proteins. *PLoS ONE, 8*, e74656.

Sarde, S. J., Bouwmeester, K., Venegas-Molina, J., David, A., Boland, W., & Dicke, M. (2019). Involvement of sweet pepper *CaLOX2* in jasmonate-dependent induced defense against western flower thrips. *Journal of Integrative Plant Biology, 61*, 1085–1098.

Schoonhoven, L., van Loon, J. J., & Dicke, M. (2005). *Insect-Plant Biology* (2nd ed.). Oxford University Press.

Schuman, M. C., & Baldwin, I. T. (2016). The layers of plant responses to insect herbivores. *Annual Review of Entomology, 61*, 373–394.

Schwartzberg, E. G., & Tumlinson, J. H. (2014). Aphid honeydew alters plant defence responses. *Functional Ecology, 28*, 386–394.

Shalileh, S., Ogada, P. A., Moualeu, D. P., & Poehling, H. M. (2016). Manipulation of *Frankliniella occidentalis* (Thysanoptera: Thripidae) by tomato spotted wilt virus (*Tospovirus*) via the host plant nutrients to enhance its transmission and spread. *Environmental Entomology, 45*, 1235–1242.

Sharma, E., Anand, G., & Kapoor, R. (2017). Terpenoids in plant and arbuscular mycorrhiza-reinforced defence against herbivorous insects. *Annals of Botany, 119*, 791–801.

Shrestha, A., Srinivasan, R., Riley, D. G., & Culbreath, A. K. (2012). Direct and indirect effects of a thrips-transmitted *Tospovirus* on the preference and fitness of its vector, *Frankliniella fusca. Entomologia Experimentalis et Applicata, 145*(3), 260–271.

Siddique, Z., Akhtar, K. P., Hameed, A., Sarwar, N., Imran-Ul-Haq, & Khan, S. A. (2014). Biochemical alterations in leaves of resistant and susceptible cotton genotypes infected systemically by cotton leaf curl Burewala virus. *Journal of Plant Interactions, 9*(1), 702–711.

Singh, A., Dilkes, B., Sela, H., & Tzin, V. (2021). The effectiveness of physical and chemical defense responses of wild emmer wheat against aphids depends on leaf position and genotype. *Frontiers in Plant Science, 12*, 667820. https://doi.org/10.3389/fpls.2021.667820.

Singh, B., Kaur, A., Chandi, R. S., & Sharma, A. (2023). Virus–vector relationship of *Squash leaf curl China virus* and its transmission through whitefly vector, *Bemisia tabaci* (Gennadius) causing yellow vein mosaic disease of pumpkin. *Indian Phytopathology.* https://doi.org/10.1007/s42360-023-00647-4 (accessed on 3 April 2025).

Singh, H. P., Kaur, S., Batish, D. R., & Kohli, R. K. (2014). Ferulic acid impairs rhizogenesis and root growth, and alters associated biochemical changes in mung bean (*Vigna radiata*) hypocotyls. *Journal of Plant Interactions, 9*, 267–274.

Smith, B. C. (1969). Effects of silica on the survival of *Coleomegilla maculata lengi* (Coleoptera: Coccinellidae) and *Leptinotarsa decemlineata* (Coleoptera: Chrysomelidae). *The Canadian Entomologist, 101*, 460–462.

Smith, M. T., & Severson, R. F. (1992). Host recognition by the black margined aphid (Homoptera: Aphididae) on pecan. *Journal of Entomological Science, 27*, 93–112.

Snoeck, S., Guayazán-Palacios, N., & Steinbrenner, A. D. (2022). Molecular tug-of-war: Plant immune recognition of herbivory. *The Plant Cell, 34*, 1497–1513. https://doi.org/10.1093/plcell/koac009.

Spiller, N. J. (1990). An ultrastructural study of the stylet pathway of the brown planthopper *Nilaparvata lugens*. *Entomologia Experimentalis et Applicata, 54*, 191–193. https://doi.org/10.1111/j.1570-7458.1990.tb01329.x.

Strauss, S. Y., & Agrawal, A. A. (1999). The ecology and evolution of plant tolerance to herbivory. *Trends in Ecology & Evolution, 14*, 179–185. https://doi.org/10.1016/S0169-5347(98)01576-6.

Sulistyo, A., & Inayati, A. (2016). Mechanism of antixenosis, antibiosis, and tolerance of fourteen soybean genotypes in response to whiteflies (*Bemisia tabaci*). *Biodiversitas, 17*(2), 447–453.

Sytykiewicz, H., Chrzanowski, G., Czerniewicz, P., Sprawka, I., Lukasik, I., & Golawska, S. (2014). Expression profiling of selected glutathione transferase genes in *Zea mays* (L.) seedlings infested with cereal aphids. *PLoS ONE, 9*, e111863. https://doi.org/10.1371/journal.pone.0111863.

Taggar, G. K., Gill, R. S., Gupta, A. K., & Singh, S. (2014). Induced changes in the antioxidative compounds of *Vigna mungo* genotypes due to infestation by *Bemisia tabaci* (Gennadius). *Journal of Environmental Biology, 35*(6), 1037–1045.

Thabet, A. F., Boraei, H. A., Galal, O. A., El Samahy, M. F. M., Mousa, K. M., Zhang, Y. Z., Tuda, M., Helmy, E. A., Wen, J., & Nozaki, T. (2021). Silica nanoparticles as pesticides against insects of different feeding types and their non-target attraction of predators. *Scientific Reports, 11*, 144–184.

Thaler, J. S., Karban, R., Ullman, D. E., Boege, K., & Bostock, R. M. (2002). Crosstalk between jasmonate and salicylate plant defense pathways: Effects on several plant parasites. *Oecologia, 131*, 227–235.

Thomas, A., Kar, A., Rebijith, K. B., Asokan, R., & Ramamurthy, V. V. (2014). *Bemisia tabaci* (Hemiptera: Aleyrodidae) species complex from cotton cultivars: A comparative study of population density, morphology, and molecular variations. *Annals of the Entomological Society of America, 107*(2), 389–398.

Tian, M., Sasvari, Z., Gonzalez, P. A., Friso, G., Rowland, E., Liu, X. M., van Wijk, K. J., Nagy, P. D., & Klessig, D. F. (2015). Salicylic acid inhibits the replication of tomato bushy stunt virus by directly targeting a host component in the replication complex. *Molecular Plant-Microbe Interactions, 28*, 379–386. https://doi.org/10.1094/MPMI09-14-0259-R.

Tjallingii, W. F. (2006). Salivary secretions by aphids interacting with proteins of phloem wound responses. *Journal of Experimental Botany, 57*, 739–745.

Tungadi, T., Simon, C., Groen, S. C., Murphy, A. M., Pate, A. E., Iqbal, J., Bruce, J. A. T., Cunniffe, N. J., & Carr, J. P. (2017). Cucumber mosaic virus and its 2b protein alter emission of host volatile organic compounds but not aphid vector settling in tobacco. *Virology Journal, 14*, 91, 1–9.

Turlings, T. C. J., & Erb, M. (2018). Tritrophic interactions mediated by herbivore-induced plant volatiles: Mechanisms, ecological relevance, and application potential. *Annual Review of Entomology, 63*, 433–452.

Vaello, T., Pineda, A., & Marcos-García, M. (2019). Role of thrips omnivory and their aggregation pheromone on multitrophic interactions between sweet pepper plants, aphids, and hoverflies. *Frontiers in Ecology and Evolution, 6*, 240.

Valle, G. E., Lourenção, A. L., & Pinheiro, J. S. (2012). Adult attractiveness and oviposition preference of *Bemisia tabaci* biotype B in soybean genotypes with different trichome density. *Journal of Pest Science, 85*(4), 431–442.

Van Doorn, A., Vries, M., Kant, M. R., & Schuurink, R. C. (2015). Whiteflies glycosylate salicylic acid and secrete the conjugate via their honeydew. *Journal of Chemical Ecology, 41*, 52–58.

Vanitha, S. C., & Umesha, S. (2008). Variations in defense-related enzyme activities in tomato during the infection with bacterial wilt pathogen. *Journal of Plant Interactions, 3*, 245–253.

Venzon, M., Janssen, A., & Sabelis, M. W. (1999). Attraction of a generalist predator towards herbivore-infested plants. *Entomologia Experimentalis et Applicata, 93*, 305–314.

Voigt, C. A. (2014). Callose-mediated resistance to pathogenic intruders in plant defense-related papillae. *Frontiers in Plant Science, 5*. https://doi.org/10.3389/fpls.2014.00168.

Walling, L. L. (2008). Avoiding effective defenses: Strategies employed by phloem-feeding insects. *Plant Physiology, 146*, 859–866.

Wan, Y., Hussain, S., Merchant, A., Xu, B., Xie, W., Wen, X., Wang, S., Zhang, Y., Zhou, X., & Wu, Q. (2020). *Tomato spotted wilt orthotospovirus* influences the reproduction of its insect vector, western flower thrips, *Frankliniella occidentalis*, to facilitate transmission. *Pest Management Science, 76*, 2406–2414.

Wang, H., Shi, S., & Hua, W. (2023). Advances of herbivore-secreted elicitors and effectors in plant-insect interactions. *Frontiers in Plant Science, 14*, 1176048. https://doi.org/10.3389/fpls.2023.1176048.

Wang, X., Liu, Q., Meissle, M., Peng, Y., Wu, K., Romeis, J., & Li, Y. (2018). Bt rice could provide ecological resistance against non-target planthoppers. *Plant Biotechnology Journal, 16*, 1748–1755.

War, A. R., Paulraj, M. G., Ahmad, T., Buhroo, A. A., Hussain, B., Ignacimuthu, S., & Sharma, H. C. (2012). Mechanisms of plant defense against insect herbivores. *Plant Signaling & Behavior, 7*(10), 1306–1320.

Ward, S. A., Leather, S. R., Pickup, J., & Harrington, R. (1998). Mortality during dispersal and the cost of host-specificity in parasites: How many aphids find hosts? *Journal of Animal Ecology, 67*, 763–773.

Watson, M. A. (1967). Epidemiology of aphid-transmitted plant-virus diseases. *Outlook on Agriculture, 5*, 155–166.

Weintraub, P. G., Pivonia, S., & Steinberg, S. (2011). How many *Orius laevigatus* are needed for effective *Frankliniella occidentalis* management in sweet pepper? *Crop Protection, 30*, 1443–1448.

Will, T., Furch, A. C. U., & Zimmermann, M. R. (2013). How phloem-feeding insects face the challenge of phloem-located defenses. *Frontiers in Plant Science, 4*, 336.

Wu, J., & Baldwin, I. T. (2010). New insights into plant responses to the attack from insect herbivores. *Annual Review of Genetics, 44*, 1–24. https://doi.org/10.1146/annurev-genet-102209-163500.

Wu, X., Xu, S., Zhao, P., Zhang, X., Yao, X., Sun, Y., Fang, R., & Ye, J. (2019). The *Orthotospovirus* nonstructural protein NSs suppresses plant MYC-regulated jasmonate signaling leading to enhanced vector attraction and performance. *PLoS Pathogens, 15*, e1007897.

Yadav, M., & Yadav, F. D. (2019). Effect of TLCV on metabolic parameter and yield of tomato. *The Scientific Temper, 11*(1, 2), 5–8.

Yadav, S. R., Yan, D., Sevilem, I., & Helariutta, Y. (2014). Plasmodesmata-mediated intercellular signaling during plant growth and development. *Frontiers in Plant Science, 5*. https://doi.org/10.3389/fpls.2014.00044.

Yakhin, O. I., Lubyanov, A. A., Yakhin, I. A., & Brown, P. H. (2016). Biostimulants in plant science: A global perspective. *Frontiers in Plant Science, 7*, 2049.

Yang, F., Zhang, Q., Yao, Q., Chen, G., Tong, H., & Zhang, J. (2020). Direct and indirect plant defenses induced by (Z)-3-hexenol in tomato against whitefly attack. *Journal of Pest Science, 93*, 1243–1254. https://doi.org/10.1007/s10340-020-01234-6.

Yasmin, F., Amin, M. R., Afroz, M., Swapon, M. A. H., & Hossain, M. M. (2022). Morphological traits and phytochemical contents of brinjal germplasm affect the incidence and infestation of aphid. *Bangladesh Journal of Entomology, 30*(1 & 2), 51–63.

Yi, S. Y., Shirasu, K., Moon, J. S., Lee, S. G., & Kwon, S. Y. (2014). The activated SA and JA signaling pathways have an influence on flg22-triggered oxidative burst and callose deposition. *PLoS ONE, 9*, e88951. https://doi.org/10.1371/journal.pone.0088951.

Yu, D., Huang, P., Chen, Y., Lin, Y., Akutse, K. S., Lan, Y., & Wei, H. (2018). Effects of flower thrips (*Thysanoptera: Thripidae*) on nutritional quality of banana (*Zingiberales: Musaceae*) buds. *PLoS ONE, 13*, e0202199.

Yuan, H. M., Liu, W. C., & Lu, Y. T. (2017). CATALASE2 coordinates SA-mediated repression of both auxin accumulation and JA biosynthesis in plant defenses. *Cell Host & Microbe, 21*, 143–155. https://doi.org/10.1016/j.chom.2017.01.007.

Zafar, Z., Rasheed, F., Atif, R. M., Maqsood, M., & Gailing, O. (2021). Salicylic acid-induced morpho-physiological and biochemical changes triggered water deficit tolerance in *Syzygium cumini* L. saplings. *Forests, 12*, 491.

Zárate, S. I., Kempema, L. A., & Walling, L. L. (2007). Silverleaf whitefly induces salicylic acid defenses and suppresses effectual jasmonic acid defenses. *Plant Physiology, 143*, 866–867.

Zavala, J. A., Mazza, C. A., Dillon, F. M., Chludil, H. D., & Ballaré, C. L. (2015). Soybean resistance to stink bugs (*Nezara viridula* and *Piezodorus guildinii*) increases with exposure to solar UV-B radiation and correlates with isoflavonoid content in pods under field conditions. *Plant, Cell & Environment, 38*, 920–928.

Zhang, P. J., Li, W. D., Huang, F., Zhang, J. M., Xu, F. C., & Lu, Y. B. (2013a). Feeding by whiteflies suppresses downstream jasmonic acid signaling by eliciting salicylic acid signaling. *Journal of Chemical Ecology, 39*, 612–619.

Zhang, P. J., Xu, C. X., Zhang, J. M., Lu, Y. B., Wei, J. N., et al. (2013b). Phloem-feeding whiteflies can fool their host plants, but not their parasitoids. *Functional Ecology, 27*, 1304–1312.

Zhang, X., Wang, C., Zhang, Y., Sun, Y., & Mou, Z. (2012). The *Arabidopsis* mediator complex subunit16 positively regulates salicylate-mediated systemic acquired resistance and jasmonate/ethylene-induced defense pathways. *The Plant Cell, 24*, 4294–4309.

Zhao, J., Zhang, X., Hong, Y., & Liu, Y. (2016). Chloroplast in plant-virus interaction. *Frontiers in Microbiology, 7*, 1565.

Zheng, X., Chen, Y., Zhao, L., Chen, Y., Zheng, L., Zheng, K., Mu, Y., Zhao, X., Gao, Y., & Zhang, J. (2019). Tripartite interactions between jasmonic salicylic acid pathways, western flower thrips, and thrips-transmitted tomato zonate spot virus infection in *Capsicum annuum*. *Arthropod-Plant Interactions, 13*, 289–297.

Zhu, J., Zhu, K., Li, L., Li, Z., Qin, W., Park, Y., & He, Y. (2020). Proteomics of the honeydew from the brown planthopper and green rice leafhopper reveal they are rich in proteins from insects, rice plant, and bacteria. *Insects, 11*(9), 582. https://doi.org/10.3390/insects11090582.

CHAPTER 4

Prey and Predators: The Functional Ecology of Insectivorous Foraging in Agroecosystems

NADIA K. TOUKEM* and ABDULLAHI YUSUF

Department of Zoology and Entomology, University of Pretoria, Pretoria, South Africa

*Corresponding author

ABSTRACT

Predation is a key ecological interaction that regulate prey populations and community dynamics. Insects, like other organisms, occupy different levels within food webs, serving as prey, predators, or both. This chapter examines the intricacies of prey-predator dynamics within insectivorous ecosystems, focusing on fundamental components of insect predation, including the functional and numerical responses of predators. Considerations of predator efficiency encompass factors such as prey defenses, predator and prey traits, and environmental conditions. Furthermore, we present case studies that elucidate functional responses demonstrated by different predatory insects targeting various insect pests in agroecosystems. The chapter delves into the implications of these interactions for biological pest control and provides insights into the complex trophic relationships between prey and predators.

4.1 INTRODUCTION

Insects play important roles at different levels of the food chain, acting as herbivores, prey, predators, and even decomposers (Goulson, 2019; Yang

Insect-Plant Interactions, Volume 1: Influence of Biotic and Abiotic Factors.
Jamin Ali, Adil Tonğa, Adnan Shakeel, and Nasheeman Ashraf (Eds.)

DOI: 10.1201/9781779644282-4

and Gratton, 2014) (Figure 4.1). Their trophic position is influenced by factors such as habitat, developmental stage, and intraspecific variability, leading to overlapping roles within food chains (Coll and Guershon, 2002). Notably, certain insects are omnivores, feeding on either prey or plants (leaves, seeds, floral resources, and homopteran honeydew) across different life stages (Coll and Guershon, 2002). A classic example is found in hoverflies (e.g., *Episyrphus* spp.), which prey on soft-bodied insects like aphids during their larval stage and later forage on floral resources as adults (Khatua et al., 2020; Van Oystaeyen et al., 2022). Species like big-eyed bugs (*Geocoris* spp.), mirid bugs (e.g., *Dicyphus errans* Wolff), and ground-beetle carabids (e.g., *Poecilus cupreus* (L.)) exhibit life-long omnivory, consuming both prey (aphids, mealybugs, leafminers, and whiteflies) and plants throughout their development (Abraços-Duarte et al., 2021; Eubanks and Styrsky, 2005; Rischen et al., 2021; Saska et al., 2007; Voigt, 2019) (Figure 4.2).

Insect predation is an important trophic interaction that regulates population dynamics and influences ecosystem processes (Dixon, 2000; Eubanks and Finke, 2014; Holling, 1961). The conservation or introduction of parasitoids and predators in agroecosystems is advocated in pest management programs, given their potential to influence trophic cascades affecting plant biomass and yield.

Insect predators can be broadly categorized into three types based on their killing process: grazers, true predators, and parasitoids (Hawkins et al., 1997) (Figure 4.1B). Grazers, like mosquitoes, consume parts of multiple prey without immediate mortality. True predators, like katydids and mantids, use physical structures, such as mandibles, to kill their prey immediately. Parasitoids, meanwhile, lay eggs within (or outside, for ectoparasitoids) a single individual host, which serves as a resource for larval development until the host's demise. Most endoparasitoids are koinobionts, allowing the host to continue its development while feeding upon it, while idiobiont parasitoids prevent further development of the host (Brodeur and Boivin, 2004). Numerous endoparasitoids, such as *Aphidius* spp., *Diaeretiella rapae* (M'Intosh), *Diadegma* spp., *Tersilochus heterocerus* Thomson, and *Phradis* spp., contribute to pest regulation in agroecosystems (Gols et al., 2008; Kos et al., 2012; Rusch et al., 2011). Insects may act as predators even without consuming herbivorous prey, altering their physiology or behavior to reduce plant damage and enhance yield (Eubanks and Finke, 2014; Fill et al., 2012; Preisser et al., 2005).

For instance, *Aphidius colemanii* Viereck induces defensive dropping behavior in green peach aphids (*Myzus persicae* Sulzer), resulting in a 50% reduction in their population (Fill et al., 2012).

The negative indirect effect of predators on crop productivity through predation on beneficial insects like pollinators and natural enemies is increasingly recognized. Many cases of bee predation by praying mantises, crab spiders, and other predators have been documented (Figure 4.1A) (Jones and Dornhaus, 2011; Llandres et al., 2013). Studies have shown that pollinators perceive danger on flowers (e.g., the presence of dead bees or crab spiders, bee hemolymph, etc.) and modify their foraging behavior by reducing the duration of their visits or visiting poorer resource areas to reduce the risk of predation (Jones and Dornhaus, 2011; Wang et al., 2013a). Predator interference can lead to intraguild predation (or hyper-parasitism in the case of parasitoid-parasitoid interactions) and alter herbivore regulation (Gontijo et al., 2015; Snyder et al., 2022). For example, the larger intraguild predator Nabis sp. may consume smaller predatory bugs (*Geocoris* spp.) in potato fields, both of which prey on aphids and thrips (Snyder et al., 2022). Interaction can also occur between parasitoids and predators, with the immature parasitoid becoming the prey. An interesting example is the predation of the parasitoid *Aphelinus mali* (Haldeman) or parasitized aphids by a guild of generalist predators, including earwigs, syrphids, bugs, and spiders, all of which prey on the aphid *Eriosoma lanigerum* (Hausmann) (Gontijo et al., 2015). Factors like prey and floral resource diversity, along with refugia availability for vulnerable prey, can mitigate interference among natural enemies (Gontijo et al., 2015; Snyder et al., 2022).

Prey-predator interactions are prey density-dependent relationships that influence prey population density and facilitate predator reproduction (Holling, 1961, 1966; Jeschke et al., 2004). The functional response of a predator, delineating the rate of prey consumption by an average predator, determines the efficiency of a predator and guides decisions in biological control programs (Holling, 1961, 1966; Jeschke et al., 2004). Over many decades, scientists have proposed more than 40 models to assess predator functional response, considering a variety of ecological factors (predator foraging tactics, prey avoidance behavior, alternative prey, habitat structure, etc.) and environmental conditions (Abraços-Duarte et al., 2021; Jeschke et al., 2002; Shah and Khan, 2013). Holling reviewed these models and classified the functional response into three main types:

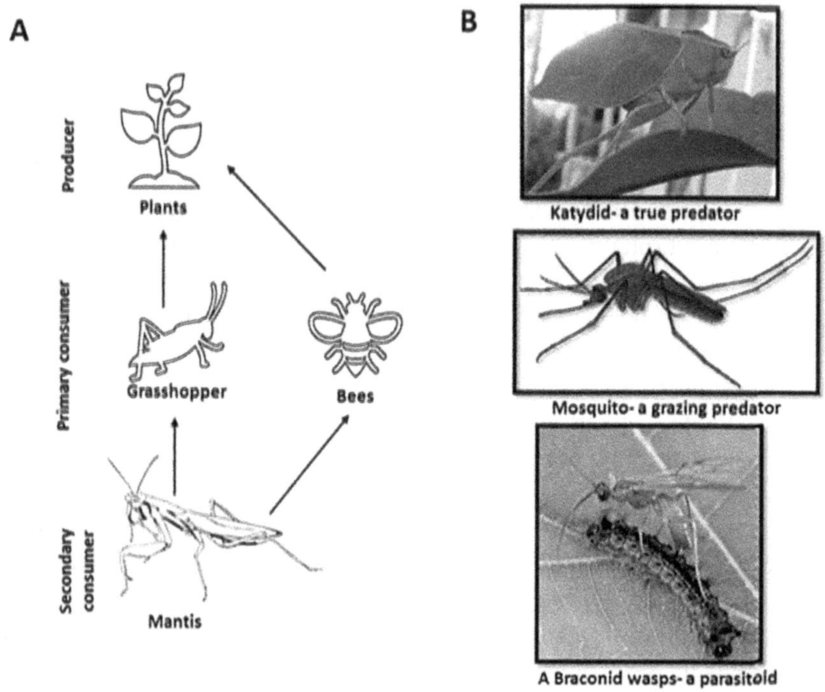

FIGURE 4.1 The trophic interactions in a food chain and main strategies of insect predation.

Note: (A) shows a trophic interaction with a mantid consuming a grasshopper, which is an antagonist, and a mutualist pollinator, which is a bee. (B) shows, from top to bottom, a common voracious predator of plant insects, which is a katydid, a common grazing predator of humans, which is a mosquito, and a wasp parasitoid laying eggs inside a caterpillar.

All photos are free to use.

type I, II, and III, all of which share a common feature in their models, the success rate, or the rate of successful search by the predator (1961, 1966). The type II functional response is commonly observed in a wide range of taxa, including invertebrates, and will therefore be the subject of most of the discussion in this chapter (Jeschke et al., 2004). However, recent studies have sparked interest in ratio-dependent functional responses, where the ratio of prey population size to predator population size is considered instead of the absolute number of either prey or predator (Abrams et al., 2000). Predator efficiency is further influenced by their numerical response, reflecting changes in reproductive rates with varying prey densities (Bayliss and Choquenot, 2002; Holling, 1966). Notably, high predation rates do not necessarily correlate with a stronger positive numerical response. One study found that *Dicyphus bolivari* Lindberg and *D. errans,* which prey on *Tuta absoluta* (Meyrick) eggs, show a type II functional response with a lower numerical response compared to *Macrolophus pygmaeus* (Rambur), which has a lower predation rate but a stronger numerical response (Ingegno et al., 2019). This chapter elucidates the dynamics of prey-predator interactions, factors shaping predator responses, and their efficacy in biological control. Furthermore, defensive strategies employed by insects to mitigate predation risks and examples of predator responses in agroecosystems are discussed.

4.2 MECHANISMS OF INSECT PREY DEFENSES AGAINST PREDATORS

Insect prey has developed a variety of defenses against predators to reduce the risk of predation. These defensive strategies can be categorized into chemical, morphological, and behavioral defenses (Sugiura, 2020). They are deployed before the attack (known as the primary defense) or in response to an attack, either during or after (the secondary defense) (Figure 4.2) (Ruxton et al., 2019; Sugiura, 2020).

4.3 CHEMICAL DEFENSES

Chemical defenses play a significant role in the defensive strategy of many insect species. For instance, formicine ants are known to secrete formic acid, not only to recruit workers but also to detoxify venoms from other ant species (Chen et al., 2013; Renyard et al., 2020). Studies have reported the toxicity of formic acid to the red imported fire ant, *Solenopsis invicta* Buren, an aggressive invasive species in the United States, China, and Australia (Chen et al., 2013; LeBrun et al., 2014; Wang and Henderson, 2016). In tropical rainforests, ants use chemical secretions to kill prey and defend their territories against territorially dominant arboreal ants. This was shown in cocoa orchards in Central and West Africa, where the native ant, *Crematogaster striatula* Emery, secretes a mixture of mono- or polyunsaturated fatty acids to paralyze prey, such as termites, and repel competing ant species like *Camponotus brutus* Forel, *Oecophylla longinoda* (Latreille), and *Tetramorium aculeatum* (Mayr) (Rifflet et al., 2011).

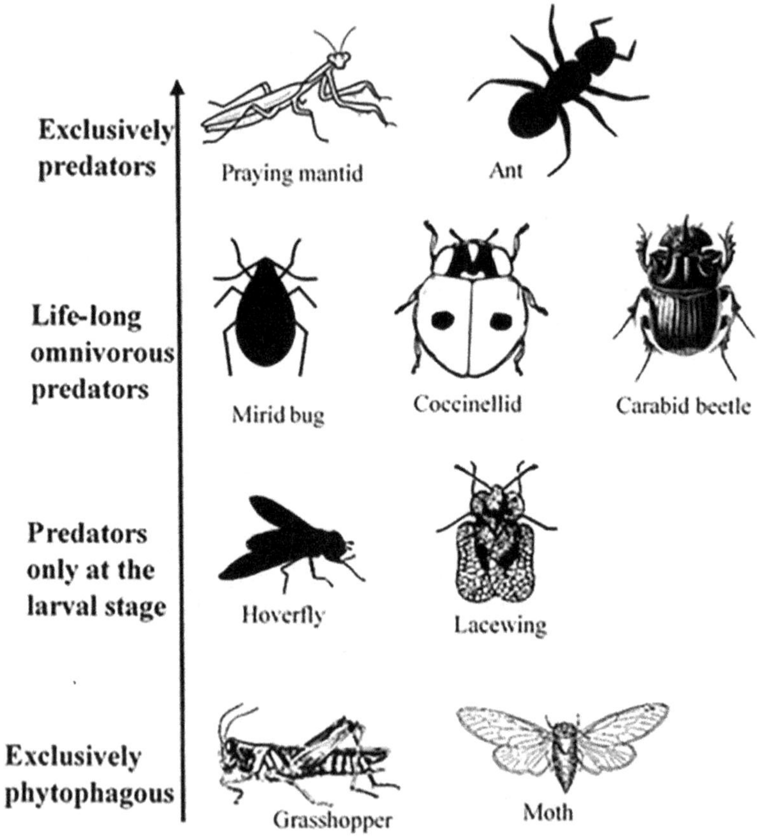

FIGURE 4.2 Food sources of insects and degree of feeding. From top to bottom are some examples of insects that are exclusively predators in the adult stage (praying mantids and ants), that prey on both plants and insects throughout their lifecycles (Dicyphine mirid bugs, coccinellids, and carabids), that prey on insects only in the larval stage and feed on plants as adults (hoverflies, lacewings), and those that feed exclusively on plants (grasshoppers and moths).

Other insects acquire defensive chemicals by sequestering secondary plant metabolites such as alkaloids, terpenoids, and phenolics, making them unpalatable to predators (Petschenka et al., 2022; Sugiura and Sato, 2018). Sequestering of plant toxins is a vehicle for plant co-evolution, which begins with the acquisition of plant defense against predatory vertebrates. For example, sawfly species of the genus *Athalia,* whose larvae feed on Brassicaceae, store glucosinolates (GSL) or iridoid glucosides in their hemolymph when feeding on Plantaginaceae to deter ant, wasp, and bird predators (Opitz et al., 2010). Specialist cruciferous aphids like *Brevycorinae brassicae* (L.) and *Lipaphis erysimi* (Kaltenbach) sequester GSLs in their hemolymph, which they hydrolyze into various volatile toxic products upon attack by predators (e.g., (iso)thiocyanates and nitriles) with the enzymatic action of aphid myrosinase (Gols et al., 2008; Kos et al., 2012; Pratt et al., 2008) (Figure 4.3). Studies have shown that larval survival, developmental time, and adult weight of hoverflies like *Episyrphus balteatus* (De Geer) and ladybirds such as *Adalia bipunctata* (L.) and *Coccinella septempunctata* L. decrease with increasing GSL concentration in *B. brassicae* (Kos et al., 2012; Pratt et al., 2008).

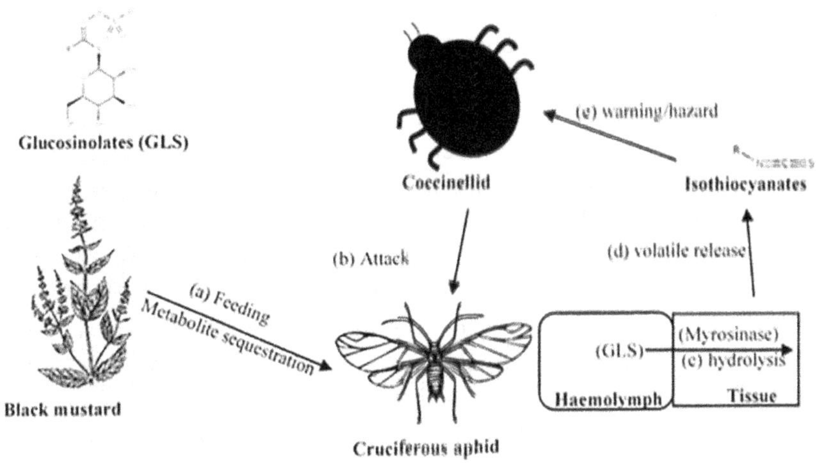

FIGURE 4.3 Steps in the acquisition of defensive chemicals by the specialist cruciferous aphid, *Brevicoryne brassicae*.

Note: The aphid feeds on the black mustard plant while sequestering glucosinolates in its hemolymph (a); upon attack by a coccinellid (b); glucosinolates are broken down by the enzyme myrosinase (c); into toxic volatiles such as isothiocyanates, which are further released (d); to warn the coccinellid predator (e).

4.3.1 *MORPHOLOGICAL DEFENSES*

In addition to chemical defenses, insects develop physical characteristics such as hairs, spines (e.g., Lepidopteran and Odonatan larvae), or a hard exoskeleton (e.g., *Coleopteran weevils*) to warn or deter access by predators (Sugiura and Yamazaki, 2014; Wang et al., 2018). For example, the larvae of the tiger moth *Lemyra imparilis* (Butler) are covered with long hairs, which act as a physical barrier against predatory carabids (Sugiura and Yamazaki, 2014). Insect prey also use crypsis to create an ambiguous image that delays the predator's response and allows the prey's escape. Many Orthopterans and Lepidopterans have evolved cryptic body colors and/or shapes to blend in with their surroundings by matching the color or texture of their environment (Shih et al., 2019). Notable examples include leaf insects (Phyllidae), leaf katydids (Tettigonidae), violin mantises, dead leaf mantises (Mantodea), leaf mimic butterflies (Nymphalidae), and moths (Notodontidae), all of which mimic angiosperm leaves (Shih et al., 2019). Another form of crypsis is masquerade, in which insects (e.g., the larva of the moth *Selenia dentaria* (Fabricius)) mimic inanimate objects such as twigs or bird droppings to be mistaken for inedible objects by predators upon detection (Skelhorn, 2015).

Mimicry is another defense strategy used by insects, and one such form is Batesian mimicry, wherein an unprotected species (e.g., hoverflies and moths) resemble or imitate the signals of protected species (e.g., hornets, wasps, and bees) (Ruxton et al., 2019). In contrast, Müllerian mimicry involves protected species such as venomous insects (e.g., bees and wasps) mimicking each other (Ruxton et al., 2019; Sherratt, 2008). This type of mimicry is most common in Lepidoptera, particularly heliconiid butterflies and burnet moths (Sherratt, 2008). However, these types of defenses can be combined in an anti-predator defense strategy, as observed in praying mantids, which defend themselves by using startle displays that combine aposematism, camouflage, and rasping sounds (Vidal-García et al., 2020).

4.3.2 BEHAVIORAL DEFENSES

Insects employ a diverse array of behaviors to escape from predators, including jumping, dropping, flight, eyespot warning, autotomy, freezing, and feigning death (Sugiura, 2020). Orthopterans have evolved well-developed legs for jumping and evading predators. Phytophagous insects, such as butterfly and moth larvae, often drop from host plants to the ground to escape from predators and return to their host by climbing silk threads (Matsubara and Sugiura, 2018; Sugiura and Yamazaki, 2006). Insects may shed or amputate part of their body to escape entrapment. This extreme form of defense, known as autotomy, has been observed in a wide range of taxa including Orthoptera, Phasmida, and Hemiptera in defense against conspecifics or nonspecific threats (Bateman and Fleming, 2008; Emberts et al., 2017; Maginnis, 2008). Nymphs of the cricket, *Gryllus bimaculatus* De Geer, escape by amputating their right hind legs without further regeneration, whereas nymphs of the plasmatic *Didymuria violescens* (Leach) regenerate lost legs (Bateman and Fleming, 2008; Maginnis, 2008). Other insects will freeze and not move in front of the predator at the first attempt, becoming immobile and feigning death if the predator does not give up (Humphreys and Ruxton, 2018). This anti-predator behavior has been extensively documented in Tribolium beetles and *S. invicta* worker ants (Asakura et al., 2022; Cassill et al., 2008).

In addition, certain insects exhibit aggressive behaviors such as biting, stinging, regurgitation, heat production, kicking, and striking, enabling insect termites and stingless bees to defend themselves in perilous situations (Sugiura, 2020). Hymenopterans, including bees, wasps, and hornets, have a modified sharp ovipositor capable of injecting venom when attacked. Some Lepidopteran larvae and Orthoptera utilize repellent fluids as a defense mechanism against predators (Bura et al., 2016; Grant, 2006). Meanwhile, larvae of the leaf beetle, *Cassida rubiginosa* Müller, and bagworm larvae, *Eumeta minuscula* (Butler), construct cryptic shelters or shields to protect themselves from predators (Sugiura, 2016).

4.4 COMPONENTS OF INSECT PREDATION

The assessment of the trophic interaction between insect prey and predators is primarily based on the number of prey attacked and the presence of predators, depending on the density of both prey and predators (Holling, 1966; Juliano, 2020). This leads to two important characteristics of insect predation that determine predator efficiency: the functional and numerical responses of the predator (Bayliss and Choquenot, 2002; Holling, 1966).

4.4.1 THE FUNCTIONAL RESPONSE OF PREDATORS

The functional response of predators refers to the change in predation rate with increasing prey density (Chan et al., 2017; Holling, 1966; Jeschke et al., 2002). There are three types of functional responses in animal predation, namely type I, II, and III (Holling, 1966) (Figure 4.4). In the type I response, the number of prey eaten per predator increases linearly with the density of prey (Holling, 1966; Juliano, 2020). Filter feeders predominantly exhibit this type I response as they capture and handle prey simultaneously (Jeschke et al., 2004). In most experimental studies, invertebrate predators, including insects, exhibit a type II functional response characterized by an increase in predation rate with increasing prey density, which levels off at a plateau where the predation rate remains constant regardless of prey density (Crookes et al., 2019; Khan et al., 2023; Wang et al., 2021). The magnitude and the character of the curve are determined by the exposure time of the predator and prey, the search or discovery rate, and the handling time involved in identifying, capturing, and consuming the prey (Holling, 1966; Juliano, 2020). The search rate is represented by a coefficient of attack and helps estimate the steepness or slope of the curve, while the handling time allows estimation of the satiation threshold, which is the peak of the curve (Holling, 1966; Juliano, 2020). Under field conditions, these two parameters are influenced by abiotic factors such as temperature, as well as biotic factors like the availability of alternative prey, prey suitability, prey size and mobility, distribution, and intraguild predation (Abraços-Duarte et al., 2021; Shah and Khan, 2013).

This type II functional response leads to unstable prey-predator dynamics because, at low prey densities, the predation rate is high, exposing the prey to the risk of extinction (Bakker and Lenteren, 1975; Murdoch and Oaten, 1975). In contrast, at higher prey densities, the control exerted by predators over the prey population is limited due to the lower predation rate. However, other factors such as temperature, host plant, alternative prey, prey species, and prey distribution may contribute to the stability of predation and shift the response to type III (Abraços-Duarte et al., 2021; Jeschke et al., 2002; Shah and Khan, 2013; Van Lenteren et al., 2016).

In the type III functional response, the curve is sigmoidal, with a predation rate initially increasing at lower prey densities and later decreasing at higher prey densities, resulting in a positive density-dependent mortality. This type of response is traditionally associated with predators that can switch between primary and alternate prey depending on primary prey density (Van Lenteren et al., 2016).

Juliano proposed a polynomial logistic regression model (Equation 4.1) to fit the response curve and determine the type of functional response

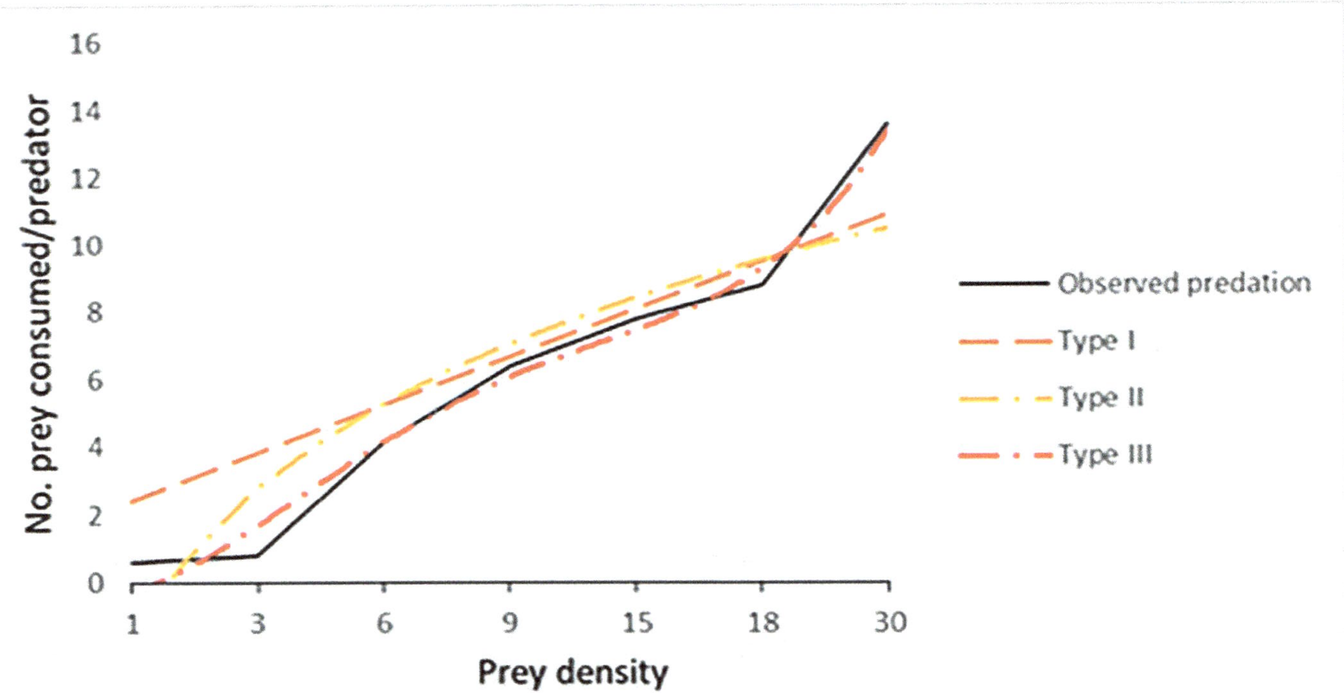

FIGURE 4.4 Functional response of the anthocorid bug *Cardiastethus affinis* Poppius preying on thrips *Frankliniella schultzei* (Trybom).

Source: Data were retrieved from Varshney et al. (2020) and different types of responses (type I, II, and III) were fitted using the polynomial logistic regression proposed by Juliano (2020). The observed predation fits the type III response better than others.

of insect predators (2020). The model uses the proportion of eaten prey (N_{ha}/N) as a function of prey density (N).

$$\frac{N_h a}{N} = \frac{(\exp(P_{\ddot{u}} + PN + PN^2)}{1 \aleph \exp(P_{\ddot{u}} \quad PN \quad PN^2)} \tag{4.1}$$

In Equation 4.1, P_0, P_1, and P_2 are the constant, linear, and quadratic coefficients, which can be estimated using the maximum likelihood method. A value of $P_1 < 0$ indicates the type II response, while $P_1 > 0$ and $P_2 < 0$ indicate a type III response (Juliano, 2020).

4.4.2 THE NUMERICAL RESPONSE OF PREDATORS

The numerical response indicates the change in the reproductive rate in response to variations in prey density (Bayliss and Choquenot, 2002). Holling has specifically categorized this response into direct or positive numerical responses and inverse numerical responses (1961). Specialist predators, such as aphidophagous coccinellids like *Hippodamia convergens* Guerin-Meneville, *Hippodamia parenthesis* (Say), and *C. septempunctata*, are known to exhibit a positive numerical response during the early growing season when aphid outbreaks occur (Elliott et al., 2002; Evans and Youssef, 1992). Conversely, a predator may display an inverse numerical response, particularly if its dispersal is limited. This phenomenon arises due to increased interference among predators under high predator densities, consequently constraining their growth even in the presence of abundant prey (Abrams et al., 2000). It is noteworthy that predators may not respond immediately to changes in prey density and may require some time to become active predators. This time delay results in oscillations in predator and prey densities, which amplify over time, eventually destabilizing the prey-predator interaction (Pandey et al., 2023). Nevertheless, intrinsic mechanisms in the attack process, such as interference among predators, prey and predator densities, and the availability of refuges, could influence prey-predator dynamics and decrease oscillations (Holt et al., 2013; Kapçak et al., 2016). Moreover, the instability of prey-predator dynamics facilitates the coexistence of specialist and generalist predators within the system, thereby limiting oscillations in the prey-predator system (Holt et al., 2013). In certain scenarios, predators may exhibit no response, especially when prey is available for only a short duration. This phenomenon is commonly observed in generalist predators like damsel bugs (Nabidae) and big-eyed bugs, whose densities showed no correlation with aphids or weevils in alfalfa fields (Evans and Youssef, 1992).

4.5 FUNCTIONAL RESPONSE OF INSECT PREDATORS IN AGRICULTURAL HABITATS AND IMPLICATIONS FOR BIOLOGICAL PEST CONTROL

4.5.1 FUNCTIONAL RESPONSE OF PREDATORS WITHIN THE COLEOPTERA TAXA: EXAMPLES FROM CARABIDS, STAPHYLINIDS, AND LADYBIRDS

Carabid and staphylinid beetles are commonly encountered epigeal predators that seek refuges in the ground layer of field boundaries (Rischen et al., 2021; Saska et al., 2007). These beetles play a crucial role in controlling insect pests but are susceptible to variations in crop type and agricultural management, which can affect their abundance and diversity (Frank and Reichhart, 2004; Shah et al., 2003). Shah et al. found a higher abundance of the carabid, *Pterostichus melanarius* (Illiger), in organic farms in England compared to conventional cropping systems, while the staphylinid, *Tachinus signatus* Gravenhorst, exhibited an opposite response (2003).

Ladybird beetles, also significant within the *Coleoptera taxa*, prey on aphids, serving as a food source during their growth and reproductive stages (Dixon, 2000; Shah and Khan, 2013). Research indicates that fourth instar larvae of ladybird beetles consume around 60% of the total aphid population during this stage (Jervis, 2007; Khan, 2010; Shah and Khan, 2013; Wang et al., 2013b). Female ladybirds exhibit the highest predation rate to meet their nutrient requirements for oviposition (Gupta et al., 2012; Khan et al., 2023; Shah and Khan, 2013; Wang et al., 2013b). Exotic ladybird species (e.g., *Cryptolaemus montrouzieri* Mulsant, *Rodolia cardinalis* (Mulsant), and *Harmonia axyridis* (Pallas)) have been introduced into new geographical areas over the past two centuries to control insect pests such as aphids, psyllids, coccids, and whiteflies (Cock et al., 2010; Rondoni et al., 2021; Soares et al., 2008). Traits such as voracity, dispersal ability, reproductive investment, and host range are crucial for the successful introduction of exotic ladybird species (Crookes et al., 2019; van Lenteren et al., 2003).

Numerous studies have demonstrated that aphidophagous ladybirds show a type II functional response in both laboratory and field settings (Gupta et al., 2012; Khan and Mir, 2008; Khan et al., 2023; Omkar and Kumar, 2013). For example, the Asian multicolored ladybird *H. axyridis* preys on the pea aphid, *Acyrthosiphon pisum* (Harris), showing a higher type II functional response compared to the native species such as *H. convergens* (Crookes et al., 2019). Similarly, *C. septempunctata* preying on aphids *L. erysimi* and *B. brassicae* in mustard and cabbage fields exhibit a type II functional response at all predatory stages (fourth instar larvae, male and female adults) (Gupta et al., 2012). Other coccinellid species are generalists and prey on both aphids and non-aphids. *Harmonia axyridis* provides natural control of the cotton leafworm *Spodoptera litura* (Fabricius), a major pest species distributed worldwide. The fourth instar larvae and adult females of *H. axyridis* exhibit increased consumption of eggs at temperatures of 30 and 35°C, showing a type II functional response, with the highest attack rate and lowest handling time (Islam et al., 2020).

4.5.2 FUNCTIONAL RESPONSE OF PREDATORS WITHIN THE HETEROPTERA TAXA: EXAMPLES FROM ANTHOCORIDS

Anthocorids, also known as flower bugs or minute pirate bugs, have the potential to serve as biological control agents in crop fields and greenhouses. They can be found across the world and feed primarily on small Lepidopteran larvae, thrips, aphids, whiteflies, psyllids, mealybugs, and scales (Panizzi, 2000; Varshney, 2023). Several biological studies have reported that certain species of flower bugs, including *Orius pumilio* (Champion), can be mass-produced for augmentative biological control in greenhouses (Kirandeep and Rabinder, 2018; Varshney, 2023). *Orius* spp. have exhibited great potential as bioagents, effectively suppressing a wide range of insect pests.

In tropical regions, *Orius albidipennis* Reuter is an efficient predator and displays different functional response types depending on prey species, their life stage, and plant traits (Gitonga et al., 2002; Rashedi et al., 2020; Salehi et al., 2016). For instance, *O. albidipennis* shows a type III functional response when preying on nymphs of *Bemisia tabaci*

(Genn.) and a type II when preying on eggs of the same species or nymphs of *Aphis fabae* Scopoli or larvae and adults of *Megalurothrips sjostedti* Trybom (Gitonga et al., 2002; Shahpouri et al., 2019). A comparable type III functional response was noted in *O. albidipennis* during predation of *T. absoluta* eggs, with a higher attack rate on low-density trichome tomato plants than on high-density trichome plants (Salehi et al., 2016). *Orius albidipennis* was found to exhibit a type I response to the predation of *M. sjostedti* at various temperatures (Gitonga et al., 2002).

Orius sauteri (Poppius) is recognized as a biological control agent with the capacity to prey on various species of thrips during nymphal and adult stages in agricultural habitats (Hemerik and Yano, 2011; Liu et al., 2018; Zhang et al., 2021). Numerous studies have demonstrated that *O. sauteri* displays a type II functional response irrespective of the specific prey species, crop type, and climatic conditions (Hemerik and Yano, 2011; Liu et al., 2018; Wang et al., 2021; Zhang et al., 2021). The species has been observed preying on eggs and larvae of the fall armyworm *Spodoptera frugiperda* (J. E. Smith), nymphs of the walnut aphids, *Chromaphis juglandicola* (Kalt.) and *Panaphis juglandis* (Goeze), consistently showing a type II functional response (Wang et al., 2021).

Orius insidiosus (Say) is an important predatory bug for controlling insect pests in vegetable and ornamental crop fields (Bernardo et al., 2017). The nymphal and adult stages of the species feed on various species of aphids, thrips, whiteflies, leafhoppers, and Lepidopterans (Baez et al., 2004). The importance of *O. insidiosus* for pest suppression like *Aphis glycines* Matsumura and thrips in soybean fields has been extensively documented (Butler and O'Neil, 2007, 2008; Harwood et al., 2007; Yoo and O'Neil, 2009). In rare instances, *O. insidiosus* may display a type I response as reported for the predation of *Helicoverpa zea* (Boddie) eggs (Parajulee et al., 2006) or a type III response during predation of eggs and larvae of corn silk flies, *Euxesta stigmatias* (Loew), *Euxesta eluta* (Loew), and *Euxesta annonae* F. (Kalsi et al., 2014).

4.5.3 FUNCTIONAL RESPONSE OF PREDATORS WITHIN THE DIPTERA TAXA: EXAMPLES FOR HOVERFLIES

Hoverflies (Diptera: Syrphidae) play a key role in the biological control of pests during the larval stage and in the pollination of crops during the adult stage (Toukem et al., 2022, 2023; Van Oystaeyen et al., 2022). These insects are specialized in preying on aphids, making them effective bioagents for aphid pest management in agricultural fields. Third instar larvae exhibit greater predatory activity compared to other instar larvae, as evidenced by shorter prey handling times and a higher prey consumption rate (Arcaya et al., 2017; Khatua et al., 2020). Additionally, different larval stages may display different functional responses depending on the species. According to Khatua et al., the first and second instars of *Eumerus albifrons* Walker showed a type II response when preying on *Aphis craccivora* Koch, whereas the third instar showed a type III response (2020). Nevertheless, *Allograpta exotica* (Wiedemann) showed a type II functional response at all larval stages when preying on the same prey, *A. craccivora* (Arcaya et al., 2017). The functional response of *E. balteatus* and *E. corollae* exhibits a type III response during the second instar larvae when consuming aphids, *Acyrtosiphon pisum* (Harris) and *A. craccivora* (Putra and Yasuda, 2006). However, a distinct response, namely type II, was recorded for *E. balteatus* at all larval stages when feeding on *A. fabae* (Amiri-Jami and Sadeghi-Namaghi, 2014; Li and Wu, 2022). In some circumstances, recent research has revealed that hoverflies can engage in bidirectional predation (Li and Wu, 2022; Li et al., 2021). This involves *S. frugiperda* fifth and sixth instar larvae preying on larvae of *E. balteatus* and *Eupeodes corollae* (Fabricius) and developing a type III response (Li and Wu, 2022; Li et al., 2021).

4.6 CONCLUSION

Insects occupy a versatile ecological role, serving as either herbivores or predators, or both, at different life stages. Their prey, far from being passive, deploy an array of mechanical, behavioral, and chemical defenses to evade predation. Predators, upon circumventing these defenses, engage in consumption and display a functional response. In the laboratory, insect predators typically demonstrate a type II functional response, in which the predation rate increases with increasing prey density until it reaches a plateau and then remains constant irrespective of prey density. Nonetheless, this response is subject to the influence of numerous biotic and abiotic factors, making insect predation a highly dynamic interaction within crop fields. Complementing the functional response, the numerical response of the predator, denoted as growth rate in relation to prey density, serves as a metric to evaluate the efficacy of biological control programs.

Prey-predator interactions extend beyond herbivores and carnivores to encompass intraguild predation, which disrupts pest regulation in agroecosystems. Studies have explored the potential of factors like prey availability and diversity to mitigate intraguild predation. Furthermore, rather than directly consuming prey, some insects modify their behavior and physiology to reduce lifetime fitness, making them more vulnerable to true predators. An emerging body of research highlights the non-lethal effects of these insects and advocates for their consideration to refine our understanding of food web dynamics.

KEYWORDS

- **anthocorids**
- **food web**
- **functional response**
- **insect predator**
- **prey defenses**
- **prey-predator density**

REFERENCES

Abraços-Duarte, G., et al. (2021). Functional response and predation rate of *Dicyphus cerastii* Wagner (Hemiptera: Miridae). *Insects, 12*(6), 530.

Abrams, P. A., et al. (2000). The nature of predation: Prey dependent, ratio dependent or neither? *Trends in Ecology & Evolution, 15*(8), 337–341.

Amiri-Jami, A. R., & Sadeghi-Namaghi, H. (2014). Responses of *Episyrphus balteatus* DeGeer (Diptera: Syrphidae) in relation to prey density and predator size. *Journal of Asia-Pacific Entomology, 17*(3), 207–211.

Arcaya, E., et al. (2017). Life table and predation rates of the syrphid fly *Allograpta exotica*, a control agent of the cowpea aphid *Aphis craccivora*. *Biological Control, 115*, 74–84.

Asakura, M., et al. (2022). Freezing or death feigning? Beetles selected for long death feigning showed different tactics against different predators. *Ecology and Evolution, 12*(2), e8533.

Baez, I., Reitz, S. R., & Funderburk, J. E. (2004). Predation by *Orius insidiosus* (Heteroptera: Anthocoridae) on life stages and species of *Frankliniella* flower thrips (Thysanoptera: Thripidae) in pepper flowers. *Environmental Entomology, 33*(3), 662–670.

Bakker, K., & van Lenteren, J. C. (1975). Functional responses in invertebrates. *Netherlands Journal of Zoology, 26*(4), 567–572.

Bateman, P. W., & Fleming, P. A. (2008). An intra- and interspecific study of body size and autotomy as a defense in Orthoptera. *Journal of Orthoptera Research, 17*(2), 315–320.

Bayliss, P., & Choquenot, D. (2002). The numerical response: Rate of increase and food limitation in herbivores and predators. *Philosophical Transactions of the Royal Society of London. Series B: Biological Sciences, 357*(1425), 1233–1248.

Bernardo, A. M. G., et al. (2017). Performance of *Orius insidiosus* on alternative foods. *Journal of Applied Entomology, 141*(9), 702–707.

Brodeur, J., & Boivin, G. (2004). Functional ecology of immature parasitoids. *Annual Review of Entomology, 49*(1), 27–49.

Bura, V. L., Kawahara, A. Y., & Yack, J. E. (2016). A comparative analysis of sonic defenses in Bombycoidea caterpillars. *Scientific Reports, 6*(1), 31469.

Butler, C. D., & O'Neil, R. J. (2007). Life history characteristics of *Orius insidiosus* (Say) fed diets of soybean aphid, *Aphis glycines* Matsumura and soybean thrips, *Neohydatothrips variabilis* (Beach). *Biological Control, 40*(3), 339–346.

Cassill, D. L., Vo, K., & Becker, B. (2008). Young fire ant workers feign death and survive aggressive neighbors. *Naturwissenschaften, 95*(7), 617–624.

Chan, K., et al. (2017). Improving the assessment of predator functional responses by considering alternate prey and predator interactions. *Ecology, 98*(7), 1787–1796.

Chen, J., et al. (2013). Defensive chemicals of tawny crazy ants, *Nylanderia fulva* (Hymenoptera: Formicidae) and their toxicity to red imported fire ants, *Solenopsis invicta* (Hymenoptera: Formicidae). *Toxicon, 76*, 160–166.

Cock, M. J. W., et al. (2010). Do new access and benefit sharing procedures under the Convention on Biological Diversity threaten the future of biological control? *BioControl, 55*(2), 199–218.

Coll, M., & Guershon, M. (2002). Omnivory in terrestrial arthropods: Mixing plant and prey diets. *Annual Review of Entomology, 47*(1), 267–297.

Crookes, S., et al. (2019). Comparative functional responses of introduced and native ladybird beetles track ecological impact through predation and competition. *Biological Invasions, 21*(2), 519–529.

Dixon, A. F. G. (2000). *Insect Predator-Prey Dynamics: Ladybird Beetles and Biological Control*. Cambridge University Press.

Elliott, N. C., et al. (2002). Predator abundance in alfalfa fields in relation to aphids, within-field vegetation, and landscape matrix. *Environmental Entomology, 31*(2), 253–260.

Emberts, Z., et al. (2017). Cut your losses: Self-amputation of injured limbs increases survival. *Behavioral Ecology, 28*(4), 1047–1054.

Eubanks, M. D., & Finke, D. L. (2014). Interaction webs in agroecosystems: Beyond who eats whom. *Current Opinion in Insect Science, 2*, 1–6.

Eubanks, M. D., & Styrsky, J. D. (2005). Effects of plant feeding on the performance of omnivorous "predators." In *Plant-Provided Food for Carnivorous Insects: A Protective Mutualism and its Applications* (pp. 148–177). Cambridge University Press.

Evans, E., & Youssef, N. (1992). Numerical responses of aphid predators to varying prey density among Utah alfalfa fields. *Journal of the Kansas Entomological Society, 65*(1), 30–38.

Fill, A., Long, E. Y., & Finke, D. L. (2012). Non-consumptive effects of a natural enemy on a non-prey herbivore population. *Ecological Entomology, 37*(1), 43–50.

Frank, T., & Reichhart, B. (2004). Staphylinidae and Carabidae overwintering in wheat and sown wildflower areas of different age. *Bulletin of Entomological Research, 94*(3), 209–217.

Gitonga, L. M., et al. (2002). Functional response of *Orius albidipennis* (Hemiptera: Anthocoridae) to *Megalurothrips sjostedti* (Thysanoptera: Thripidae). *Biological Control, 24*(1), 1–6.

Gols, R., et al. (2008). Performance of generalist and specialist herbivores and their endoparasitoids differs on cultivated and wild *Brassica* populations. *Journal of Chemical Ecology, 34*(2), 132–143.

Gontijo, L. M., Beers, E. H., & Snyder, W. E. (2015). Complementary suppression of aphids by predators and parasitoids. *Biological Control, 90*, 83–91.

Goulson, D. (2019). The insect apocalypse, and why it matters. *Current Biology, 29*(19), R967–R971.

Grant, J. B. (2006). Diversification of gut morphology in caterpillars is associated with defensive behavior. *Journal of Experimental Biology, 209*(15), 3018–3024.

Gupta, R. K., et al. (2012). Stage-specific functional response of an aphidophagous ladybird, *Coccinella septempunctata* (Coleoptera: Coccinellidae), to two aphid species. *International Journal of Tropical Insect Science, 32*(3), 136–141.

Harwood, J. D., et al. (2007). Tracking the role of alternative prey in soybean aphid predation by *Orius insidiosus*: A molecular approach. *Molecular Ecology, 16*(20), 4390–4400.

Hawkins, B. A., Cornell, H. V., & Hochberg, M. E. (1997). Predators, parasitoids, and pathogens as mortality agents in phytophagous insect populations. *Ecology, 78*(7), 2145–2152.

Hemerik, L., & Yano, E. (2011). Scaling up from individual behavior of *Orius sauteri* foraging on *Thrips palmi* to its daily functional response. *Population Ecology, 53*(4), 563–572.

Holling, C. S. (1961). Principles of insect predation. *Annual Review of Entomology, 6*(1), 163–182.

Holling, C. S. (1966). The functional response of invertebrate predators to prey density. *The Memoirs of the Entomological Society of Canada, 98*(S48), 5–86.

Holt, R. D., Roy, M., & Barfield, M. (2013). Unstable predator–prey dynamics permits the coexistence of generalist and specialist predators, and the maintenance of partial preferences. *Israel Journal of Ecology and Evolution, 59*(1), 27–36.

Humphreys, R. K., & Ruxton, G. D. (2018). A review of thanatosis (death feigning) as an anti-predator behavior. *Behavioral Ecology and Sociobiology, 72*(2), 22.

Ingegno, B. L., et al. (2019). Functional response of the mirid predators *Dicyphus bolivari* and *Dicyphus errans* and their efficacy as biological control agents of *Tuta absoluta* on tomato. *Journal of Pest Science, 92*(4), 1457–1466.

Islam, Y., et al. (2020). Temperature-dependent functional response of *Harmonia axyridis* (Coleoptera: Coccinellidae) on the eggs of *Spodoptera litura* (Lepidoptera: Noctuidae) in laboratory. *Insects, 11*(9), 583.

Jervis, M. A. (2007). *Insects as Natural Enemies: A Practical Perspective.* Springer Science & Business Media.

Jeschke, J. M., Kopp, M., & Tollrian, R. (2002). Predator functional responses: Discriminating between handling and digesting prey. *Ecological Monographs, 72*(1), 95–112.

Jeschke, J. M., Kopp, M., & Tollrian, R. (2004). Consumer-food systems: Why type I functional responses are exclusive to filter feeders. *Biological Reviews, 79*(2), 337–349.

Jones, E. I., & Dornhaus, A. (2011). Predation risk makes bees reject rewarding flowers and reduce foraging activity. *Behavioral Ecology and Sociobiology, 65*(8), 1505–1511.

Juliano, S. A. (2020). Nonlinear curve fitting: Predation and functional response curves. In *Design and Analysis of Ecological Experiments* (pp. 159–182). Chapman and Hall/CRC.

Kalsi, M., et al. (2014). Distribution of arthropod predators and their responses to *Euxesta* spp. (Diptera: Ulidiidae) in the laboratory and in corn fields in South Florida. *The Florida Entomologist, 97*(3), 911–920.

Kapçak, S., Elaydi, S., & Ufuktepe, Ü. (2016). Stability of a predator–prey model with refuge effect. *Journal of Difference Equations and Applications, 22*(7), 989–1004.

Khan, M. H., Yoldaş, Z., & Madahi, K. (2023). High prey density affects the functional response of variegated ladybird beetles against pea aphids. *Gesunde Pflanzen*. https://doi.org/10.1007/s10343-023-00913-5.

Khatua, P., Mohapatra, L. N., & Nayak, P. (2020). Feeding potential and functional response of syrphid fly, *Eumerus albifrons* Walker to cowpea aphid, *Aphis craccivora* Koch. *Journal of Entomology and Zoology Studies, 8*, 2324–2327.

Kos, M., et al. (2012). Herbivore-mediated effects of glucosinolates on different natural enemies of a specialist aphid. *Journal of Chemical Ecology, 38*(1), 100–115.

LeBrun, E. G., Jones, N. T., & Gilbert, L. E. (2014). Chemical warfare among invaders: A detoxification interaction facilitates an ant invasion. *Science, 343*(6174), 1014–1017.

Li, H., & Wu, K. (2022). Bidirectional predation between larvae of the hoverfly *Episyrphus balteatus* (Diptera: Syrphidae) and the fall armyworm *Spodoptera frugiperda* (Lepidoptera: Noctuidae). *Journal of Economic Entomology, 115*(2), 545–555.

Li, H., et al. (2021). Two-way predation between immature stages of the hoverfly *Eupeodes corollae* and the invasive fall armyworm (*Spodoptera frugiperda* J. E. Smith). *Journal of Integrative Agriculture, 20*(3), 829–839.

Liu, P., et al. (2018). Predation functional response and life table parameters of *Orius sauteri* (Hemiptera: Anthocoridae) feeding on *Megalurothrips usitatus* (Thysanoptera: Thripidae). *Florida Entomologist, 101*(2), 254–259.

Llandres, A. L., De Mas, E., & Rodríguez-Gironés, M. A. (2012). Response of pollinators to the tradeoff between resource acquisition and predator avoidance. *Oikos, 121*(5), 687–696.

Maginnis, T. L. (2008). Autotomy in a stick insect (Insecta: Phasmida): Predation versus molting. *Florida Entomologist, 91*(1), 126–127.

Matsubara, S., & Sugiura, S. (2018). Host plant architecture affects the costs of dropping behavior in *Phaedon brassicae* (Coleoptera: Chrysomelidae). *Applied Entomology and Zoology, 53*(4), 501–508.

Murdoch, W. W., & Oaten, A. (1975). Predation and population stability. In *Advances in Ecological Research* (pp. 1–131). Elsevier.

Omkar, & Kumar, G. (2013). Responses of an aphidophagous ladybird beetle, *Anegleis cardoni*, to varying densities of *Aphis gossypii*. *Journal of Insect Science, 13*(1), 24.

Opitz, S. E. W., Jensen, S. R., & Müller, C. (2010). Sequestration of glucosinolates and iridoid glucosides in sawfly species of the genus *Athalia* and their role in defense against ants. *Journal of Chemical Ecology, 36*(2), 148–157.

Pandey, S., et al. (2023). Dynamics of a delay-induced prey-predator system with interaction between immature prey and predators. *International Journal of Biomathematics, 17*(2), 2350016.

Panizzi, C. W. S., & Antonio Ricardo. (2000). *Heteroptera of Economic Importance*. CRC Press.

Parajulee, M. N., et al. (2006). Evaluation of the functional response of selected arthropod predators on bollworm eggs in the laboratory and effect of temperature on their predation efficiency. *Environmental Entomology, 35*(2), 379–386.

Petschenka, G., et al. (2022). Sequestration of defenses against predators drives specialized host plant associations in preadapted milkweed bugs (Heteroptera: Lygaeinae). *The American Naturalist, 199*(6), E211–E228.

Pratt, C., et al. (2008). Accumulation of glucosinolates by the cabbage aphid *Brevicoryne brassicae* as a defense against two coccinellid species. *Journal of Chemical Ecology, 34*(3), 323–329.

Preisser, E. L., Bolnick, D. I., & Benard, M. F. (2005). Scared to death? The effects of intimidation and consumption in predator–prey interactions. *Ecology, 86*(2), 501–509.

Putra, N. S., & Yasuda, H. (2006). Effects of prey species and its density on larval performance of two species of hoverfly larvae, *Episyrphus balteatus* de Geer and *Eupeodes corollae* Fabricius (Diptera: Syrphidae). *Applied Entomology and Zoology, 41*(3), 389–397.

Rashedi, A., et al. (2020). Prey stage preference and functional response of *Orius albidipennis* (Heteroptera, Anthocoridae) to *Aphis fabae* (Homoptera, Aphididae). *International Journal of Tropical Insect Science, 40*(1), 13–19.

Renyard, A., Gries, R., & Gries, G. (2020). A blend of formic acid, benzoic acid, and aliphatic alkanes mediates alarm recruitment responses in western carpenter ants, *Camponotus modoc*. *Entomologia Experimentalis et Applicata, 168*(4), 311–321.

Rifflet, A., et al. (2011). Paralyzing action from a distance in an arboreal African ant species. *PLOS ONE, 6*(12), e28571.

Rischen, T., Frenzel, T., & Fischer, K. (2021). Biodiversity in agricultural landscapes: Different non-crop habitats increase diversity of ground-dwelling beetles (Coleoptera) but support different communities. *Biodiversity and Conservation, 30*(13), 3965–3981.

Rondoni, G., et al. (2021). Exotic ladybirds for biological control of herbivorous insects – A review. *Entomologia Experimentalis et Applicata, 169*(1), 6–27.

Rusch, A., et al. (2011). Multi-scale effects of landscape complexity and crop management on pollen beetle parasitism rate. *Landscape Ecology, 26*(4), 473–486.

Ruxton, G. D., et al. (2019). *Avoiding Attack: The Evolutionary Ecology of Crypsis, Aposematism, and Mimicry*. Oxford University Press.

Salehi, Z., et al. (2016). Functional responses of *Orius albidipennis* Reuter (Hemiptera, Anthocoridae) to *Tuta absoluta* Meyrick (Lepidoptera, Gelechiidae) on two tomato cultivars with different leaf morphological characteristics. *Entomologia Generalis, 36*(2).

Saska, P., et al. (2007). The significance of a grassy field boundary for the spatial distribution of carabids within two cereal fields. *Agriculture, Ecosystems & Environment, 122*(4), 427–434.

Shah, M. A., & Khan, A. A. (2013). Functional response—a function of predator and prey species. *The Bioscan, 8*(3), 751–758.

Shah, P. A., et al. (2003). Diversity and abundance of the coleopteran fauna from organic and conventional management systems in southern England. *Agricultural and Forest Entomology, 5*(1), 51–60.

Shahpouri, A., Yarahmadi, F., & Zandi Sohani, N. (2019). Functional response of the predatory species *Orius albidipennis* Reuter (Hemiptera: Anthocoridae) to two life stages of *Bemisia tabaci* (Genn.) (Hemiptera: Aleyrodidae). *Egyptian Journal of Biological Pest Control, 29*(1), 14.

Sherratt, T. N. (2008). The evolution of Müllerian mimicry. *Naturwissenschaften, 95*(8), 681–695.

Shih, C., Wang, Y., & Ren, D. (2019). Camouflage, mimicry or eyespot warning. In D. Ren et al. (Eds.), *Rhythms of Insect Evolution* (1st ed., pp. 651–665). Wiley.

Skelhorn, J. (2015). Masquerade. *Current Biology, 25*(15), R643–R644.

Snyder, G. B., et al. (2022). Alternative prey mediate intraguild predation in the open field. *Pest Management Science, 78*(9), 3939–3946.

Soares, A. O., et al. (2008). *Harmonia axyridis*: What will stop the invader? *BioControl, 53*(1), 127–145.

Sugiura, S. (2016). Bagworm bags as portable armor against invertebrate predators. *PeerJ, 4*, e1686.

Sugiura, S. (2020). Predators as drivers of insect defenses. *Entomological Science, 23*(3), 316–337.

Sugiura, S., & Sato, T. (2018). Successful escape of bombardier beetles from predator digestive systems. *Biology Letters, 14*(2), 20170647.

Sugiura, S., & Yamazaki, K. (2006). The role of silk threads as lifelines for caterpillars: Pattern and significance of lifeline-climbing behavior. *Ecological Entomology, 31*(1), 52–57.

Sugiura, S., & Yamazaki, K. (2014). Caterpillar hair as a physical barrier against invertebrate predators. *Behavioral Ecology, 25*(4), 975–983.

Toukem, N. K., et al. (2022). Interactions between integrated pest management, pollinator introduction, and landscape context on avocado *Persea americana* productivity. *Entomologia Generalis, 42*(4), 579–587.

Toukem, N. K., et al. (2023). The effect of annual flower strips on pollinator visitation and fruit set of avocadoes (*Persea americana* Mill.) in Kenya. *Arthropod-Plant Interactions, 17*(1), 19–29.

Van Lenteren, J. C., et al. (2003). Environmental risk assessment of exotic natural enemies used in inundative biological control. *BioControl, 48*(1), 3–38.

Van Lenteren, J. C., et al. (2016). Functional responses of three neotropical mirid predators to eggs of *Tuta absoluta* on tomato. *Insects, 7*(3), 34.

Van Oystaeyen, A., et al. (2022). Dual purpose: Predatory hoverflies pollinate strawberry crops and protect them against the strawberry aphid, *Chaetospihon fragaefolii*. *Pest Management Science, 78*(7), 3051–3060.

Varshney, C. R. B., Tripti, & Gupta, R. (2023). The flower bugs (Anthocoridae). In *Insect Predators in Pest Management*. CRC Press.

Varshney, R., Budhlakoti, N., & Ballal, C. R. (2020). Functional response of three species of predatory pirate bugs to different densities of blossom thrips, *Frankliniella schultzei* Trybom (Thysanoptera: Thripidae). *Current Science, 118*(5).

Vidal-García, M., et al. (2020). The evolution of startle displays: A case study in praying mantises. *Proceedings of the Royal Society B: Biological Sciences, 287*(1934), 20201016.

Voigt, D. (2019). Foothold matters: Attachment on plant surfaces promotes the vitality of omnivorous mirid bugs *Dicyphus errans*. *Arthropod-Plant Interactions, 13*(6), 819–834.

Wang, C., & Henderson, G. (2016). Repellent effect of formic acid against the red imported fire ant (Hymenoptera: Formicidae): A field study. *Journal of Economic Entomology, 109*(2), 779–784.

Wang, L. Y., et al. (2018). Too hard to swallow: A secret secondary defense of an aposematic insect. *Journal of Experimental Biology, 221*(2), jeb172486.

Wang, M. Y., et al. (2013a). Can bees simultaneously engage in adaptive foraging behavior and attend to cryptic predators? *Animal Behavior, 86*(4), 859–866.

Wang, S., et al. (2013b). Effect of temperature and photoperiod on the development, reproduction, and predation of the predatory ladybird *Cheilomenes sexmaculata* (Coleoptera: Coccinellidae). *Journal of Economic Entomology, 106*(6), 2621–2629.

Wang, T., et al. (2021). Is *Orius sauteri* Poppius a promising biological control agent for walnut aphids? An assessment from the laboratory to field. *Insects, 12*(1), 25.

Yang, L. H., & Gratton, C. (2014). Insects as drivers of ecosystem processes. *Current Opinion in Insect Science, 2*, 26–32.

Yoo, H. J. S., & O'Neil, R. J. (2009). Temporal relationships between the generalist predator, *Orius insidiosus*, and its two major preys in soybean. *Biological Control, 48*(2), 168–180.

Zhang, Q., et al. (2021). Functional response and control potential of *Orius sauteri* (Hemiptera: Anthocoridae) on tea thrips (*Dendrothrips minowai* Priesner). *Insects, 12*(12), 1132.

CHAPTER 5

Nematodes and Their Role in Insect-Plant Interactions

AASHAQ HUSSAIN BHAT[1,2,*] and VASEEM RAJA[2]

[1]*Department of Biosciences, Chandigarh University, Mohali, Punjab, India*

[2]*University Center for Research and Development, Chandigarh University, Mohali, Punjab, India*

*Corresponding author

ABSTRACT

Nematodes are microscopic roundworms widely distributed in soil, where they play vital roles in terrestrial ecosystems and influence both agriculture and ecosystem health. This chapter explores the intricate interactions among nematodes, insects, and plants, focusing on the ecological and functional mechanisms involved. Nematodes act as predators, parasites, or vectors, directly suppressing insect populations through predation or indirectly by transmitting entomopathogenic species. Their plant interactions are equally multifaceted; they can enhance plant health by suppressing insect herbivores, activating plant defense responses, and influencing plant physiology. However, some nematodes also facilitate the spread of plant pathogens, adding complexity to plant-insect-nematode dynamics. The chapter highlights the need to understand these multi-trophic relationships and the molecular mechanisms underpinning them, aiming to apply this knowledge to sustainable pest management and biodiversity conservation. It also considers how anthropogenic changes, such as land-use shifts and climate variability, may affect these interactions, emphasizing

Insect-Plant Interactions, Volume 1: Influence of Biotic and Abiotic Factors.
Jamin Ali, Adil Tonğa, Adnan Shakeel, and Nasheeman Ashraf (Eds.)

DOI: 10.1201/9781779644282-5

the importance of integrated strategies. By examining the roles nematodes play within soil ecosystems, this chapter provides insights into their contributions to ecological balance and proposes ways to harness their functions to improve crop protection, boost productivity, and maintain ecosystem resilience in the face of global environmental changes.

5.1 INTRODUCTION TO NEMATODES AND INSECT–PLANT INTERACTIONS

Nematodes play a pivotal role in the complex interactions among ecosystems, particularly between insects and plants, despite receiving limited attention in general discussions about ecological systems. These microscopic organisms significantly affect the dynamics within terrestrial ecosystems (Li et al., 2024; Pires et al., 2023). Their diverse feeding behaviors and roles within these environments directly influence insect populations and indirectly affect the health and efficiency of plants (Lazarova et al., 2021; Topalović and Geisen, 2023; Yeates et al., 2009). The subtle yet profound connections between nematodes, insects, and plants play a crucial role not only in natural settings but also in agricultural landscapes, influencing either the challenges or opportunities for managing pests sustainably. A deeper understanding of these intricate ecological relationships is essential for the effective stewardship and preservation of ecosystems, underscoring the need to recognize and study the less visible constituents of ecological networks.

Nematodes are microscopic worms that inhabit a vast array of environments, from terrestrial soils to aquatic ecosystems. Their remarkable diversity and adaptation abilities are highlighted in research, revealing that they are ancient, varied, and present in nearly all habitats, yet their existence often remains obscure due to their concealed way of life (Okamura et al., 2018). These creatures have an elongated, cylindrical shape, tapering at both ends and resembling thin threads or eels. Their anatomy is essentially composed of two tubes; the outer tube consists of the body wall and muscles, while the inner tube includes the digestive system. A fluid-filled cavity called the pseudocoelom, which lies between these tubes, contains the reproductive and other vital organs. This cavity acts as a hydrostatic skeleton that provides structural support to the nematodes, which lack a rigid body structure (Harris and Crofton, 1957). Nematodes possess an intricate nervous system and an advanced excretory-secretory system, although they do not have a circulatory system for transporting blood. The sexes are distinct, with males usually smaller and having a curved tail and a specialized reproductive organ for mating (Mehlhorn, 2015). Nathan A. Cobb, a seminal figure in nematology, described nematodes from an incredible ecological range, noting their presence in extreme environments from arid deserts to the cold, deep waters of the polar seas. He recounted how nematodes had been revived from Antarctic ice, showcasing their extreme resilience, a discovery made by Shackleton's exploration team. These creatures are also found in the depths of Alpine lakes and deep-sea environments (Chen et al., 2004). Cobb famously speculated that if all matter except nematodes were erased from the universe, the contours of our planet—including mountains, valleys, and bodies of water—would remain faintly discernible, outlined by a thin layer of nematodes.

Some nematode species exist independently as free-living organisms, while others have evolved into specialized parasites impacting plants, vertebrates, and invertebrates (Vlaar et al., 2021). Human infections by certain nematodes such as *Trichinella spiralis, Anisakis simplex*, and *Pseudoterranova decipiens*, often originating from livestock and pets, have led to severe health issues globally. According to the WHO, nematode infections affect more than 2 million people worldwide (http://www.who.int/wormcontrol/statistics/). Additionally, nematodes that target plants are responsible for substantial economic damage, contributing to an estimated annual loss of over $157 billion, or 12.3% of potential output (Holden-Dye, 2007; McCarthy and Moore, 2000). Nematodes that attack plants also cause significant economic losses, estimated at 12.3% annually, which translates to over 157 billion dollars lost each year (Mendoza-de Gives, 2022; Singh, 2015). Conversely, nematodes that parasitize only insects are valuable for biological pest control (Bhat et al., 2023). These EPNs are increasingly recognized as potent alternatives to chemical pesticides due to their effectiveness in managing pest populations (Bhat et al., 2021a; Campos-Herrera, 2015).

Parasitic nematodes display specialized adaptations that allow them to thrive in complex interactions involving their insect and plant hosts, often forming elaborate ecological bonds (Bharti et al., 2020). Nematodes significantly influence the dynamics between plants and insects. For example, while some nematodes harm plants, others protect them by targeting insect pests, attracted by specific volatile emissions from plant roots during insect feeding (Rasmann et al., 2005). Research by Hiltpold et al. (2015) showed that certain stable compounds in pea root exudates could trigger dormancy in both types of nematodes at high levels. This finding is crucial for Integrated Pest and Disease Management (IPDM), suggesting that storing nematodes in root exudates preserves their qualities better than storage in water. Furthermore, when root exudates are present at normal rhizosphere concentrations, they increase the vigor of entomopathogenic nematodes while suppressing the activity of plant-parasitic nematodes. This enhanced vigor of entomopathogenic nematodes also leads to more effective infection of *Galleria mellonella* larvae (Sikder and Vestergård, 2020). These studies indicate that metabolites in root exudates can deter detrimental nematodes and bolster beneficial ones, underscoring the complex coevolution between plant chemicals and the diverse life forms in the rhizosphere.

5.2 NEMATODES AS PLANT PATHOGENS

Nematodes play a pivotal role in the soil ecosystem, but they are often notorious for their harmful impacts on other living organisms. A significant number of these microscopic worms are parasites, causing problems for both plants and animals. Researchers have cataloged over 4100 species of nematodes that attack plants, commonly referred to as phytophagous nematodes (Poveda et al., 2020; Pulavarty et al., 2021). These parasites typically invade plant roots and create numerous challenges for their hosts. With specialized mouthparts shaped like needles, known as stylets, these nematodes pierce plant cells and drain their nutrients, leading to significant destruction in both agriculture and natural habitats. Research indicates that about 15% of all nematode species are plant parasites (Watkins et al., 2012). The harm they cause is significant, with their detrimental effects on plant health and crop yields contributing to global economic losses that reach as much as U.S. $173 billion each year (Elling, 2013; Pires et al., 2022). Beyond agriculture, their presence also threatens the balance of natural ecosystems.

PPNs employ a variety of feeding behaviors, with some residing in the soil around plant roots, known as the rhizosphere, like ectoparasitic nematodes, while others, termed endoparasitic nematodes, inhabit the internal tissues of plants for most of their lifecycle (Poveda et al., 2020; Sikder and Vestergård, 2020). These strategies allow them to evade natural predators and significantly harm their plant hosts. The range of plants affected by these nematodes is broad, including fruits, nuts, vegetables, foliage plants, and trees (Poveda et al., 2020). According to the American Society of Phytopathology, nematodes cause approximately 14% of global agricultural yield losses annually, amounting to around $125 billion. Specifically, PPNs are responsible for global annual agricultural losses estimated at U.S. $157 billion, impacting roughly 15% of agricultural production, highlighting their significant role as pathogens (Khanna et al., 2021). These minute organisms represent a substantial economic threat due to their destructive capabilities.

Various well-known species of plant-parasitic nematodes are recognized for their specific host preferences and methods of infestation. These include the soybean cyst nematode (Heterodera), potato cyst nematodes (*Globodera* spp. and Ditylenchus), root-knot nematodes (RKN) (*Meloidogyne* spp. and *Longidorus* spp.), root-lesion nematodes (*Nacobbus* spp. and *Pratylenchus* spp.), stunt nematodes (*Tylenchorhynchus* spp., *Merlinius* spp., and *Amplimerlinius* spp.), and dagger nematodes (*Trichodorus* spp. and *Xiphinema* spp.) (Ali et al., 2014; Bharti et al., 2020; Kiewnick et al., 2008; Price et al., 2021a; Shakeel et al., 2022). Their feeding strategies vary significantly; for example, *Helicotylenchus* uses its sharp, short stylet to feed externally on plant tissues (Soomro et al., 2022), whereas *Belonolaimus* and *Longidorus* penetrate deeper into root tissues using their longer stylets (Kutsuwa et al., 2015). The first recorded plant-parasitic nematode, *Anguina tritici*, known for causing seed galls, was discovered in England in 1743 by Needham. The presence of nematodes is typically identified by symptoms like root cysts, galls, or nematode wool on corms and bulbs (Lambert and Bekal, 2002). Additionally, nematodes also pose threats beyond terrestrial plant environments; for instance, *Bursaphelenchus xylophilus*, also known as the pinewood nematode, has been reported damaging bark and forest trees in the Americas, Asia, and more recently in Europe (Pimentel, 2022).

Nematodes that parasitize plants not only damage crops directly but also amplify their threat by spreading plant viruses. Notably, the nematode *X. index* carries the grapevine fanleaf virus, while Arabis mosaic virus is transmitted by *Xiphinema diversicaudatum* (Chariou and Steinmetz, 2017). Such complexities underscore the intertwined nature of nematode-related plant diseases and emphasize the challenges in developing effective disease control measures. Among the most harmful plant-parasitic nematodes, the list is topped by RKN (*Meloidogyne* spp.), followed by CN (*Heterodera* and *Globodera* spp.), root-lesion nematodes (*Pratylenchus* spp.), and *Radopholus similis,* the burrowing nematode. Also listed are *Ditylenchus dipsaci*, known for its broad range of crop hosts; *Bursaphelenchus xylophilus*, infamous for causing pine wilt disease; and other significant pests like *Rotylenchulus reniformis, Xiphinema index, Nacobbus aberrans*, and *Aphelenchoides besseyi* (Abad et al., 2009; Pimentel, 2022). These nematodes are linked to substantial financial losses in agriculture due to their vast host range, diverse feeding habits, and virus transmission roles. Addressing the devastation wrought by these nematodes requires a deep understanding of their biology, breeding plants that can resist them, and employing comprehensive pest management approaches. These measures are vital for maintaining agricultural productivity and preserving ecosystem health in regions plagued by nematode pests.

5.3 UNDERSTANDING THE ENDOPARASITIC NEMATODE LIFE CYCLE

The developmental cycle of endoparasitic nematodes starts when J2, which have just hatched from eggs previously termed J1, emerge into the soil. These juveniles employ specialized sensory organs known as amphids to detect chemical cues present in the exudates of plant roots, aiding them in locating appropriate host plants (Grundler et al., 1991; Teillet et al., 2013). Their primary objective is to secure host tissue before their energy reserves, mainly composed of lipids, deplete. For host penetration, J2 nematodes utilize a stylet to apply mechanical force and breach the plant cell walls. Concurrently, they secrete enzymes that assist in breaking down these walls during both penetration and subsequent migration through the host tissue. Depending on the species, this migration might occur inter-cellularly (as seen in RKN) or intracellularly (as in CN), aiming to reach the plant's vascular tissues.

Hatching from their eggs, J2 CNs penetrate the host plant's roots and move through the cortex towards the vascular bundle. At this site, they release substances that alter plant cell metabolism, causing nearby cells to merge and form a nutrient-dense structure known as a syncytium (Sobczak and Golinowski, 2011, 2019). Conversely, RKNs make their entry near the base of the root and travel intercellularly towards the growing root tip (Holbein et al., 2019). Shifting direction, RKNs head towards young vascular tissues where they manipulate several surrounding pre-vascular cells into feeding sites (Jones and Goto, 2011). Through their stylet, RKNs inject specific substances that prompt these cells to undergo extensive nuclear divisions and DNA replication, leading to the formation of large, multinucleated, and expanded cells called giant cells. These cells, which are critical for the nematodes' nourishment, can swell to up to 100 times their original size, typically forming clusters of five to seven giant cells.

Nematodes penetrate plant vascular bundles and focus on certain xylem parenchyma cells to create unique NFS. Once these sites are established, the nematode stops moving and undergoes several developmental stages, evolving from the J2 juvenile phase through J3 and J4 stages, culminating in adulthood. Adult male nematodes eventually become slender, mobile creatures and exit the host plant. In contrast, adult females take on a distinct, swollen pear-shaped configuration and maintain a small structure for nutrient intake. In the case of CNs, females engage in sexual reproduction, generating a multitude of eggs (Price et al., 2021b). After the death of a female, her exoskeleton morphs into a cyst that protects the eggs, which are subsequently released from the plant roots, thus deriving the name "cyst nematodes." On the other hand, RKN reproduce asexually via parthenogenesis, leading to the formation of gelatinous egg masses (Castagnone-Sereno, 2006). These masses are expelled externally on the root knots while the female stays embedded within, setting the stage for another infection cycle as the eggs hatch into J2 juveniles (Abad and Williamson, 2010).

Large, metabolically active multinucleated NFS play a crucial role in feeding embryonic nematodes with their dense, granular cytoplasm. In CNs, these units are called syncytia (Jones and Payne, 1977), while in RKNs, they are known as colossal cells (Jones and Goto, 2011). The formation of colossal cells involves karyokinesis without cytokinesis in about 5–7 cells of the vascular parenchyma. This cellular activity, coupled with hyperplasia and hypertrophy in the adjacent cortical areas, leads to noticeable swellings called root-knots (Rodiuc et al., 2014). Similarly, giant cells arise from the activation of karyokinesis without cytokinesis in a similar number of vascular parenchymal cells (Figure 5.1). This process, alongside the hyperplastic and hypertrophic responses of the surrounding cortical tissue, results in the emergence of distinctive galls, also termed root-knots (Rodiuc et al., 2014). Nematodes have developed complex strategies to interact with their host plants, modifying these interactions to optimize their growth and reproductive success. Although previous research has explored various dimensions of nematode behavior, morphology, physiology, genetics, and life cycle, recent studies have focused more on understanding the molecular underpinnings of the interactions between plants and nematodes.

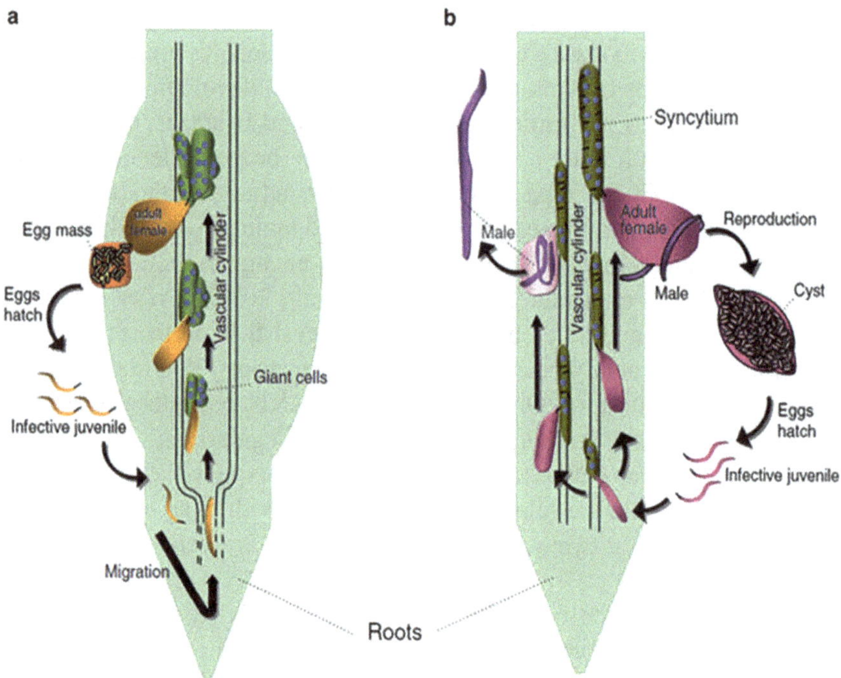

FIGURE 5.1 Life cycle of (a) RKN and (b) CN.

Source: Reprinted with permission from Abad et al., 2020. Copyright © 2010 Elsevier Ltd.

5.4 EVOLUTION OF INSECT–NEMATODE SYMBIOSIS

The interactions between nematodes and arthropods vary widely, from mutualistic to detrimental, and are grouped into four main types.

First, there is phoresy, where nematodes use insects simply for transport or shelter, without gaining any nutritional value from the host. Second, necromancy involves nematodes consuming deceased insects, yet they do not cause the death of these insects. Third, facultative parasitism occurs when nematodes feed on live insects, potentially harming or killing them. Finally, in obligate entomopathogeny, nematodes, together with their bacterial endosymbionts, rapidly terminate their insect hosts, relying on both the insects and bacterial by-products for their reproductive cycle (Sudhaus, 2008). It is suggested that these parasitic relationships evolved progressively from non-parasitic origins (Poinar, 2011; Sudhaus, 2008).

The interplay between nematodes and bacteria has evolved through several distinct phases: (i) trophism, where nematodes feed on bacteria; (ii) parasitism, in which nematodes are adversely affected by bacterial pathogens and must evolve defenses; and (iii) mutualism, a symbiotic relationship where both nematodes and bacteria gain advantages and engage in cooperative interactions (Kaya and Gaugler, 1993). These evolutionary steps showcase an increasing complexity and dependency in the interactions between nematodes and bacteria, highlighting a key aspect of evolutionary biology where mutual advantages promote the co-evolution of species. The dynamics of nematode-arthropod relationships range from beneficial to hostile (Gaugler and Kaya, 1990), encompassing four main forms: (i) phoresy, involving nematodes using insects simply for transportation or as a habitat without feeding on them; (ii) necromancy, a scenario where nematodes consume dead insects but do not cause their death; (iii) facultative parasitism, with nematodes feeding on living insects, which may harm or kill them; and (iv) obligate entomopathogeny, where nematodes, along with their bacterial symbionts, swiftly exterminate insects and use both insect and bacterial resources to complete their life cycles (Sudhaus, 2008). It is theorized that parasitic tendencies likely evolved from non-parasitic origins in this sequence (Poinar, 1983; Sudhaus, 2008). These relationships between nematodes and bacteria are believed to have developed through the aforementioned stages: trophism, involving nematode consumption of bacteria; parasitism, where nematodes endure bacterial infections unless they have protective adaptations; and mutualism, where both entities experience reciprocal benefits and engage in close cooperation.

5.5 ENTOMOPATHOGENIC NEMATODES AS BIOINSECTICIDES

India, the world's second most populous country and seventh largest by area, relies heavily on agriculture, which is central to its economy. Agriculture is a major contributor to India's GDP and serves as the main source of income for over 58% of its rural households (*Source:* IBEF). Despite advances in farming techniques, Indian agriculture is challenged by an extensive array of pests. This includes around 10,000 insect species, 30,000 weed species, and over 100,000 disease-causing pathogens such as fungi, viruses, and bacteria, along with around 1000 types of nematodes (Dhaliwal et al., 2007; Hall, 1995). Annually, the global expenditure on pesticides to combat these threats is around $40 billion, which includes the cost of 3 million metric tons of pesticides as well as various biological and non-chemical pest controls (Pimentel, 2007). Of the 70,000 pests identified, insects alone are responsible for approximately 14% of crop damage worldwide (Peshin and Zhang, 2014). In India, pests lead to yearly agricultural losses estimated at around ₹6000 crore, with insects causing about 20% of these losses. Weeds, diseases, birds, and other factors account for 33%, 26%, 10%, and 11% of the damage, respectively (Janardhan et al., 2023). The increasing severity of this issue underscores the need for more effective and environmentally friendly pest control strategies in the future. India is the world's second-largest producer of vegetables, only behind China, generating 101.43 million tons annually from 6.76 million hectares. The country tops global production of cauliflowers and ranks second in eggplant production. Key pests affecting Indian crops include *Spodoptera litura, Plutella xylostella L.* (Lepidoptera: Plutellidae), *Crocidolomia binotalis* Zeller (Lepidoptera: Pyralidae), *Helicoverpa armigera* Hubner (Lepidoptera: Noctuidae), *Leucinodes orbonalis* Guenee (Lepidoptera: Crambidae), *Pieris brassicae* L. (Lepidoptera: Pieridae), *Hellula undalis* Fab. (Lepidoptera: Crambidae), *Spilosoma obliqua* Walker (Lepidoptera: Erebidae), and *Brevicoryne brassicae* L. (Hemiptera: Aphididae) (Askary et al., 2020, 2022; Bhat et al., 2021a, 2021b, 2022, 2023; Janardhan et al., 2023; Rana et al., 2020; Sharma et al., 2022). These pests are prevalent across different agricultural zones in India and pose significant threats to crop yields.

Pest management, particularly targeting insects, represents a critical challenge within the realms of agriculture, entomology, and ecology. As these fields search for methods to control pests in a sustainable and environmentally friendly manner, the focus has shifted toward biocontrol strategies that limit ecological damage and promote biodiversity preservation. One such emerging strategy involves employing nematodes as biological control agents, a method supported by recent research (Askary et al., 2022; Campos-Herrera, 2015; Janardhan et al., 2023; Loulou et al., 2023). Nematodes, a phylum comprising a wide range of species found in nearly every global ecosystem, mostly contribute positively to nutrient cycling and soil structure. However, some species have adapted to a parasitic way of life, targeting insects and other invertebrates. These predatory nematodes are now being recognized for their capacity to naturally suppress insect pest populations, making them valuable allies in integrated pest management (IPM) efforts.

The use of nematodes as biological control agents isn't a new concept; their predatory habits were noted centuries ago. However, it is only in the past few decades that there has been a deliberate focus on effectively utilizing their abilities. Recent scientific advances, technological developments, and improvements in farming methods have placed nematodes at the forefront of eco-friendly pest control strategies (Campos-Herrera, 2015; Pires et al., 2022). Nematodes are particularly valued as biocontrol agents (BCA) due to their selectivity, effectiveness, and safety for the environment. In contrast to chemical pesticides, which typically affect a broad spectrum of organisms, nematodes target specific pests, a trait developed through thousands of years of natural evolution alongside their insect hosts. Additionally, they employ a precise method of attacking pests, which includes entering their bodies, releasing harmful toxins, and utilizing their symbiotic bacteria to incapacitate or kill the pest (Campos-Herrera, 2015).

EPNs, specifically from the *Steinernema* and *Heterorhabditis* genera, are widely recognized for their use in biological control (Bhat et al., 2017, 2019, 2020). These nematodes have a symbiotic association with the *Xenorhabdus* and *Photorhabdus* bacteria, respectively, which are lethal to many insects (Machado et al., 2021, 2022). The EPNs deploy infective juveniles (IJ) that hunt for insect hosts in the soil, entering through natural body openings or by piercing the skin. Once inside, they release their bacterial partners, causing the host's demise typically within 24–48 hours (Machado et al., 2023a; Yadav et al., 2020). EPNs are noted for their safety towards non-target species and environmental compatibility, ease of application, and compatibility with conventional agricultural chemicals (Kaya and Gaugler, 1993). They exhibit a wide range of hosts, possess strong pest-seeking capabilities, and can multiply after being deployed (Kaya and Gaugler, 1993).

EPNs are chosen for biological pest control based on their distinctive host-seeking abilities, which include "ambush" tactics on the soil surface and "cruise" strategies deeper within the soil (Koppenhöfer and Fuzy, 2008). Such knowledge about their foraging behavior is crucial for optimizing their use in various settings where pest control is required, alongside the development of suitable application technologies (Dillon et al., 2006; Shapiro-Ilan et al., 2006). Due to their non-toxic nature to humans and vertebrates, EPNs are considered vital and effective adversaries against challenging insect pests (Athanassiou et al., 2010; Campos-Herrera et al., 2012; Kergunteuil et al., 2016; Lacey et al., 2015; Peters et al., 2017). For example, the nematode species *Heterorhabditis bacteriophora* has shown effective control over the Mediterranean fruit fly, *Ceratitis capitata* (Askary et al., 2022), and *Heterorhabditis zealandica* has been evaluated for control of the citrus mealybug, *Planococcus citri* (Van Niekerk and Malan, 2014).

EPNs are increasingly recognized globally for their role as biological control agents against various insect pests, marketed under numerous brand names such as Ecomask, Savoir Weevil larvae, Guardian, and others. These agents have demonstrated significant efficacy in pest management (Bhat et al., 2020). In the UK and Europe, companies like Bionema and E-nema have successfully commercialized EPN formulations, generating substantial revenue annually. However, in India, initial attempts to harness EPNs like *S. carpocapsae, S. glaseri, S. feltiae*, and *H. bacteriophora*—species originally imported from more developed nations—were unsuccessful, largely due to their limited adaptability to the diverse local agricultural climates (Bhat et al., 2020). Research on a global scale has extensively focused on non-native species such as *S. carpocapsae* and *H. bacteriophora*, which were commercialized by Ecomax Company in the 1980s under the names Green Commandos and Soil Commandos, respectively. Despite the initial enthusiasm, these products did not yield consistent results under field conditions in India, likely due to adaptability issues related to local environmental conditions, or perhaps problems with production and formulation techniques (Bhat et al., 2020). The ongoing challenge underscores the potential benefits of utilizing local, indigenous EPN species which might adapt better and perform more effectively as bio-insecticides. Despite the limited research and trials involving indigenous EPN species against various pests in India, the successes have been modest in practical field applications. Currently, no firms in India are engaged in the commercial marketing of entomopathogenic nematodes (Bhat et al., 2020; Kaya and Gaugler, 1993).

Nematode species like those in the *Oscheius* genus, particularly from the Insectivora group, along with *Caenorhabditis briggsae,* demonstrate effectiveness in eradicating insect pests such as *Galleria mellonella* larvae. They do this by employing a method called soil baiting, which is commonly used to detect juvenile nematodes capable of infection (Bedding and Akhurst, 1975; Hunt and Nguyen, 2016; Loulou et al., 2023; Torres-Barragan et al., 2010; Zhang et al., 2008). The lethality exhibited by these nematodes against insects is largely due to their symbiotic relationships with Serratia bacteria. Specifically, *O. carolinensis* is noted for its association with Serratia as well as other symbiotic bacteria (Abebe et al., 2010; Torres-Barragan et al., 2010; Zhang et al., 2008). Both *O. chongmingensis* and *C. briggsae* rely on their symbiotic Serratia to facilitate their insect-killing process, which is crucial for their lifecycle and the survival of their offspring (Abebe et al., 2010; Torres-Barragan et al., 2010). While some specialists consider these nematodes to be entomopathogenic nematodes (EPNs), debates continue regarding their classification, as discussed at various academic forums including the Nematode-Bacteria Symbioses National Science Foundation Research Coordination Network meeting in November 2010 and at the Society of Nematologists meeting in July 2011 (Abebe et al., 2010; Rae and Sommer, 2011; Stock et al., 2011).

Using nematodes as biological agents to control pests has significant benefits over traditional pest management techniques. These microscopic worms are largely harmless to humans, animals, and beneficial insects, thus minimizing environmental harm and helping to maintain ecological integrity. Additionally, nematodes align well with organic agricultural practices and can be easily incorporated into current pest management protocols (Van Niekerk and Malan, 2014). They are effective against a broad spectrum of insect pests, impacting both those that reside in the soil and those that are found above ground, which enhances their utility in a variety of farming environments (Bhat et al., 2020). However, the deployment of nematodes in biocontrol faces several hurdles. There is a pressing need to gain a better grasp of the ecological dynamics of nematodes, including the variables that affect their performance in different farming systems, and ways to improve their production, formulation, and delivery. Moreover, issues such as the development of resistance by pests and the long-term survival of nematodes in the environment need more research and solutions.

5.6 ENTOMOPATHOGENIC NEMATODE LIFE CYCLE

Entomopathogenic nematodes (EPNs) exhibit a complex life cycle that includes two adult stages and four larval stages, identified as J1 through J4. Among these, only the third-stage larvae, termed IJs, possess the ability to infect and kill insect hosts. Unique in their life cycle, IJs represent the sole free-living stage, with all other stages developing internally within the host insect. Analogous to the dauer stage in *Caenorhabditis elegans*, IJs are adapted for survival under stress. They do not consume food and are equipped with bacteria that they transport to new hosts. The primary mode of entry for IJs into insect hosts involves penetration through natural body openings such as the spiracles, mouth, and anus, or less commonly, through the cuticle's inter-segmental membranes (Bedding and Molyneux, 1982; Hunt and Nguyen, 2016; Koppenhöfer and Fuzy, 2008). Notably, Heterorhabditis species IJs have the additional ability to breach the host's skin directly. Once inside, IJs expel the symbiotic bacteria harbored within their digestive systems into the host's internal cavity, or hemocoel. These bacteria rapidly multiply, releasing toxins and substances that suppress the host's immune response, ultimately causing death within 24–48 hours (Adams et al., 2006; Shairra, 2009).

Photorhabdus and *Xenorhabdus* bacteria, which are symbiotic with *Steinernema* and *Heterorhabditis* nematodes, respectively, not only kill their host insects but also block other microbes from utilizing the nutrients in the host's body. This strategy ensures that these nutrients remain available exclusively for the bacteria and their nematode partners. While these bacteria are key agents in killing the insects, the nematodes themselves also contribute significantly to this process. They can produce harmful substances that impair the insect's immune response and damage its tissues (Bedding and Molyneux, 1982; Kaya and Gaugler, 1993; Machado et al., 2021). So far, researchers have not identified any EPNs that operate without these bacterial partners, even though some nematodes are intrinsically harmful to hosts. Following the host's demise, nematodes consume the bacteria and the dissolved tissues in the cadaver, developing into reproductive adult males and females. When sufficient resources exist, these nematodes can even produce a second generation of adults within the same host body, and this cycle can repeat several times based on nutrient availability. When the resources are exhausted, a large number of IJ nematodes IJ leave the cadaver in search of new hosts, thus continuing their lifecycle. These IJs have the capability to survive for extended periods, relying on stored nutrients or entering a near-dormant state when conditions are unfavorable, with their survival and ability to reproduce being heavily influenced by the availability of new hosts and the success of their offspring.

Reproduction in heterorhabditid and steinernematid nematodes shows significant differences. In steinernematid nematodes, each generation consistently develops into amphimictic males and females (a process known as gonochorism). In contrast, heterorhabditid nematodes start their life cycle with self-fertilizing hermaphrodites, followed by later generations that include a combination of males, females, and hermaphrodites (Grewal et al., 2005). Additionally, the coloration of the insect cadaver can serve as an indicator of the species: heterorhabditids typically result in red-colored cadavers, whereas steinernematids cause the cadaver to turn brown or tan. This variation in color is due to the specific pigments produced by the bacterial monoculture present within the host.

5.7 ROLE OF SOIL NEMATODES IN CONTROLLING PLANT-PARASITIC NEMATODES

Entomopathogenic nematodes (EPNs), primarily from the Steinernematidae and Heterorhabditidae families, are increasingly recognized for their capacity to manage PPNs, thus offering an effective biological control method in agricultural settings (Shapiro-Ilan et al., 2006). EPNs, which are naturally occurring insect parasites, have a remarkable capability to not only attack a broad spectrum of insect pests but also to suppress PPNs, thereby enhancing their applicability in IPM strategies. Species such as *Steinernema* and *Heterorhabditis* have a mutualistic association with bacteria from the genera *Xenorhabdus* and *Photorhabdus,* respectively (Bhat et al., 2020; Machado et al., 2023b). These bacteria reside within the nematodes' gastrointestinal tract and are released into the insect host upon infection, leading to septicemia and eventual host death. The deceased host then serves as a resource for nematode reproduction and further infection events. Research has shown that EPNs are effective against a variety of significant PPNs, such as RKNs (*Meloidogyne* spp.), CNs (*Heterodera* and *Globodera* spp.), and lesion nematodes (*Pratylenchus* spp.), highlighting their potential as a biocontrol agent (Bird and Bird, 1986; Caccia et al., 2018; El Aimani et al., 2022; Grewal et al., 2005; Kenney and Eleftherianos, 2016; Li et al., 2023; Shapiro-Ilan et al., 2006).

A promising method for controlling PPNs involves using EPNs and their symbiotic bacteria. This technique leverages the interaction between EPNs and their mutualistic bacteria, such as *Xenorhabdus* spp., within an infected insect host. The bacteria released by EPNs generate substances that boost the infection process and have been shown to have adverse effects on PPNs. Research indicates that the supernatants from cultures of EPN-associated *Xenorhabdus* spp. have nematicidal properties, causing significant mortality in PPN juveniles (Caccia et al., 2018; Li et al., 2023; Shapiro-Ilan et al., 2006). These bacteria produce various compounds like ammonia, indole, and stilbene derivatives, known for their selective nematicidal capacities (Caccia et al., 2018). These bacterial supernatants contain various chemical compounds, including ammonia, indole, and derivatives of stilbene, which are particularly toxic to nematodes. Among EPNs, *Steinernema feltiae*, which hosts *Xenorhabdus bovienii,* has been highlighted as an effective agent in PPN biocontrol strategies, consistently producing substances that deter PPNs (Gulcu et al., 2012). However, application techniques such as immersing plants in these bacterial culture filtrates have shown only temporary effectiveness (Kenney and Eleftherianos, 2016).

The stability of soil environments when using biological control agents is not guaranteed, as there are issues with bacterial reversion and a rapid decrease in the numbers of nematodes following their application. It is crucial to develop methods that can maintain or enhance the effectiveness of EPNs for the control of PPNs (Gulcu et al., 2012). One approach could be the creation of genetically stable *Xenorhabdus* strains that work across different plant types or the innovative technique of simultaneously planting both EPNs and insect hosts into the soil. Another important area of research is understanding the extent to which the substances produced by *Xenorhabdus* specifically target PPNs (Grewal, 2005). Increasing the specificity of these interactions could improve the effectiveness of biocontrol methods while reducing unwanted side effects (Kenney and Eleftherianos, 2016). The intricate relationships among EPNs, PPNs, and their host species, however, require more detailed study to fully understand the potential benefits and downsides over the long term. Nevertheless, identifying compounds that specifically deter or kill PPNs could lead to more efficient and targeted biocontrol solutions, thus reducing unintended impacts on the ecosystem.

Biological control has recently gained traction as an effective method for nematode management, particularly within integrated systems that aim to minimize ecological impacts. This approach is not only eco-friendly but also cost-efficient. Predatory nematodes, when properly supported, can play a pivotal role in controlling nematode populations. In nature, these predatory species are key in regulating the numbers of plant-eating nematodes, thus maintaining the health of soil ecosystems. The study of these predator-prey relationships is critical for devising successful biocontrol strategies. Predatory nematodes are important members of the soil community, influencing its structure based on their feeding groups (Devi and George, 2018; Wardle and Yeates, 1993). The groups such as Mononchida, Aphelenchida, Dorylaimida, and Diplogasterida are noted for their predatory capabilities, with Mononchida being exclusively predatory (Ciancio and Mukerji, 2008; Devi and George, 2018; Mulvey, 1961). Although they represent a minor portion of the total soil biomass, predatory nematodes are crucial for their role in targeting various organisms, including plants, fungi, and bacteria. These nematodes come equipped with a variety of feeding structures, exhibit specific dietary preferences, and utilize unique mechanisms to consume a broad range of prey, each contributing distinctly to the ecological equilibrium of their habitats. Research on Mononchida, in particular, has been scant, suggesting room for further investigation. The importance of predatory nematodes in controlling plant-parasitic nematodes has been supported by various researchers (Keith et al., 2009). For example, studies indicate that *Prionchulus punctatus* can effectively reduce numbers of both PCNs (*G. rostochiensis*) and RKNs (*M. incognita*) (Khan and Kim, 2007; Small, 1979).

The initial suggestion to deploy predatory nematodes as a method for controlling plant-parasitic nematodes was made by Cobb (1917). This idea was later supported by Steiner and Heinly (1922), who proposed the use of *Clarkus papillatus* for managing nematode infestations in sugar beet crops. Thorne (1927) also highlighted the effectiveness of predatory nematodes, notably *Clarkus papillatus*, in subduing *Heterodera radicicola*. It was noted that under ideal circumstances, *Iotonchus brachylaimus* could be instrumental in controlling harmful nematode populations. Subsequent research expanded the list of beneficial predatory nematode species to include *Sectonema, Nygolaimus, Labronema, Aporcelaimus, Dorylaimus, Actinolaimus*, and *Dorylaimoides*, indicating a rich variety in this biological group (Keith et al., 2009; Peña-Santiago and Álvarez-Ortega, 2014). Numerous studies have consistently recommended integrating predatory nematodes into strategies aimed at mitigating phytophagous nematode threats.

The renewed interest in deploying nematodes for biocontrol initiatives against plant parasites and pests has been invigorated by the identification of predatory species. Research has shown a clear inverse relationship between the population levels of *Mylonchulus sigmaturus*, a predatory nematode, and the numbers of citrus nematodes (*Tylenchulus semipenetrans*). Additionally, the predatory capabilities of *Prionchulus punctatus*, particularly in comparison with predatory arthropods, were proven effective in controlled environment experiments targeting pests such as *M. incognita* and *G. rostochiensis* (Small, 1979). Numerous studies have consistently observed strong negative correlations between the populations of plant-feeding nematodes (*Trichodorus, Hemicycliophora*) and the predatory nematode *Parahadronchus shakili* (Khan and Kim, 2007). The predatory nature of what is now known as *Mononchoides potohikus* (formerly *Diplenteron colobocercus*) towards PPNs was first noted by Yeates in 1969, highlighting the complex interactions of predator-prey dynamics in soil environments. Furthermore, research indicates that predatory behaviors in Mononchids vary with age, adding another layer of complexity to their ecological roles (Khan and Kim, 2007; Sidhu and Kanwar, 2020).

Studies involving agar plate experiments have shown the significance of predatory nematodes in understanding their dietary and hunting practices. Developing effective biocontrol strategies hinges on how susceptible the target pests are to these predatory organisms. *Diplogasterids* are specifically noted for their selective eating habits (Chitambar and Noffsinger, 1989), a critical factor for effective biological control measures. The dynamics between these predators and their prey are crucial for optimizing biocontrol approaches (Steel et al., 2010). *Mononchus potohikus,* a species of diplogasterid nematodes, highlighted their potential as BCA in research by Yeates (1969) due to their predation capabilities. These nematodes thrive in nutrient-rich settings such as decomposed manure and are well-suited for laboratory cultivation. They exhibit advantageous characteristics like versatile feeding phases, high predation and reproduction rates, short life spans, and a keen response to prey emissions (Bilgrami et al., 2005), underscoring their suitability for biocontrol purposes. Studies have confirmed that diplogasterids effectively control populations of RKNs (Bajaj and Kanwar, 2015). Among them, *Fictor composticola*, known for its vigorous predatory behavior, typically inhabits compost and similar nutrient-dense substrates. It targets a range of nematode groups (Bajaj and Kanwar, 2015) and is prevalent in compost used for cultivating button mushrooms in regions such as Haryana and Bihar, India. Laboratory tests have demonstrated *F. composticola's* ability to manage various nematode densities that impact mushrooms, including species like *Aphelenchus avenae, Aphelenchoides swarupi, Ditylenchus myceliophagous*, and *Panagrolaimus* sp. Nevertheless, further investigation is required to fully grasp the effectiveness of predatory nematodes in biocontrol applications.

Studies by Bajaj and Kanwar (2015) and Khan and Kim (2005) shed light on the predation behaviors and effectiveness of nematodes such as *Neoactinolaimus* and *M. longicaudatus* in controlling nematode populations and reducing root galling. Bilgrami (2008) examined the broader implications and future prospects of nematode biocontrol, while Bilgrami et al. (2005) investigated predator-prey interactions for the purpose of biological pest control. Bilgrami and Brey (2005) studied the feeding habits and prey preferences of the Diplogasterid predator *Mononchoides gaugleri*, contributing to a better understanding of its role in pest management. Furthermore, the studies by Khan and Kim (2005) and Bilgrami (2008) emphasized the practical applications of predatory nematodes in controlling PPN populations, with *Mononchoides gaugleri* being released for field management. Certain predatory nematodes display a wide range of prey, including borers, grubs, thrips, and beetles, thus earning the title of entomopathogenic nematodes. The distinctive feeding structure of Mononchid nematodes, characterized by a buccal cavity with a sharp dorsal tooth and associated hardened denticles, highlights their effectiveness as predators. Investigations into *Mononchus aquaticus, Prionchulus punctatus*, and *Mylonchulus dentatus* have elucidated their feeding behaviors, emphasizing the importance of prey density and encounter rates in predation dynamics (Grootaert and Wyss, 1979). The utilization of IJ nematodes for pest management offers an environmentally friendly alternative. Overall, this body of research underscores the significance of predatory nematodes in maintaining ecological balance and their potential role in IPM strategies. Through ongoing exploration and refinement of these techniques, predatory nematodes present promising opportunities for sustainable pest control and ecosystem management.

5.8 ECOLOGICAL IMPLICATIONS AND FUTURE DIRECTIONS

Nematodes have a diverse impact on both natural and agricultural environments due to their interactions with insects and plants, with effects that range from positive to negative, or even neutral. Some nematodes, especially entomopathogenic types like *Steinernema* and *Heterorhabditis,* are crucial in controlling insect pests naturally (Campos-Herrera et al., 2012; Shapiro-Ilan et al., 2006). These nematodes hunt and kill various insect pests, which lessens dependence on synthetic pesticides. This shift towards natural pest control methods supports more sustainable farming by helping maintain ecological balance and minimizing the adverse effects associated with chemical treatments (Bhat et al., 2020). Nematodes engage with plants in contrasting manners: certain types promote plant development and increase resistance to pests, whereas others harm plants, leading to substantial reductions in crop yields. This ambivalence in their roles—as both benefactors and adversaries in plant health—complicates the management of farming ecosystems. Beneficial nematodes can enhance soil quality and nutrient distribution, thereby indirectly influencing the interactions between plants and insects.

Future investigations in this domain are anticipated to concentrate on incorporating new schemes that address their dual roles as both pests and beneficial organisms. This involves creating techniques to boost the effectiveness and sustainability of EPNs in combating insect pests while also reducing the adverse impacts of plant-pathogenic nematodes. Emerging genetic and molecular technologies could permit precise alterations to nematode behaviors and functionalities, facilitating more specific strategies in managing pests and diseases. Furthermore, elucidating the ecological functions of nematodes across different habitats will aid in forecasting how climate change might alter nematode behaviors and their effects on interactions between plants and insects. Such comprehensive strategies will necessitate collaborative efforts across multiple disciplines, including ecology, molecular biology, agronomy, and environmental science, to forge sustainable agricultural methods. Given their crucial role in the soil ecosystem, effectively managing nematodes in agricultural settings will not only enhance crop yields but also affect broader ecological health, impacting biodiversity, soil integrity, and the control of greenhouse gases. Therefore, exploring nematodes at the nexus of plant and insect interactions remains an essential and growing area of research with profound implications for both ecosystem understanding and the practical management of farming systems.

5.9 CONCLUSION

Nematodes, which are microscopic worms, play a pivotal role in the relationship between plants and insects, affecting both natural ecosystems and agricultural settings. These organisms can be either detrimental, acting as pests that damage crops, or beneficial, helping control insect populations that are harmful to plants. Their contribution is not limited to direct interaction; nematodes also facilitate the spread of fungi that kill insects, adding layers to their impact on both managed and wild environments. The broader changes in the environment, such as shifts in climate and land use, further influence nematode activities and the overall balance of ecosystems. It is vital to expand our understanding of these intricate relationships at both molecular and ecological levels to devise effective IPM strategies that protect the health of the environment. Continued research into the dynamic interactions among nematodes, insects, and plants will enhance our comprehension of ecosystem functions and lead to advancements in agricultural practices. By leveraging the positive attributes of nematodes and curbing their negative effects, we can improve agricultural yields, sustain ecological harmony, and promote biodiversity, particularly as global environmental changes persist. This expanding knowledge base not only enriches our understanding of nematodes in terrestrial ecosystems but also highlights the necessity for interdisciplinary efforts in ecological studies and the management of natural resources.

KEYWORDS

- **biocontrol agents**
- **biodiversity conservation**
- **ecological significance**
- **insect-plant interactions**
- **nematodes**
- **pest management**
- **sustainable agriculture**
- **trophic interactions**

REFERENCES

Abad, P., & Williamson, V. M. (2010). Plant-nematode interaction: A sophisticated dialogue. In *Advances in Botanical Research, 53*, 147–192. Elsevier. https://doi.org/10.1016/S0065-2296(10)53005-2.

Abad, P., Castagnone-Sereno, P., Rosso, M. N., Engler, J. D. A., & Favery, B. (2009). Invasion, feeding and development. In R. N. Perry, M. Moens, & J. L. Starr (Eds.), *Root-Knot Nematodes* (1st ed., pp. 163–181). CABI. https://doi.org/10.1079/9781845934927.0163.

Abebe, E., Jumba, M., Bonner, K., Gray, V., Morris, K., & Thomas, W. K. (2010). An entomopathogenic *Caenorhabditis briggsae*. *Journal of Experimental Biology, 213*(18), 3223–3229. https://doi.org/10.1242/jeb.043109.

Adams, B. J., Fodor, A., Koppenhöfer, H. S., Stackebrandt, E., Stock, P. S., & Klein, M. G. (2006). Biodiversity and systematics of nematode–bacterium entomopathogens. *Biological Control, 37*(1), 32–49. https://doi.org/10.1016/j.biocontrol.2005.11.008.

Ali, M. A., Wieczorek, K., Kreil, D. P., & Bohlmann, H. (2014). The beet cyst nematode *Heterodera schachtii* modulates the expression of WRKY transcription factors in syncytia to favor its development in *Arabidopsis* roots. *PLOS ONE, 9*(7), e102360. https://doi.org/10.1371/journal.pone.0102360.

Askary, T. H., Bhat, A. H., Machado, R. A. R., Ahmad, M. J., Abd-Elgawad, M. M. M., Khan, A. A., & Gani, M. (2022). Virulence and reproductive potential of Indian entomopathogenic nematodes against the larvae of the rice meal moth. *Archives of Phytopathology and Plant Protection, 55*(19), 2237–2249. https://doi.org/10.1080/03235408.2022.2161293.

Athanassiou, C. G., Kavallieratos, N. G., Menti, H., & Karanastasi, E. (2010). Mortality of four stored product pests in stored wheat when exposed to doses of three entomopathogenic nematodes. *Journal of Economic Entomology, 103*(3), 977–984. https://doi.org/10.1603/EC09202.

Bajaj, H. K., & Kanwar, R. S. (2015). Biology and predatory attributes of a diplogasterid nematode, *Fictor composticola Khan et al.,* 2008. *Helminthologia, 52*(1), 50–57. https://doi.org/10.1515/helmin-2015-0009.

Bedding, R. A., & Akhurst, R. J. (1975). A simple technique for the detection of insect-parasitic rhabditid nematodes in soil. *Nematologica, 21*(1), 109–110. https://doi.org/10.1163/187529275X00419.

Bedding, R. A., & Molyneux, A. S. (1982). Penetration of insect cuticle by infective juveniles of *Heterorhabditis* spp. (Heterorhabditidae: Nematoda). *Nematologica, 28*(3), 354–359. https://doi.org/10.1163/187529282X00402.

Bharti, L., Bhat, A. H., Chaubey, A. K., & Abolafia, J. (2020). Morphological and molecular characterisation of *Merlinius brevidens* (Allen, 1955) Siddiqi, 1970 (Nematoda: Rhabditida: Merlinidae) from India. *Journal of Natural History, 54*(23, 24), 1477–1498. https://doi.org/10.1080/00222933.2020.1810352.

Bhat, A. H., Chaubey, A. K., & Askary, T. H. (2020). Global distribution of entomopathogenic nematodes, *Steinernema* and *Heterorhabditis*. *Egyptian Journal of Biological Pest Control, 30*(1), 31. https://doi.org/10.1186/s41938-020-0212-y.

Bhat, A. H., Chaubey, A. K., & Půža, V. (2019). The first report of *Xenorhabdus indica* from *Steinernema pakistanense*: Co-phylogenetic study suggests co-speciation between *X. indica* and its steinernematid nematodes. *Journal of Helminthology, 93*(1), 81–90. https://doi.org/10.1017/S0022149X17001171.

Bhat, A. H., Chaubey, A. K., Hartmann, J., Nermut, J., & Půža, V. (2021a). Notes on the morphology, bionomics, distribution, and efficacy of *Steinernema siamkayai* (Rhabditida: Steinernematidae) from western Uttar Pradesh, India. *Nematology, 23*(7), 817–836. https://doi.org/10.1163/15685411-bja10079.

Bhat, A. H., Gautum, S., Rana, A., Chaubey, A. K., Abolafia, J., & Půža, V. (2021b). Morphological, morphometrical and molecular characterization of *Oscheius siddiqii Tabassum and Shahina, 2010* (Rhabditida, Rhabditidae) from India with its taxonomic consequences for the subgenus *Oscheius Andrássy, 1976. Biology, 10*(12), 1239. https://doi.org/10.3390/biology10121239.

Bhat, A. H., Istkhar, Chaubey, A. K., Puža, V., & San-Blas, E. (2017). First report and comparative study of *Steinernema surkhetense* (Rhabditida: Steinernematidae) and its symbiont bacteria from subcontinental India. *Journal of Nematology, 49*(1), 92–102. https://doi.org/10.21307/jofnem-2017-049.

Bhat, A. H., Loulou, A., Abolafia, J., Machado, R. A. R., & Kallel, S. (2022). Comparative morphological and molecular analyses of *Acrobeloides bodenheimeri* and *A. tricornis Cobb, 1924* (Rhabditida, Cephalobidae) from Tunisia. *Nematology, 25*(2), 207–226. https://doi.org/10.1163/15685411-bja10215.

Bhat, A. H., Machado, R. A. R., Abolafia, J., Ruiz-Cuenca, A. N., Askary, T. H., Ameen, F., & Dass, W. M. (2023). Taxonomic and molecular characterization of a new entomopathogenic nematode species, *Heterorhabditis casmirica* n. sp., and whole genome sequencing of its associated bacterial symbiont. *Parasites & Vectors, 16*(1), 383. https://doi.org/10.1186/s13071-023-05990-z.

Bilgrami, A. L. (2008). Biological control potentials of predatory nematodes. In A. Ciancio & K. G. Mukerji (Eds.), *Integrated Management and Biocontrol of Vegetable and Grain Crops Nematodes* (pp. 3–28). Springer Netherlands. https://doi.org/10.1007/978-1-4020-6063-2_1.

Bilgrami, A. L., & Brey, C. (2005). Potential of predatory nematodes to control plant-parasitic nematodes. In P. S. Grewal, R. U. Ehlers, & D. I. Shapiro-Ilan (Eds.), *Nematodes as Biocontrol Agents* (1st ed., pp. 447–464). CABI Publishing. https://doi.org/10.1079/9780851990170.0447.

Bilgrami, A., Gaugler, R., & Brey, C. (2005). Prey preference and feeding behavior of the diplogastrid predator *Mononchoides gaugleri* (Nematoda: Diplogastrida). *Nematology, 7*(3), 333–342. https://doi.org/10.1163/156854105774355563.

Bird, A. F., & Bird, J. (1986). Observations on the use of insect-parasitic nematodes as a means of biological control of root-knot nematodes. *International Journal for Parasitology, 16*(5), 511–516. https://doi.org/10.1016/0020-7519(86)90086-X.

Caccia, M., Marro, N., Dueñas, J. R., Doucet, M. E., & Lax, P. (2018). Effect of the entomopathogenic nematode-bacterial symbiont complex on *Meloidogyne hapla* and *Nacobbus aberrans* in short-term greenhouse trials. *Crop Protection, 114*, 162–166. https://doi.org/10.1016/j.cropro.2018.07.016.

Campos-Herrera, R. (2015). *Nematode Pathogenesis of Insects and Other Pests: Ecology and Applied Technologies for Sustainable Plant and Crop Protection.* Springer Berlin Heidelberg.

Campos-Herrera, R., Barbercheck, M., Hoy, C. W., & Stock, S. P. (2012). Entomopathogenic nematodes as a model system for advancing the frontiers of ecology. *Journal of Nematology, 44*(2), 162–176.

Castagnone-Sereno, P. (2006). Genetic variability and adaptive evolution in parthenogenetic root-knot nematodes. *Heredity, 96*(4), 282–289. https://doi.org/10.1038/sj.hdy.6800794.

Chariou, P. L., & Steinmetz, N. F. (2017). Delivery of pesticides to plant-parasitic nematodes using *Tobacco mild green mosaic virus* as a nanocarrier. *ACS Nano, 11*(5), 4719–4730. https://doi.org/10.1021/acsnano.7b00823.

Chen, Z. X., Chen, S. Y., & Dickson, D. W. (2004). *Nematology: Advances and Perspectives.* Tsinghua University Press; CABI Pub.

Chitambar, J. J., & Noffsinger, E. M. (1989). Predaceous behavior and life history of *Odontopharynx longicaudata* (Diplogasterida). *Journal of Nematology, 21*(2), 284–291.

Ciancio, A., & Mukerji, K. G. (2008). *Integrated Management and Biocontrol of Vegetable and Grain Crops Nematodes.* Springer Netherlands. https://doi.org/10.1007/978-1-4020-6063-2.

Devi, G., & George, J. (2018). Predatory nematodes as bio-control agents against plant-parasitic nematodes—A review. *Agricultural Reviews, 38*(1). https://doi.org/10.18805/ag.R-1715.

Dhaliwal, G., Dhawan, A., & Singh, R. (2007). Biodiversity and ecological agriculture: Issues and perspectives. *Indian Journal of Ecology, 34*(2), 100–109.

Dillon, A. B., Ward, D., Downes, M. J., & Griffin, C. T. (2006). Suppression of the large pine weevil *Hylobius abietis* (L.) (Coleoptera: Curculionidae) in pine stumps by entomopathogenic nematodes with different foraging strategies. *Biological Control, 38*(2), 217–226. https://doi.org/10.1016/j.biocontrol.2006.03.004.

El Aimani, A., Houari, A., Laasli, S. E., Mentag, R., Iraqi, D., Diria, G., Khayi, S., Lahlali, R., Dababat, A. A., & Mokrini, F. (2022). Antagonistic potential of Moroccan entomopathogenic nematodes against root-knot nematodes, *Meloidogyne javanica*, on tomato under greenhouse conditions. *Scientific Reports, 12*(1), 2915. https://doi.org/10.1038/s41598-022-07039-0.

Elling, A. A. (2013). Major emerging problems with minor *Meloidogyne* species. *Phytopathology, 103*(11), 1092–1102. https://doi.org/10.1094/PHYTO-01-13-0019-RVW.

Gaugler, R., & Kaya, H. K. (1990). *Entomopathogenic Nematodes in Biological Control*. CRC Press.

Grewal, P. S., Ehlers, R. U., & Shapiro-Ilan, D. I. (2005). *Nematodes as Biocontrol Agents*. CABI Pub.

Grootaert, P., & Wyss, U. (1979). Ultrastructure and function of the anterior feeding apparatus in *Mononchus aquaticus*. *Nematologica, 25*(2), 163–173. https://doi.org/10.1163/187529279X00181.

Grundler, F., Schnibbe, L., & Wyss, U. (1991). *In vitro* studies on the behavior of second-stage juveniles of *Heterodera schachtii* (Nematoda: Heteroderidae) in response to host plant root exudates. *Parasitology, 103*(1), 149–155. https://doi.org/10.1017/S0031182000059394.

Gulcu, B., Hazir, S., & Kaya, H. K. (2012). Scavenger deterrent factor (SDF) from symbiotic bacteria of entomopathogenic nematodes. *Journal of Invertebrate Pathology, 110*(3), 326–333. https://doi.org/10.1016/j.jip.2012.03.014.

Hall, R. (1995). Challenges and prospects of integrated pest management. In *Novel Approaches to Integrated Pest Management* (pp. 1–99).

Harris, J. E., & Crofton, H. D. (1957). Structure and function in the nematodes: Internal pressure and cuticular structure in *Ascaris*. *Journal of Experimental Biology, 34*(1), 116–130. https://doi.org/10.1242/jeb.34.1.116.

Hiltpold, I., Jaffuel, G., & Turlings, T. C. J. (2015). The dual effects of root-cap exudates on nematodes: From quiescence in plant-parasitic nematodes to frenzy in entomopathogenic nematodes. *Journal of Experimental Botany, 66*(2), 603–611. https://doi.org/10.1093/jxb/eru345.

Holbein, J., Franke, R. B., Marhavý, P., Fujita, S., Górecka, M., Sobczak, M., Geldner, N., Schreiber, L., Grundler, F. M. W., & Siddique, S. (2019). Root endodermal barrier system contributes to defence against plant-parasitic cyst and root-knot nematodes. *The Plant Journal, 100*(2), 221–236. https://doi.org/10.1111/tpj.14459.

Holden-Dye, L. (2007). Anthelmintic drugs. *Worm Book*, 1–13. https://doi.org/10.1895/wormbook.1.143.1.

Hunt, D. J., & Nguyen, K. B. (2016). *Advances in Entomopathogenic Nematode Taxonomy and Phylogeny*. Brill.

Janardhan, H. N., Askary, T. H., Bhat, A. H., Rana, A., Ahad, I., & Al-Qahtani, W. H. (2023). Morphological and molecular profiling of an entomopathogenic nematode *Steinernema feltiae*: Unlocking its biocontrol potential against vegetable insect pests. *Zootaxa, 5351*(2), 202–220. https://doi.org/10.11646/zootaxa.5351.2.2.

Jones, M. G. K., & Goto, D. B. (2011). Root-knot nematodes and giant cells. In J. Jones, G. Gheysen, & C. Fenoll (Eds.), *Genomics and Molecular Genetics of Plant-Nematode Interactions* (pp. 83–100). Springer Netherlands. https://doi.org/10.1007/978-94-007-0434-3_5.

Jones, M. G. K., & Payne, H. L. (1977). The structure of syncytia induced by the phytoparasitic nematode *Nacobbus aberrans* in tomato roots, and the possible role of plasmodesmata in their nutrition. *Journal of Cell Science, 23*(1), 299–313. https://doi.org/10.1242/jcs.23.1.299.

Kaya, H. K., & Gaugler, R. (1993). Entomopathogenic nematodes. *Annual Review of Entomology, 38*(1), 181–206. https://doi.org/10.1146/annurev.en.38.010193.001145.

Keith, A. M., Griffin, C. T., & Schmidt, O. (2009). Predatory soil nematodes (Nematoda: Mononchida) in major land-use types across Ireland. *Journal of Natural History, 43*(41, 42), 2571–2577. https://doi.org/10.1080/00222930903219970.

Kenney, E., & Eleftherianos, I. (2016). Entomopathogenic and plant pathogenic nematodes as opposing forces in agriculture. *International Journal for Parasitology, 46*(1), 13–19. https://doi.org/10.1016/j.ijpara.2015.09.005.

Kergunteuil, A., Bakhtiari, M., Formenti, L., Xiao, Z., Defossez, E., & Rasmann, S. (2016). Biological control beneath the feet: A review of crop protection against insect root herbivores. *Insects, 7*(4), 70. https://doi.org/10.3390/insects7040070.

Khan, Z., & Kim, Y. H. (2005). The predatory nematode, *Mononchoides fortidens* (Nematoda: Diplogasterida), suppresses the root-knot nematode, *Meloidogyne arenaria*, in potted field soil. *Biological Control, 35*(1), 78–82. https://doi.org/10.1016/j.biocontrol.2005.05.015.

Khan, Z., & Kim, Y. H. (2007). A review on the role of predatory soil nematodes in the biological control of plant-parasitic nematodes. *Applied Soil Ecology, 35*(2), 370–379. https://doi.org/10.1016/j.apsoil.2006.07.007.

Khanna, K., Kohli, S. K., Ohri, P., & Bhardwaj, R. (2021). Plants-nematodes-microbes crosstalk within soil: A trade-off among friends or foes. *Microbiological Research, 248*, 126755. https://doi.org/10.1016/j.micres.2021.126755.

Kiewnick, S., Karssen, G., Brito, J. A., Oggenfuss, M., & Frey, J. E. (2008). First report of root-knot nematode *Meloidogyne enterolobii* on tomato and cucumber in Switzerland. *Plant Disease, 92*(9), 1370–1370. https://doi.org/10.1094/PDIS-92-9-1370A.

Koppenhöfer, A. M., & Fuzy, E. M. (2008). Attraction of four entomopathogenic nematodes to four white grub species. *Journal of Invertebrate Pathology, 99*(2), 227–234. https://doi.org/10.1016/j.jip.2008.05.003.

Kutsuwa, K., Dickson, D. W., Brito, J. A., Jeyaprakash, A., & Drew, A. (2015). *Belonolaimus longicaudatus*: An emerging pathogen of peanut in Florida. *Journal of Nematology, 47*(2), 87–96.

Lacey, L. A., Grzywacz, D., Shapiro-Ilan, D. I., Frutos, R., Brownbridge, M., & Goettel, M. S. (2015). Insect pathogens as biological control agents: Back to the future. *Journal of Invertebrate Pathology, 132*, 1–41. https://doi.org/10.1016/j.jip.2015.07.009.

Lambert, K., & Bekal, S. (2002). Introduction to plant-parasitic nematodes. *The Plant Health Instructor.* https://doi.org/10.1094/PHI-I-2002-1218-01.

Lazarova, S., Coyne, D., Rodríguez, M. G. G., Peteira, B., & Ciancio, A. (2021). Functional diversity of soil nematodes in relation to the impact of agriculture—A review. *Diversity, 13*(2), 64. https://doi.org/10.3390/d13020064.

Li, G., Liu, T., Whalen, J. K., & Wei, Z. (2024). Nematodes: An overlooked tiny engineer of plant health. *Trends in Plant Science, 29*(1), 52–63. https://doi.org/10.1016/j.tplants.2023.06.022.

Li, J., Li, Y., Wei, X., Cui, Y., Gu, X., Li, X., Yoshiga, T., Abd-Elgawad, M. M., Shapiro-Ilan, D., Ruan, W., & Rasmann, S. (2023). Direct antagonistic effect of entomopathogenic nematodes and their symbiotic bacteria on root-knot nematodes migration toward tomato roots. *Plant and Soil, 484*(1, 2), 441–455. https://doi.org/10.1007/s11104-022-05808-4.

Loulou, A., Mastore, M., Caramella, S., Bhat, A. H., Brivio, M. F., Machado, R. A. R., & Kallel, S. (2023). Entomopathogenic potential of bacteria associated with soil-borne nematodes and insect immune responses to their infection. *PLOS ONE, 18*(1), e0280675. https://doi.org/10.1371/journal.pone.0280675.

Machado, R. A. R., Bhat, A. H., Abolafia, J., Muller, A., Bruno, P., Fallet, P., Arce, C. C. M., Turlings, T. C. J., Bernal, J. S., Kajuga, J., Waweru, B., & Toepfer, S. (2021). Multi-locus phylogenetic analyses uncover species boundaries and reveal the occurrence of two new entomopathogenic nematode species, *Heterorhabditis ruandica* n. sp. and *Heterorhabditis zacatecana* n. sp. *Journal of Nematology, 53*(1), 1–42. https://doi.org/10.21307/jofnem-2021-089.

Machado, R. A. R., Bhat, A. H., Abolafia, J., Shokoohi, E., Fallet, P., Turlings, T. C. J., Tarasco, E., Půža, V., Kajuga, J., Yan, X., & Toepfer, S. (2022). *Steinernema africanum* n. sp. (*Rhabditida, Steinernematidae*), a new entomopathogenic nematode species isolated in the Republic of Rwanda. *Journal of Nematology, 54*(1), 20220049. https://doi.org/10.2478/jofnem-2022-0049.

Machado, R. A. R., Bhat, A. H., Castaneda-Alvarez, C., Askary, T. H., Půža, V., Pagès, S., & Abolafia, J. (2023a). *Xenorhabdus aichiensis* sp. nov., *Xenorhabdus anantnagensis* sp. nov., and *Xenorhabdus yunnanensis* sp. nov., isolated from *Steinernema* entomopathogenic nematodes. *Current Microbiology, 80*(9), 300. https://doi.org/10.1007/s00284-023-03373-2.

Machado, R. A. R., Bhat, A. H., Castaneda-Alvarez, C., Půža, V., & San-Blas, E. (2023b). *Photorhabdus aballayi* sp. nov. and *Photorhabdus luminescens subsp. venezuelensis* subsp. nov., isolated from *Heterorhabditis amazonensis* entomopathogenic nematodes. *International Journal of Systematic and Evolutionary Microbiology, 73*(5). https://doi.org/10.1099/ijsem.0.005872.

Machado, R. A. R., Bhat, A. H., Fallet, P., Turlings, T. C. J., Kajuga, J., Yan, X., & Toepfer, S. (2023c). *Xenorhabdus bovienii* subsp. *africana* subsp. nov., isolated from *Steinernema africanum* entomopathogenic nematodes. *International Journal of Systematic and Evolutionary Microbiology, 73*(4). https://doi.org/10.1099/ijsem.0.005795.

McCarthy, J., & Moore, T. A. (2000). Emerging helminth zoonoses. *International Journal for Parasitology, 30*(12, 13), 1351–1359. https://doi.org/10.1016/S0020-7519(00)00122-3.

Mehlhorn, H. (2015). Nematodes. In H. Mehlhorn (Ed.), *Encyclopedia of Parasitology* (pp. 1–37). Springer Berlin Heidelberg. https://doi.org/10.1007/978-3-642-27769-6_2086-2.

Mendoza-de Gives, P. (2022). Soil-borne nematodes: Impact in agriculture and livestock and sustainable strategies of prevention and control with special reference to the use of nematode natural enemies. *Pathogens, 11*(6), 640. https://doi.org/10.3390/pathogens11060640.

Mulvey, R. H. (1961). The Mononchidae: A family of predaceous nematodes: I. Genus *Mylonchulus* (*Enoplida: Mononchidae*). *Canadian Journal of Zoology, 39*(5), 665–696. https://doi.org/10.1139/z61-070.

Okamura, B., Hartigan, A., & Naldoni, J. (2018). Extensive uncharted biodiversity: The parasite dimension. *Integrative and Comparative Biology.* https://doi.org/10.1093/icb/icy039.

Peña-Santiago, R., & Álvarez-Ortega, S. (2014). Re-description of three species of the genus *Sectonema* Thorne, 1930 (*Nematoda: Dorylaimida: Aporcelaimidae*) originally studied by E. Altherr. *Zootaxa, 3881*(1). https://doi.org/10.11646/zootaxa.3881.1.5.

Peshin, R., & Zhang, W. (2014). Integrated pest management and pesticide use. In D. Pimentel & R. Peshin (Eds.), *Integrated Pest Management* (pp. 1–46). Springer Netherlands. https://doi.org/10.1007/978-94-007-7796-5_1.

Peters, A., Han, R., Yan, X., & Leite, L. G. (2017). Production of entomopathogenic nematodes. In *Microbial Control of Insect and Mite Pests* (pp. 157–170). Elsevier. https://doi.org/10.1016/B978-0-12-803527-6.00010-X.

Pimentel, C. (2022). *Bursaphelenchus xylophilus* (pinewood nematode) (p. 10448) [Dataset]. https://doi.org/10.1079/cabicompendium.10448.

Pimentel, D. (2007). Area-wide pest management: Environmental, economic and food issues. In M. J. B. Vreysen, A. S. Robinson, & J. Hendrichs (Eds.), *Area-Wide Control of Insect Pests* (pp. 35–47). Springer Netherlands. https://doi.org/10.1007/978-1-4020-6059-5_2.

Pires, D., Orlando, V., Collett, R. L., Moreira, D., Costa, S. R., & Inácio, M. L. (2023). Linking nematode communities and soil health under climate change. *Sustainability, 15*(15), 11747. https://doi.org/10.3390/su151511747.

Pires, D., Vicente, C. S. L., Menéndez, E., Faria, J. M. S., Rusinque, L., Camacho, M. J., & Inácio, M. L. (2022). The fight against plant-parasitic nematodes: Current status of bacterial and fungal biocontrol agents. *Pathogens, 11*(10), 1178. https://doi.org/10.3390/pathogens11101178.

Poveda, J., Abril-Urias, P., & Escobar, C. (2020). Biological control of plant-parasitic nematodes by filamentous fungi inducers of resistance: *Trichoderma*, mycorrhizal and endophytic fungi. *Frontiers in Microbiology, 11*, 992. https://doi.org/10.3389/fmicb.2020.00992.

Price, J. A., Coyne, D., Blok, V. C., & Jones, J. T. (2021). Potato cyst nematodes *Globodera rostochiensis* and *G. pallida*. *Molecular Plant Pathology, 22*(5), 495–507. https://doi.org/10.1111/mpp.13047.

Pulavarty, A., Egan, A., Karpinska, A., Horgan, K., & Kakouli-Duarte, T. (2021). Plant-parasitic nematodes: A review on their behavior, host interaction, management approaches and their occurrence in two sites in the Republic of Ireland. *Plants, 10*(11), 2352. https://doi.org/10.3390/plants10112352.

Rae, R., & Sommer, R. J. (2011). Bugs don't make worms kill. *Journal of Experimental Biology, 214*(6), 1053. https://doi.org/10.1242/jeb.052480.

Rana, A., Bhat, A. H., Bhargava, S., Chaubey, A. K., & Abolafia, J. (2020). Morphological and molecular characterization of *Acrobeloides saeedi* Siddiqi, De Ley and Khan, 1992 (Rhabditida, Cephalobidae) from India and comments on its status. *Journal of Nematology, 52*(1), 1–21. https://doi.org/10.21307/jofnem-2020-027.

Rasmann, S., Köllner, T. G., Degenhardt, J., Hiltpold, I., Toepfer, S., Kuhlmann, U., Gershenzon, J., & Turlings, T. C. J. (2005). Recruitment of entomopathogenic nematodes by insect-damaged maize roots. *Nature, 434*(7034), 732–737. https://doi.org/10.1038/nature03451.

Rodiuc, N., Vieira, P., Banora, M. Y., & De Almeida Engler, J. (2014). On the track of transfer cell formation by specialized plant-parasitic nematodes. *Frontiers in Plant Science, 5*. https://doi.org/10.3389/fpls.2014.00160.

Shairra, S. (2009). Parasitism of locust by entomopathogenic nematode in relation to insect microaggregation inhibitor. *Egyptian Academic Journal of Biological Sciences. A, Entomology, 2*(2), 221–230. https://doi.org/10.21608/eajbsa.2009.15445.

Shakeel, A., Bhat, A. H., Bhat, A. A., & Khan, A. A. (2022). Interactive effect of *Meloidogyne incognita* and fly ash on the growth, physiology, and antioxidant properties of carrot (*Daucus carota* L.). *Environmental Science and Pollution Research, 29*(5), 7661–7677. https://doi.org/10.1007/s11356-021-16160-y.

Shapiro-Ilan, D. I., Gouge, D. H., Piggott, S. J., & Fife, J. P. (2006). Application technology and environmental considerations for use of entomopathogenic nematodes in biological control. *Biological Control, 38*(1), 124–133. https://doi.org/10.1016/j.biocontrol.2005.09.005.

Sharma, H., Rana, A., Bhat, A. H., & Chaubey, A. K. (2022). Entomopathogenic nematodes: Their characterization, bio-control properties and new perspectives. In C. Bellé & T. Edu Kaspary (Eds.), *Nematodes—Recent Advances, Management and New Perspectives*. IntechOpen. https://doi.org/10.5772/intechopen.99319.

Sidhu, H. S., & Kanwar, R. S. (2020). Effect of prey density of *Meloidogyne incognita* on the predation efficiency of the predator, *Fictor composticola* (Nematoda: Diplogasteridae). *Egyptian Journal of Biological Pest Control, 30*(1), 72. https://doi.org/10.1186/s41938-020-00274-5.

Sikder, M. M., & Vestergård, M. (2020). Impacts of root metabolites on soil nematodes. *Frontiers in Plant Science, 10*, 1792. https://doi.org/10.3389/fpls.2019.01792.

Singh, R. (2015). *Plant Parasitic Nematodes: The Hidden Enemies of Farmers*. https://doi.org/10.13140/RG.2.1.2994.7763.

Small, R. W. (1979). The effects of predatory nematodes on populations of plant-parasitic nematodes in pots. *Nematologica, 25*(1), 94–103. https://doi.org/10.1163/187529279X00424.

Sobczak, M., & Golinowski, W. (2011). *Genomics and Molecular Genetics of Plant-Nematode Interactions*. Scholars Portal.

Soomro, M. H., Iqbal, E., & Kazi, F. (2022). *Textbook of Plant Nematology* (First ed.). National Nematological Research Center, University of Karachi. https://doi.org/10.33804/978.969.23704.0.0.

Spiridonov, S. E., Chaubey, A. K., Ahmad, M. J., Bhat, A. H., & Askary, T. H. (2020). *Steinernema feltiae* (Rhabditida: Steinernematidae) from hilly areas of Kashmir Valley, India, with a note on its geographical distribution.

Steel, H., De La Peña, E., Fonderie, P., Willekens, K., Borgonie, G., & Bert, W. (2010). Nematode succession during composting and the potential of the nematode community as an indicator of compost maturity. *Pedobiologia, 53*(3), 181–190. https://doi.org/10.1016/j.pedobi.2009.09.003.

Teillet, A., Dybal, K., Kerry, B. R., Miller, A. J., Curtis, R. H. C., & Hedden, P. (2013). Transcriptional changes of the root-knot nematode *Meloidogyne incognita* in response to *Arabidopsis thaliana* root signals. *PLOS ONE, 8*(4), e61259. https://doi.org/10.1371/journal.pone.0061259.

Topalović, O., & Geisen, S. (2023). Nematodes as suppressors and facilitators of plant performance. *New Phytologist, 238*(6), 2305–2312. https://doi.org/10.1111/nph.18925.

Torres-Barragan, A., Ye, W., & Cardoza, Y. (2010). *Oscheius carolinensis* n. sp. (Nematoda: Rhabditidae), a potential entomopathogenic nematode from vermicompost. *Nematology, 12*(1), 121–135. https://doi.org/10.1163/156854109X458464.

Van Niekerk, S., & Malan, A. P. (2014). Compatibility of biological control agents and agrochemicals to entomopathogenic nematodes, *Steinernema yirgalemense* and *Heterorhabditis zealandica*. *African Entomology, 22*(1), 49–56. https://doi.org/10.4001/003.022.0132.

Vlaar, L. E., Bertran, A., Rahimi, M., Dong, L., Kammenga, J. E., Helder, J., Goverse, A., & Bouwmeester, H. J. (2021). On the role of dauer in the adaptation of nematodes to a parasitic lifestyle. *Parasites & Vectors, 14*(1), 554. https://doi.org/10.1186/s13071-021-04953-6.

Wardle, D. A., & Yeates, G. W. (1993). The dual importance of competition and predation as regulatory forces in terrestrial ecosystems: Evidence from decomposer food-webs. *Oecologia, 93*(2), 303–306. https://doi.org/10.1007/BF00317685.

Watkins, P. R., Huesing, J. E., Margam, V., Murdock, L. L., & Higgins, T. J. V. (2012). Insects, nematodes, and other pests. In *Plant Biotechnology and Agriculture* (pp. 353–370). Elsevier. https://doi.org/10.1016/B978-0-12-381466-1.00023-7.

Yadav, K., Bhat, A. H., Rana, A., & Chaubey, A. K. (2020). Biochemical and molecular characterization of *Photorhabdus akhurstii* associated with *Heterorhabditis indica* from Meerut, India. *Pakistan Journal of Nematology, 38*(1), 15–24. https://doi.org/10.18681/pjn.v38.i01.p15-24.

Yeates, G. (1969). Predation by *Mononchoides potohikus* (Nematoda: Diplogasteridae) in laboratory culture. *Nematologica, 15*(1), 1, 2. https://doi.org/10.1163/187529269X00010.

Yeates, G. W., Ferris, H., Moens, T., & Van der Putten, W. H. (2009). The role of nematodes in ecosystems. In M. J. Wilson & T. Kakouli-Duarte (Eds.), *Nematodes as Environmental Indicators* (1st ed., pp. 1–44). CABI. https://doi.org/10.1079/9781845933852.0001.

Zhang, C., Liu, J., Xu, M., Sun, J., Yang, S., An, X., Gao, G., Lin, M., Lai, R., He, Z., Wu, Y., & Zhang, K. (2008). *Heterorhabditidoides chongmingensis* gen. nov., sp. nov. (Rhabditida: Rhabditidae), a novel member of the entomopathogenic nematodes. *Journal of Invertebrate Pathology, 98*(2), 153–168. https://doi.org/10.1016/j.jip.2008.02.011.

CHAPTER 6

Urban Green Spaces: Insect-Plant Interaction Under Urban Stress

MOHSIN IKRAM,[1] MANSI SAHU,[2] RAJENDRA PRASAD,[1] and MAHIPAL SINGH[1]

[1]*School of Agriculture, Uttaranchal University, Dehradun, Uttarakhand, India*

[2]*School of Applied and Life Sciences, Uttaranchal University, Dehradun, Uttarakhand, India*

*Corresponding author

ABSTRACT

As urbanization increases, green spaces play a vital role in supporting wildlife. While evidence suggests that specific green space designs influence biodiversity, the most effective approaches for sustaining and attracting biodiversity are still uncertain. Traditional views in urban ecology suggest that simply providing green spaces will attract animals, but taxon-specific responses to plant resources reveal the complexity of this relationship. The ability of animals to utilize resources in green spaces determines how well these areas promote biodiversity. Rapid urbanization alters ecosystems, impacting local biodiversity. Urban green spaces like parks, gardens, and green roofs provide critical habitats for insect species, which often go underappreciated despite their roles as pollinators, prey, and decomposers. Protected green spaces support biological processes and enhance insect diversity. This chapter explores the interactions between plant and insect species in these spaces, highlighting their importance for shelter and food. It covers urban insect diversity, factors influencing these communities,

Insect-Plant Interactions, Volume 1: Influence of Biotic and Abiotic Factors.
Jamin Ali, Adil Tonğa, Adnan Shakeel, and Nasheeman Ashraf (Eds.)

DOI: 10.1201/9781779644282-6

conservation strategies, and future directions for research. The chapter emphasizes the need for proactive urban planning to conserve these ecosystems and the essential ecological services insects provide. Recommendations for further research and conservation efforts are also discussed.

6.1 INTRODUCTION

The need for more urban green space ranks highly on city policy agendas worldwide. Urban land that is covered in grass, trees, shrubs, or other vegetation, either entirely or partially, is referred to as urban green space, or UGS. Parks, community gardens, cemeteries, rooftop and vertical gardens, meadows, and woodlands are all considered forms of urban green space. Since ditches, canals, inland waterways, rivers, and riverbanks are all considered to constitute urban water, UGS is also known as the "blue-green zone." UGS supports biodiversity, cools cities in the summer, and lowers the risk of floods in cities (Haase et al., 2014). According to Harttig et al. (2014) and Hunter et al. (2019), UGS promotes social cohesiveness, improved health, and the well-being of cities and their citizens. Cities are now becoming an attractive place to reside. Cities have been able to recover because of the variety and quantity of occupations as well as enhancing options for leisure. This kind of urbanization is intended and planned, as opposed to the industrial era. As a result, parks, public lounges, and recreational areas have been reconnected and are now given more importance in urban and city development. Modern conveniences that are easily available greatly improve quality of life (Addas, 2021). A vibrant, socially engaged, and healthful city is the result of an active public life. Developing needs and changing habits and lifestyles of urban people, which influence the value and use of public spaces, also serve as an illustration of this understanding. Because most of the employment used to be manual, standing jobs before evolving to largely seated office work, people are moving considerably less than they used to. Hence, deliberate exercise needs to make up for the lack of regular exercise. Exercises like gymnastics, jogging, athletics, and at-home training demand endurance and willpower. Still, a lot of folks do not have the time, money, or energy to do these kinds of things multiple times a week (Cheng, 2023).

The significant action of human activities in the environment has led to the degradation of many ecosystems worldwide, obstructing the improvement of living standards despite the recent rapid economic development and significant social progress (Mata et al., 2017). Due to the increase of building land, population density, and economic growth, urbanization has resulted in ecological and environmental issues such as resource depletion, pollution, and deterioration of the environment, which has weakened the functions of ecosystems (Tamara et al., 2021). For instance, a significant part of ecological land, such as forest land, farmland, and grassland, has been destroyed by land expansion in urban areas, resulting in structural changes in the use of land, which have drastically minimized the capacity of ecosystems for carbon sequestration, hydrological regulation, and net primary productivity. This has led to the release of pollutants into the ecosystem, including waste, exhaust gases, and wastewater, as well as an increase in the consumption of natural resources like water and minerals (Dylewski et al., 2019).

Le Roux et al. (2014) and Threlfall et al. (2016) reported that urban green spaces have reduced habitats, such as those devoid of large hollow-bearing trees, decaying logs, native ground, and mid-story vegetation. For the management and preservation of these complex ecosystems, experts require knowledge on how to effectively manage vegetation.

6.1.1 THE ROLE OF GREEN SPACES IN URBAN ECOSYSTEMS

An important aspect of urban biodiversity is insects, which carry out a variety of ecological tasks that result in both benefits and drawbacks for the environment. Insects also serve an important ecological role as food for other species, including microbats, reptiles, and bird species that consume insects (Mata et al., 2021). The impact of plant changes during and after urbanization on insect diversity in urban settings is not well studied. The best ways to protect these functional groups and the ecosystem benefits they offer may be discovered by looking into how important functional groups, such as herbivores and predators, react to changes in vegetation layout and diversity (Enssle and Kabaisch, 2020). In recent decades, the environment has undergone a fundamental transformation due to urbanization, which is the swift growth of cities and the conversion of natural landscapes into urban regions. The disturbance of wildlife habitats and the change of local ecosystems are two of the most important effects of this urban development. But in the midst of all the steel and concrete of cities, green areas like gardens, parks, and green roofs are essential to maintaining and even improving urban ecosystems (Daniels et al., 2020). According to estimates, more than half of the world's population lived in urban settlements in 2018, and by 2030, that number is expected to rise to almost 60% (United Nations, 2018). More urbanization processes are expected to occur in the megacities that are already in existence or in the cities that are just starting to emerge. However, more people who live in these cities are directly affected by the effects of climate change, which include an ongoing rise in extreme weather events, the loss of biodiversity and terrestrial ecosystems, as well as the ecosystem services (ES) that go along with them (IPCC, 2019). Certain adjustments along sustainable developments, such as physical and social infrastructure, should be implemented in order to lessen vulnerability and increase the resilience of urban settlements (IPCC, 2018; Revi et al., 2014). One of the most important adaptation strategies to improve societies' capacity for adaptation is the preservation of ecological services (IPCC, 2018). Both sustainable urban growth and the preservation of ES can be aided by the design of land adaptation-related policies and governance frameworks, such as integrated landscape planning and land use (IPCC, 2019). ES are defined as "the advantages of the ecology and its functioning (in conjunction with various variables) to human well-being" (Burkhard and Maes, 2017). Human well-being is mostly dependent on ES, or the ES that nature provides to human society (Burkhard and Maes, 2017). However, the ability and availability of ES by ecosystems is predominantly dependent on their state. Thus, the evaluation of ecosystem conditions (EC) in conjunction with the mapping and quantification of the ES at different spatial scales in biophysical, social, and economic contexts would be a possible tool for policy and decision-making in the development of human society, including the development of urban green infrastructure (GI) strategies (Burkhard and Maes, 2017). Urban green spaces are well-liked components in the research on urban ES, with an emphasis on the urban scale. The significance or contribution of urban green spaces and GI has been explained by a number of research studies using the ES assessment technique for planning purposes. Ronchi, Arcidiacono, and Pogliani have looked into the need to connect ES, GI, and nature-based solutions.

6.2 CONSERVATION OF BIODIVERSITY AND ENVIRONMENT

Biodiversity encompasses the rich variety of life forms, from the tiniest microorganisms to the towering trees, from the colorful coral reefs to the vast array of animal species. Each element, no matter how small or seemingly insignificant, plays a unique role in the web of life. Biodiversity ensures the stability of ecosystems, making them more resilient to disturbances such as climate change, diseases, and natural disasters. It also underpins the functioning of ecosystems, influencing processes like pollination, water purification, and nutrient cycling (Dou et al., 2017). Urban green spaces are essential havens for a wide variety of plant and animal species, including insects. Urban ecosystems depend on insects for various ecological functions, including pollination and pest management. Many bug species would find it difficult to survive in urban settings without appropriate habitats (Semeraro et al., 2021). These habitats are provided by green spaces, which support biodiversity even in the middle of the metropolis. The plant life found in green spaces provides food and shelter for a variety of insects, including bees, butterflies, ants, and beetles. A wide variety of insect species are supported by the diversity of plant species in these environments, and this diversity is crucial to the insects' survival. The consistency of urban areas without green spaces would result in a fall in insect diversity (Rocha and Fellowes, 2020).

6.2.1 ECOSYSTEM PRESERVATION: NURTURING BIODIVERSITY AND ENVIRONMENTAL BALANCE

Maintaining a healthy planet is ethically important; it is not just an environmental goal. Entrusted with the responsibility of protecting Earth's precious assets, we are its protectors. Future generations should inherit a planet that is just as rich in biodiversity and natural beauty as the one we currently live on; hence, it is ethically our duty to protect the environment. In addition to preserving plants and animals, conservation efforts also aim to preserve the vital services that ecosystems offer, such as fresh water, clean air, and fertile soil for all. Large tracts of forest cover may be found in many Indian states. These woods serve as a repository for biodiversity and must be shielded from the ongoing pressures of growing urbanization. Nowadays, conservation of biodiversity while also working to bring the damaged landscapes back to their original beauty is essential. The numerous efforts of both national and state governments on that front, as well as their collaboration with the international community in pursuit of more comprehensive and long-lasting solutions, are encouraging.

1. **Ecosystem Services:** The incredible benefits that nature offers—often without our knowledge or awareness—are known as ES. These services can be divided into four categories: services that are provisional (like food, water, and medicine); services that are regulated (like disease control and climate regulation); supporting services (like crop pollination and nutrient cycling); and cultural services (like leisure and spiritual enrichment). According to Feng (2022), biodiversity is the fundamental component of these services, contributing to their sustainability and improving quality.

2. **Pollination and Agriculture:** Pollination depends on biodiversity, especially when it comes to the variety of plant and insect species. Many plants, including those that provide fruits and vegetables, are able to reproduce through the transport of pollen between blossoms by bees, butterflies, and other pollinators. Agriculture relies on this mechanism to provide abundant harvests and a varied supply of food. Our agricultural systems would be severely harmed by biodiversity loss, which would result in a shortage of food and financial difficulties (Meacham et al., 2022).

3. **Climate Regulation:** Forests, wetlands, and oceans are rich in biodiversity and operate as carbon sinks, absorbing carbon dioxide from the atmosphere and helping to mitigate climate change. Trees, in particular, help to regulate the climate by absorbing CO_2 during photosynthesis. Furthermore, varied ecosystems influence local weather patterns, mitigating extremes like droughts and floods. Preserving biodiversity is vital for sustaining these natural climate-regulating mechanisms, safeguarding our world from the severe effects of climate change (Ndong et al., 2022).

4. **Medicine and Biodiversity:** Many of the medicines we use are derived from plants, animals, and microorganisms found in various habitats. Biodiversity is a massive pharmaceutical treasure trove, containing molecules that have been transformed into medications for a variety of ailments, including cancer, malaria, and bacterial infections. As biodiversity declines, we risk losing possible cures and therapies for a variety of illnesses, emphasizing the importance of conservation efforts.

5. **Conservation Challenges and Solutions:** Despite the crucial importance of biodiversity and ES, human activities like deforestation, pollution, overfishing, and climate change pose serious challenges to these fragile systems. Habitat destruction, fueled by urbanization and agricultural growth, disturbs ecosystems, resulting in the extinction of species and the services they provide. Pollution, which includes plastic waste and chemical toxins, affects aquatic life, disrupts food chains, and degrades water quality. Overfishing depletes marine biodiversity, threatening coastal populations' livelihoods and the supply of fish for human consumption. Climate change, caused by greenhouse gas (GHG) emissions, affects ecosystems and disturbs the natural balance of biodiversity, resulting in species extinctions and ecosystem deterioration (Cord et al., 2017). To deal with these difficulties and promote biodiversity for a healthy planet, conservation activities at the local, national, and global levels must be increased. Protected areas, such as national parks and marine reserves, are critical to biodiversity conservation, offering safe havens for endangered species and allowing ecosystems to thrive. Agroforestry and organic farming are examples of sustainable land use methods that increase biodiversity while still serving human food and resource demands. Reducing pollution through strict legislation and public awareness campaigns protects ecosystems and ensures the quality of ES. International collaboration is critical for combating climate change, mitigating its effects, and preserving biodiversity for future generations.

6.3 ENVIRONMENTAL RESOURCES

Urban ecosystems rely on a variety of ecological services provided by green spaces. Pollination is one of the most important functions that insects perform in these situations. A wide variety of plant species, including many crops, can be effectively pollinated by insects (Theodorou, 2022). For both urban agriculture and the upkeep of green spaces, this pollination service is essential. The absence of pollinators would cause the local flora to diminish, which would have a domino effect on the ecosystem as a whole (Figure 6.1). Insects are essential for pollination as well as pest management. Insect predators, such as parasitoid wasps and ladybugs, aid in controlling pest populations. Without these natural barriers, pests may spread widely and harm urban agriculture and green areas (Wooster, 2022).

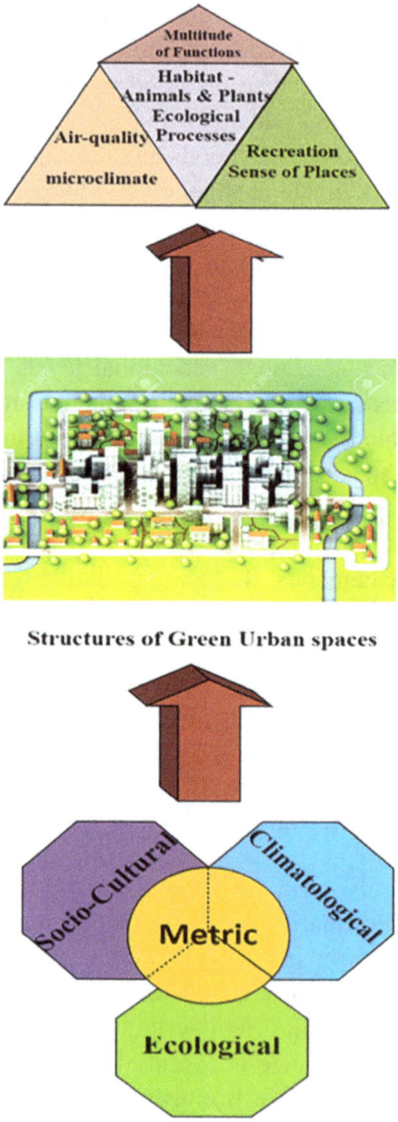

FIGURE 6.1 Structure of green urban spaces.

6.3.1 WATER AND AIR PURITY

In cities, green spaces also help to improve the quality of the air and water. In parks and gardens, trees and other vegetation aid in filtering air pollutants, which is especially helpful in highly populated areas with high air pollution levels. Furthermore, by offering shade and transpiration-based cooling, these green spaces might lessen the urban heat island effect, thus reducing the effects of climate change in metropolitan areas. Additionally, green areas contribute to better water quality (Fenoglio, 2021). When rainwater seeps through the soil, they can serve as organic buffers, filtering and cleansing the water. As a result, aquatic ecosystems become cleaner and healthier, and the number of pollutants entering urban water bodies is decreased. Thus, green areas directly affect the quality of urban lakes, ponds, and rivers (Samways, 2020).

6.3.2 WELFARE AND THE HEALTH OF HUMANS

In addition to their ecological value, green spaces significantly improve the well-being of city dwellers. Having access to green spaces has been linked to better mental and physical health, as well as lower levels of stress. These areas allow city residents opportunities for socializing, working out, and unwinding. In summary, green areas improve urban quality of life by making cities more sustainable and livable (Wilson and Fox, 2021).

6.4 URBAN INSECT SPECIES DIVERSIFICATION

Urban environments, with their concrete jungles and human activity, might not seem like the best places for a variety of insect groups to live. Nonetheless, a surprisingly wide range of insect species, each tailored to specific urban needs, can be found in cities (Adams et al., 2020). The variety of urban insect communities is evidence of the flexibility and tenacity of these microscopic but essential organisms. Growing throughout the Earth's surface, the urban habitat is having an impact on natural ecosystems. UN estimates state that by 2030, there will be a 200% rise in the urban population. In light of the impending urban expansion, it is critical to expand our knowledge of how urbanization modifies natural ecosystems in order to lessen its effects on ecological functioning and species diversity.

Given that insects are commonly used as bioindicators of habitat quality and that they play a crucial role in maintaining a variety of ecosystem functions, researching the diversity of insects in urban environments can provide important insights into the sustainability of these regions. Still, there isn't much study done on urban insects. With two perspectives: (i) as an altered landscape necessitating an understanding of its ecological functioning; and (ii) emphasizing the potential of green-blue infrastructures for urban development and insect diversity conservation, this research topic seeks to advance our knowledge of insects in the urban environment.

Pollutants from human-made environments, both urban and agricultural (although also Wang et al., 2023), frequently have an adverse effect on wildlife. Given that soil arthropods play a critical role in nutrient recycling and soil formation, researchers undertook experimental assessments to better understand the impacts of nitrogen inputs from fossil fuel emissions on soil arthropods. According to their findings, there may be a nitrogen threshold at which nitrogen levels below 25 kg N/ha positively impact biodiversity. It has been acknowledged that the idea of "green infrastructure," which combines urban development and green space conservation, has the potential to address a number of environmental issues, including lowering pollution, improving stormwater management, and mitigating the heat island effect. Ponds and their green environs are examples of blue infrastructures that support biodiversity, slow down climate change, and offer ES.

6.4.1 HABITAT DIVERSITY

Habitat heterogeneity is one of the main factors influencing insect diversity in urban areas. In contrast to natural ecosystems, which may exhibit a degree of homogeneity, urban landscapes present a diverse array of conditions. Various surfaces such as rooftops, brownfields, cracks in the pavement, and green spaces offer unique microhabitats that insects can take advantage of (Zverev et al., 2018). For example, a variety of plant species are found in parks and gardens, and these plant species, in turn, draw a wide variety of herbivorous insects. In addition to pollinators like bees and butterflies, a wide variety of other insects that play different ecological roles are drawn to these verdant havens. Similarly, specialist insects that have adapted to the circumstances present in these urban wastelands can be discovered living in abandoned buildings and industrial sites (Vilarroya et al., 2021).

The local environment can serve as a filter to shape the structure of the local community. The distribution and abundance of certain species can be affected by the microclimate in some local habitats, which can be advantageous to some but harmful to others. Buildings can change the lighting in an urban setting by obstructing sunlight, which can have an impact on moisture and temperature. Eliminating leaf litter, where insects may hibernate, is a common management strategy. The organization of plant-feeding insect communities can be significantly impacted by such heterogeneity in local environmental conditions. Movement across habitats and, consequently, local community structure can be significantly impacted by the spatial connectedness of local habitats, which depends on the distance between plants and the obstacles that separate them. Rates of colonization and extinction have an impact on the number of species that are present in a certain local habitat at any one moment; a patch with reduced connectedness will often have a lower rate of colonization than a habitat with better connectivity. Higher dispersal rates may also have an impact on the rescue effect, which maintains species richness by allowing colonization to offset local extinctions. However, not every species in a community is impacted by connectivity in the same way. Depending on the ecological guild, trophic level, and level of specialization, limits on dispersal may vary. Host plants can be viewed as habitat islands by specialized plant-feeding insects that are encircled by a matrix of structures and other non-habitat locations. Recolonization of a remnant patch by an insect is less likely to happen if it is surrounded by these more urbanized areas, and species composition may change as a result of variations in dispersal capacity across species (Vilarroya et al., 2021).

6.4.2 IMPACT OF DIVERSE PLANTS

Insect communities in urban green spaces are significantly shaped by the range of plants found there. In urban areas with well-planned green spaces featuring a diverse range of plant species, there is typically a greater diversity of insects. This is due to the fact that different plants support a wider variety of insect species by offering diverse resources, such as nectar, pollen, and leaves. For instance, native flora is essential for promoting variety (Figure 6.2). They have frequently developed alongside native insect species, giving them the resources they require to survive. Therefore, native plants play a crucial role in preserving native insect groups, which are essential to the health of nearby ecosystems (Wintergerst et al., 2021).

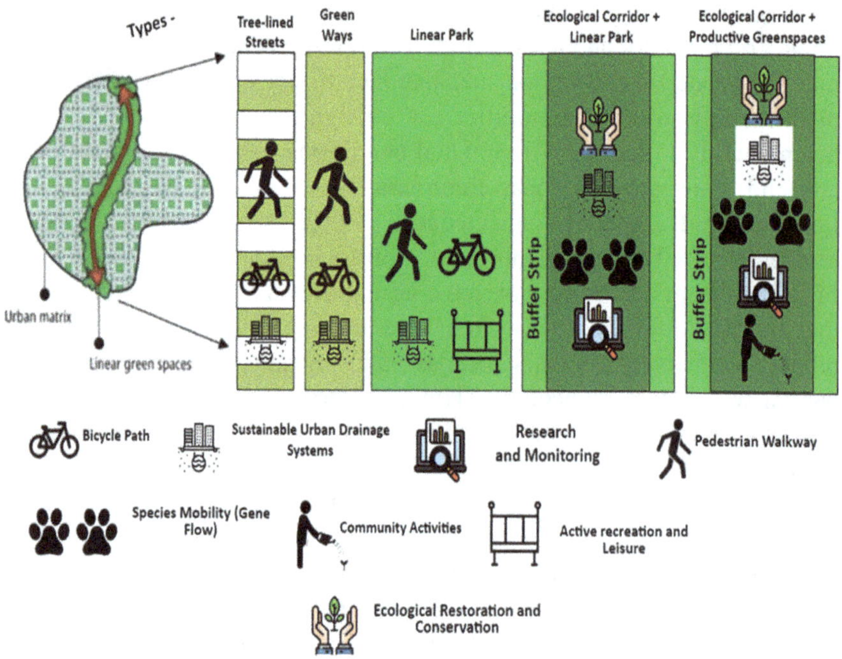

FIGURE 6.2 Types of green urban spaces.

6.4.3 THE IMPACT OF CLIMATE CHANGE

The urban environment is comprised of entirely or partially anthropogenically altered habitats formed by natural areas, sometimes via agricultural land (Robinson, 2005). In the context of the urban environment, "urban" refers not just to the city center, but also to the surrounding suburbs, urban green spaces, industrial regions, and the rural-urban borders. Urban regions have a distinct structure that sets them apart from natural environments. Human construction and behavior are the primary factors of an urban ecosystem. Urbanity consists of a dense human population, collected resources such as food and water, a range of niches such as microhabitats, flora and animal diversity, climate control options, different theories, and so on. However, one of the most noticeable elements in metropolitan environments is the heat island effect. Because of their reliance on humans, cities around the world are known to harbor common pests. Roy et al. (2009) anticipated the impact of climate change on nuisance bug species in the UK. *Blattella germanica* (German cockroach), *Cimex lectularius* (bed bug), *Monomorium pharaonis* (Pharaoh ant), *Anobium punctatum* (woodworm), *Ctenocephalides felis* (cat fleas), *Lyctus brunneus* (powderpost beetle), *Hylotrupes bajulus* (house longhorn), *Tineola bisselliella* (common clothes moth), *Dolichovespula media* (media wasp), and *Vespa crabro* (European hornet) are among the nuisance insect species highlighted in the report. *Tinea ria alternata* (moth fly) is one of the 10 species that are most likely to increase with climate change. *Lasius neglectus* (invasive garden ant), *Thaumetopoea processionea* (oak processionary moth), *Linepithema humile* (Argentine ant), *Reticulitermes grassei* (Mediterranean termite), *Culex pipiens* molestus (urban mosquito), *Culex pipiens* (mosquito), *Aedes vexans* (wetland mosquito), *Ochlerotatus cantans* (woodland mosquito), and *Musca domestica* (house fly) are also included. The study also projected that the 10 species most likely to increase with changes in precipitation patterns would be the same as those with rising temperatures, with the exception of *Musca domestica*, although it did include *Phlebotomus mascittii* (sand fly). Another aspect affecting urban insect communities is climate change. Certain insect species are spreading into cities as a result of rising temperatures. This may change the dynamics of the community by introducing new contacts and rivalry between urban and non-urban insect species (Vilenica et al., 2022).

6.5 CHALLENGES TO BIODIVERSITY OF URBAN INSECTS

Urban environments present many challenges to biodiversity, especially for insects, which comprise a significant part of urban species. These problems arise as a result of urban areas' unique features, including habitat fragmentation, pollution, loss of habitat, and the introduction of non-native species. Understanding these challenges is crucial to developing effective conservation strategies that combat biodiversity loss in urban insect populations (Ameen and Mourshed, 2017). One of the most significant threats to urban insect biodiversity is habitat fragmentation. Urbanization frequently involves the fragmentation of natural ecosystems into small pockets of green space intermingled with buildings and infrastructure. This fragmentation affects insect populations by isolating them in narrow, isolated habitats, limiting gene flow and raising the possibility of local extinction. Furthermore, fragmented habitats may lack critical resources such as food, shelter, and breeding places, reducing insect numbers. Pollution is another major concern for urban insects. Pollution in urban areas is significant due to a variety of causes, including automobile emissions, industrial operations, and household chemicals. Pollutants such as nitrogen oxides, heavy metals, and pesticides can affect insect populations by fouling their habitats and food supplies. Furthermore, pollution can have an indirect effect on insects by disturbing their reproductive cycles, weakening their immune systems, and affecting their behavior. Habitat degradation poses a significant danger to urban insect biodiversity. Natural ecosystems are frequently lost or damaged when cities grow to make space for buildings, roads, and other infrastructure. This loss of habitat limits the amount of space available for insects to live and develop, driving them into smaller and more fragmented areas.

In addition to direct habitat loss, urban expansion can degrade remaining green spaces by causing soil compaction, habitat fragmentation, and the introduction of invasive plant species. The introduction of non-native species has a considerable impact on urban insect biodiversity. Many non-native species thrive in urban contexts, outcompeting native insects for resources. These invasive species can harm native ecosystems by feeding on native insects, competing for food and habitat, and spreading diseases. In rare circumstances, non-native insects may hybridize with native species, resulting in genetic contamination and loss of genetic diversity. Climate change exacerbates many of the difficulties confronting urban insect biodiversity. Rising temperatures, altered precipitation patterns, and severe weather events associated with climate change can directly impact insect populations by affecting their physiology, behavior, and distribution. Furthermore, climate change can indirectly influence insect biodiversity by altering habitat conditions and disrupting ecological interactions. Addressing the challenges to urban insect biodiversity requires a multifaceted approach that considers the complex interactions between urbanization, pollution, habitat destruction, invasive species, and climate change. Conservation efforts should focus on preserving and restoring green spaces within cities, reducing pollution through sustainable practices, managing invasive species, and promoting climate resilience in urban ecosystems.

6.6 FACTORS AFFECTING GREEN SPACE INSECT COMMUNITIES

Urban green spaces are not independent ecosystems; rather, they are inextricably linked to the surrounding metropolitan landscape. Pollution, habitat fragmentation, and climate change are some of the variables affecting the health and composition of insect communities in these places (Maher et al., 2022). In order to manage and preserve urban insect communities, it is essential to comprehend these influences.

6.6.1 *VARIATIONS IN HABITAT*

The split of natural habitats into smaller, isolated pieces is referred to as habitat fragmentation. This is a serious problem for insect communities in green spaces in urban settings. Green spaces are becoming more dispersed as cities grow, creating smaller, more isolated habitat pockets (Lundquist et al., 2022). This may negatively impact insect communities in a number of ways, including:

1. **Decreased Gene Flow:** Fragmentation can restrict insect species' ability to migrate between patches, which lowers genetic diversity and raises the possibility of inbreeding. The long-term viability of the local population may suffer as a result. Smaller green areas may have modified microclimates, which include differences in humidity and temperature. This may have an impact on the kinds of insects that can live in certain environments.
2. **Effects at the Edge:** The surroundings surrounding green spaces frequently differ from those inside. The composition of communities might change as a result of these edge effects, which can benefit some insect species while disadvantaging others (Donkersley, 2022).

6.6.2 *ENVIRONMENTAL CONTAMINATION*

Pollution from urban areas can come from a variety of sources, such as water and air pollution. Insect communities in green areas may be impacted by these contaminants both directly and indirectly.

1. **Air Pollution:** Due to industrial operations and vehicle emissions, many metropolitan areas experience high levels of air pollution. Pollutants like nitrogen oxides and sulfur dioxide can affect insects directly or indirectly by causing harm to the plants they depend on. Furthermore, pollutants such as particulate matter can build up on plant surfaces, making them less suitable for insects to live in (Beaujour, 2022).
2. **Water Pollution:** Pollutants are frequently carried into green spaces' water bodies by urban runoff. Numerous species' larvae and aquatic insects can be poisoned by contaminated water, which will lower their numbers. Additionally, the aquatic plants that insects depend on may suffer due to pollution in the water (Soler et al., 2022).
3. **Non-Native Animals:** Non-native plant species frequently overrun urban green spaces. Because these invasive plants have the ability to outcompete native flora, fewer resources are available for native insect species. Non-native plants sometimes lack the coevolutionary history required to maintain local insect groups. The diversity of native insects may decrease as a result (Hodjat, 2022).

6.6.3 GLOBAL WARMING

The worldwide problem of climate change also has an impact on urban insect communities. Extreme weather events, changes in precipitation patterns, and rising temperatures can all affect the behavior, distribution, and life cycles of insects (Watanabe, 2022). The following are a few effects of climate change on populations of urban insects:

1. **Range Shifts:** Certain bug species are relocating to higher altitudes or latitudes in response to rising temperatures. In metropolitan settings, this may lead to new relationships and rivalries.
2. **Phenological Changes:** The timing of insect life cycles and their synchronization with plant blossoming and resource availability can be impacted by climate change. Pollination patterns and bug species' access to food may be affected by this.

3. **Increased Pest Pressure:** In urban green spaces, warmer temperatures may result in an increase in pest pressure. Increased pest management efforts would be necessary as a result, which might have unexpected effects on insects that are not the intended targets (Harvey et al., 2023).

6.6.4 TECHNIQUES FOR URBAN MANAGEMENT

The management of urban green spaces has a significant influence on insect communities. For instance, using herbicides and insecticides to eradicate weeds and other unwanted vegetation may directly harm insect populations (Vanderstock, 2022). In a similar vein, clearing out dead wood and plant detritus might remove crucial insect habitat.

6.6.5 FUNCTION FOR URBAN GREEN SPACES LIKE POLLINATION

For many plant species, pollination—the transfer of genetic material from one plant to another through reproduction—is an essential part of the process of continuously generating new generations. For pollinators, which are mostly bees (*Anthophila*), flies (*Diptera*), butterflies, and moths (*Lepidoptera*), urban trees provide an important habitat resource. For instance, urban Mead et al. (2021) have observed that wild bee communities in remnant forests are vertically stratified in the forest canopy and comprise distinct species that are unable to survive in the nearby built-up areas. According to urban parks with remaining forest patches, they support a higher diversity of butterfly species, including those that are specific to the woodland, than parks with only planted flora. But planted trees—even exotic ones—can likewise be verified by the barcoding of pollen samples taken from four different kinds of bees in five different EU cities (Müller, 2021), which are significant to pollination. Furthermore, pollination services are rendered to nearby urban and rural ecosystems by pollinators that are supported by urban forests.

Social and isolated wild bees have substantial nesting resources in urban woodlands. Social bumblebees, for example, will browse in private or communal gardens, but the less damaged soils of urban parks and woodland margins are where queens build their eggs. Many solitary bees that nest in cavities use logs, snags, and stumps as their nesting sites. Some, like leaf-cutting bees, gather leaves from various trees and bushes to divide their brood cells within the nest. Numerous bees, including members of the genus *Xylocopa*, may be restricted in their ability to find dead wood in urban areas, where they are necessary for breeding, because they are almost exclusively found in remaining urban forests. In general, a wide range of urban land coverings is necessary for many pollinators to finish their intricate life cycles; they depend on flower-rich urban green spaces for foraging and residual forests for breeding. As a result, maintaining sufficient nesting supplies in urban woods will enhance pollination in neighboring urban green spaces that value it (such as community and residential gardens).

Many weedy herbaceous plants can be found in degraded urban forests, and these species frequently offer generalist pollinators, such as non-native bees, with which they can communicate with native bees of ecological importance. But weeds can flower later or for longer periods of time than native plants, extending the food-gathering season for many pollinator groups or possibly "filling the gap" caused by climate change if fruiting and flowering phenologies change and consequently produce times when resources are scarce. Therefore, even those urban forest types that are thought to be of inferior quality need to be valued and managed effectively.

The organization of the forest supply in the landscape is significant for these taxa. Urban areas also contain a large number of tree and shrub genera that are needed by butterflies and moths for oviposition and the subsequent development of progeny.

6.7 METHODS OF CONSERVATION FOR URBAN INSECT POPULATIONS

The preservation of insect communities in urban green spaces is becoming a more crucial project as urbanization continues to change the environment. Urban insect colonies play a vital role in pollination and pest management, as well as in preserving ecological equilibrium (Diamond et al., 2023). Many conservation methods need to be put into practice in order to maintain and improve these communities.

6.7.1 CONSERVATION OF NATIVE PLANTS

The restoration and propagation of native plant species in green spaces is one of the main tactics for maintaining urban insect ecosystems. Nectar, pollen, and host plants for herbivorous species are among the vital resources that native plants offer to native insects (Forister et al., 2023). Urban green spaces can better support local insect communities by giving preference to native flora.

1. **Community Activity:** Getting the neighborhood involved in the creation and upkeep of native gardens helps promote a sense of accountability and ownership. Participation in the community can also help spread knowledge about the value of native plants and how they help sustain urban insects (Edwards, 2023).
2. **Educational Initiatives:** Workshops and public education campaigns can teach locals about the value of protecting indigenous insect species and the advantages of using native plants. The utilization of native plants in both public and private gardens can be promoted by these programs (Chowdhary, 2023).

6.8 IMPROVING THE HABITAT

In order to make urban green spaces more hospitable for insects, habitat improvements are frequently required. This may entail making structural modifications to green areas and incorporating elements that encourage insect-friendly surroundings (Gasca Alvarez and Costa, 2022).

1. **Synthesis of Ecosystems:** Meadows, wetlands, and hedgerows are just a few of the many habitats that urban planners and environmentalists can incorporate into their designs for green areas. Numerous insect species can be supported by these varied habitats, which also offer vital resources including food, shelter, and breeding grounds.
2. **Insect-Friendly Landscaping:** By creating gardens with a variety of plant species that bloom throughout the year, pollinators are guaranteed a steady supply of nectar and pollen. Furthermore, using a variety of plant heights and structures opens up more niches for insects to occupy (Dilts et al., 2023).

6.9 INTEGRATED PEST CONTROL OF URBAN INSECTS

Integrated pest management (IPM) techniques can also help insect-friendly green spaces by lowering the use of chemical pesticides (Abajue, 2023). The goal of IPM is to prevent pest issues and manage them as little as possible.

6.9.1 THE BIOLOGICAL CONTROLS

Keeping ladybugs, parasitoid wasps, birds, and other naturally occurring predators in green spaces can help manage pest populations. Utilizing biological controls increases insect diversity overall and lessens the demand for chemical insecticides (Anderson, 2022). When using pesticides, it is important to be focused and selective in order to minimize damage to insects that are not the intended target. In some situations, organic and biopesticides might be better.

6.9.2 MONITORING AND CITIZEN SCIENCE

Insect conservation can greatly benefit from community participation in scientific endeavors. Initiatives in citizen science can gather important information about the distribution, behavior, and population size of insects.

1. **Programs for Tracking Insect Populations:** Local communities can take part in these programs, which track insect populations and submit their findings to the appropriate authorities (Nsevolo, 2022). These initiatives give researchers access to useful data that they can use to spot patterns and possible dangers to populations of urban insects.

2. **Community Scientific Initiatives:** Projects involving citizen science can increase public awareness of the need to conserve urban insects. Volunteers are frequently involved in data gathering, advocacy, and teaching for these programs, which fosters in participants a sense of ownership (Palmersheim et al., 2022).

6.9.3 ECOLOGICAL PATHWAYS

Insect migration between isolated patches of green space can be facilitated by establishing and maintaining green corridors inside urban areas.

The detrimental consequences of habitat fragmentation can be mitigated with the aid of these corridors (Raghavendra et al., 2022).

1. **Bridges and Tunnels for Wildlife:** Insects and other wildlife may be able to cross busy roadways and developed areas securely through the use of bridges and tunnels. The interconnectivity of green spaces is preserved by these structures.
2. **Biodiverse Planting in Linear Green Spaces:** A variety of native plant species can be used to provide corridors for insects to travel through urban environments, even in linear green spaces such as roadside plantings (Hane and Korfmacher, 2022).

6.9.4 URBAN GREEN MANAGEMENT

Encouraging urban planning is essential to the preservation of insects. It is possible to guarantee that green spaces are established and maintained in a way that benefits urban insect communities by incorporating insect-friendly design principles into urban planning (Wingnall et al., 2023). Policies relevant to green roofs and vertical gardens: In metropolitan regions with high population densities, policies that promote the building of green roofs and vertical gardens can help insects find more resources and habitats.

6.9.5 PRESERVATION OF NATURALLY SITES

Regulations to preserve natural areas and ensure that urban growth does not encroach on vital habitats for insects and other species should be incorporated into urban planning.

6.10 CHALLENGES

Urban insect communities face many difficulties in their conservation, but by identifying these challenges and making plans for the future, we can endeavor to guarantee that insects will continue to flourish in our urban settings. Here, we discuss some of the major issues facing urban insect conservation as well as potential future paths.

1. **Loss and Fragmentation of Habitat:** The effects of urbanization are still being felt in the loss and fragmentation of habitat. Green spaces become smaller and more isolated as cities grow. Insect populations are disturbed, which may result in genetic bottlenecks that lower genetic diversity and increase the possibility of local populations becoming extinct.
2. **Environmental Pollution:** Air and water pollutants, among others, are frequently associated with urban areas. In addition to directly harming insects, this pollution can also destroy the plants that they depend on and taint green space water bodies, which will impact aquatic insects.
3. **Alien Species:** Non-native invasive plants have the power to displace native vegetation and reduce the resources that native insects have access to. The variety of nearby insect groups can be diminished, and established ecological interactions might be disrupted by invasive species.
4. **Changes in the Climate:** The behavior, distribution, and phenology of insects are being impacted by climate change. Pollination and other ecological interactions may be affected, as well as changes in the distribution of insect species and the synchronization between insects and their host plants.
5. **Application of Chemicals:** Pesticides can kill beneficial insects that are not targets, as well as target pests in urban green spaces. It remains challenging to ensure that harmful chemical use is kept to a minimum and that IPM techniques are widely implemented.

6.11 DISCUSSION FOR THE CONSERVATION OF URBAN INSECTS

1. **Sustainability in Urban Development:** The establishment of green corridors, native plant restoration, and green areas should be given top priority in future urban development. Insects require space to flourish, and sustainable urban development can help preserve the connections between green spaces and safeguard natural habitats.

2. **Citizen Science and Assessment:** Participating in citizen science projects and insect monitoring programs can help communities maintain their valuable supply of insect population data. Increased public education campaigns are needed to emphasize the value of urban conservation and insects.
3. **Analysis and Advancement:** To gain a deeper understanding of the dynamics of urban insect communities and the outcomes of different conservation measures, ongoing research is necessary. Technological advancements can help us better understand and protect urban insect populations. One such example is the monitoring of insects using environmental DNA (eDNA).
4. **Learning and Communication:** Raising public awareness of the value of insects in urban ecosystems should be the main goal of educational programs and outreach campaigns. A stronger sense of community involvement can result from encouraging stewardship and accountability for the conservation of insects.
5. **Management and Legislation:** National, regional, and local policies ought to encourage the protection of insects in urban settings. This involves laws that restrict the use of dangerous pesticides, incentives for the development of green areas that are conducive to insects, and safeguards for natural habitats.
6. **Integration:** Effective urban insect conservation requires cooperation between governmental bodies, nonprofits, academic institutions, and nearby communities. To exchange resources, talent, and knowledge, these groups ought to cooperate.

6.12 CONCLUSION

Urban insect community preservation is a significant effort with wide consequences for the well-being and sustainability of our cities. Insects play a vital role in urban ecosystems, despite the limitations presented by pollution, invasive species, habitat loss, climate change, and pesticide use. The review regarding prospective possibilities and varied strategies provides a way forward for effectively addressing these difficulties. Insect-friendly green spaces are developed and preserved through the use of sustainable urban planning, native plant restoration, and green corridor development. Integrating citizen science and learning opportunities into towns and cities cultivates a sense of responsibility and awareness for insect conservation. The efficiency of conservation efforts and our understanding of urban insect communities are enhanced by ongoing study and technological advancements. To ensure that insect conservation is prioritized in urban development, policy changes, regulation, and cooperation among numerous groups are essential. Ultimately, our ability to strike a balance between human requirements and the preservation of these little but invaluable creatures will determine the resilience and vibrancy of urban ecosystems as well as the services rendered by insects. We can create greener, more biodiverse, and sustainable cities if we work together and are dedicated to living in harmony with urban insects.

ACKNOWLEDGEMENT

This research is supported by Division of Research and Innovation, Uttaranchal University, Dehradun, India under the seed money grant reference no. UU/DRI/SM/2023-24/004.

KEYWORDS

- **biodiversity**
- **climate change**
- **decomposers**
- **green roofs**
- **green spaces**
- **urbanization**

REFERENCES

Abajue, M. C. (2023). Environmental sustainability: Relevance of forensic insects and other ecosystem services in Africa. In *Sustainable Utilization and Conservation of Africa's Biological Resources and Environment* (pp. 603–634). Singapore: Springer Nature.

Adams, B. J., Li, E., Bahlai, C. A., Meineke, E. K., McGlynn, T. P., & Brown, B. V. (2020). Local-and landscape-scale variables shape insect diversity in an urban biodiversity hotspot. *Ecological Applications, 30*(4), e02089.

Addas, A. (2023). The importance of urban green spaces in the development of smart cities. *Frontiers in Environmental Science, 11*.

Anderson, A. G. (2022). Evaluating the attractiveness of Pacific Northwest native plants to insects and gardeners.

Beaujour, P. M., & Cézilly, F. (2022). The importance of urban green spaces for pollinating insects: The case of the metropolitan area of Port-au-Prince, Haiti. *Caribbean Journal of Science, 52*(2), 238–249.

Burkhard, B., & Maes, J. (2017). Mapping ecosystem services. *Advanced Books, 1*, e12837.

Cheng, Y., et al. (2023). Impact of urbanization on ecosystem service value from the perspective of spatio-temporal heterogeneity: A case study from the Yellow River Basin. *Land, 12*.

Chowdhury, S., et al. (2023). Protected areas and the future of insect conservation. *(Trends in Ecology & Evolution, 38*(7), 681. https://doi.org/10.1016/j.tree.2023.04.006.

Cord, A. F., Bartkowski, B., Beckmann, M., & Dittrich, A., et al. (2017). Towards systematic analyses of ecosystem service trade-offs and synergies: Main concepts, methods, and the road ahead. *Ecosystem Services, 28*, 264–272.

Daniels, B., Jedamski, J., Ottermanns, R., & Ross-Nickoll, M. (2020). A "plan bee" for cities: Pollinator diversity and plant-pollinator interactions in urban green spaces. *PLoS One, 15*(7), e0235492. https://doi.org/10.1371/journal.pone.0235492.

Diamond, S. E., Bellino, G., & Deme, G. G. (2023). Urban insect bioarks of the 21st century. *Current Opinion in Insect Science, 57*, 101028. https://doi.org/10.1016/j.cois.2023.101028.

Dilts, T. E., et al. (2023). Agricultural margins could enhance landscape connectivity for pollinating insects across the Central Valley of California, USA. *PLOS ONE, 18*(2).

Donkersley, P., et al. (2022). Global insect decline is the result of willful political failure: A battle plan for entomology. *Ecology and Evolution, 12*(10), e9417. https://doi.org/10.1002/ece3.9417.

Dou, Y., Zhen, L., De Groot, R., Du, B., & Yu, X. (2017). Assessing the importance of cultural ecosystem services in urban areas of Beijing municipality. *Ecosystem Services, 24*, 79–90. https://doi.org/10.1016/j.ecoser.2017.02.011.

Dylewski, Ł., Maćkowiak, Ł., & Banaszak-Cibicka, W. (2019). Are all urban green spaces a favorable habitat for pollinator communities? Bees, butterflies, and hoverflies in different urban green areas. *Ecological Entomology, 44*(5), 678–689. https://doi.org/10.1111/een.12744.

Edwards, F. (2023). Making urban natures visible. *Urban Natures: Living the More-than-Human City, 1*.

Enssle, F., & Kabisch, N. (2020). Urban green spaces for social interaction, health, and well-being of older people—An integrated view of urban ecosystem services and socio-environmental justice. *Environmental Science & Policy, 109*, 36–44.

Feng, X. (2022). A review of the heterogeneous landscape of biodiversity databases: Opportunities and challenges for a synthesized biodiversity knowledge base. *Global Ecology and Biogeography, 31*, 1242–1260.

Fenoglio, M. S., et al. (2021). Urbanisation drivers and underlying mechanisms of terrestrial insect diversity loss in cities. *Ecological Entomology, 46*(4), 757–771. https://doi.org/10.1111/een.13041.

Forister, M. L., et al. (2023). Missing the bigger picture: Why insect monitoring programs are limited in their ability to document the effects of habitat loss. *Conservation Letters, 16*(3). https://doi.org/10.1111/conl.12951.

Gasca-Álvarez, H. J., & Costa-Neto, E. M. (2022). Insects as a food source for indigenous communities in Colombia: A review and research perspectives. *Journal of Insects as Food and Feed, 8*(6), 593–603. https://doi.org/10.3920/jiff2021.0148.

Haase, D., Frantzeskaki, N., & Elmqvist, T. (2014). Ecosystem services in urban landscapes: Practical applications and governance implications. *AMBIO, 43*(4), 407–412. https://doi.org/10.1007/s13280-014-0503-1.

Hane, E. N., & Korfmacher, K. F. (2022). Insect 'Bee&Bees' and pollinator penthouses: Teaching students about pollinators and their services in an urban environment. *Urban Ecosystems, 25*(4), 1057–1064. https://doi.org/10.1007/s11252-021-01186-4.

Hartig, T., Mitchell, R., De Vries, S., & Frumkin, H. (2014). Nature and health. *Annual Review of Public Health, 35*, 207–228. https://doi.org/10.1146/annurev-publhealth-032013-182443.

Harvey, J. A., Tougeron, K., Gols, R., Heinen, R., Abarca, M., Abram, P. K., & Chown, S. L. (2023). Scientists' warning on climate change and insects. *Ecological Monographs, 93*(1), e1553.

Hodjat, S. H. (2022). Trait in insects and trends for their control.

Hunter, R. F., Cleland, C., Cleary, A., Droomers, M., Wheeler, B. W., Sinnett, D., et al. (2019). Environmental, health, wellbeing, social, and equity effects of urban green space interventions: A meta-narrative evidence synthesis. *Environment International, 130*, 104923. https://doi.org/10.1016/j.envint.2019.104923.

Le Roux, D. S., Ikin, K., Lindenmayer, D. B., Manning, A. D., & Gibbons, P. (2014). The future of large old trees in urban landscapes. *PLOS ONE, 9*(6), e99403. https://doi.org/10.1371/journal.pone.0099403.

Lundquist, M. J., Weisend, M. R., & Kenmore, H. H. (2022). Insect biodiversity in urban tree pit habitats. *Urban Forestry & Urban Greening, 78*, 127788. https://doi.org/10.1016/j.ufug.2022.127788.

Maher, G. M., Johnson, G. A., & Burdine, J. D. (2022). Impervious surface and local abiotic conditions influence arthropod communities within urban greenspaces. *PeerJ, 10*, e12818.

Mata, L., et al. (2017). Conserving herbivorous and predatory insects in urban green spaces. *Scientific Reports, 7*(1). https://doi.org/10.1038/srep40970.

Mata, L., et al. (2021). Indigenous plants promote insect biodiversity in urban greenspaces. *Ecological Applications, 31*(4), e02309. https://doi.org/10.1002/eap.2309.

Meacham, M., Norström, A. V., Peterson, G. D., & Andersson, E., et al. (2022). Advancing research on ecosystem service bundles for comparative assessments and synthesis. *Ecosystem People, 18*, 99–111.

Ndong, G. O., Therond, O., & Cousin, I. (2020). Analysis of relationships between ecosystem services: A generic classification and review of the literature. *Ecosystem Services, 43*, 101120.

Nsevolo, M. P., et al. (2022). Reviewing entomophagy in the Democratic Republic of Congo: Species and host plant diversity, seasonality, patterns of consumption, and challenges of the edible insect sector. *Journal of Insects as Food and Feed*, 1–20. https://doi.org/10.3920/jiff2022.0024.

Palmersheim, M. C., Schürch, R., O'Rourke, M. E., Slezak, J., & Couvillon, M. J. (2022). If you grow it, they will come: Ornamental plants impact the abundance and diversity of pollinators and other flower-visiting insects in gardens. *Horticulturae, 8*(11), 1068.

Raghavendra, K. V., et al. (2022). Insects: Biodiversity, threat status, and conservation approaches. *Current Science, 122*(12), 1374. https://doi.org/10.18520/cs/v122/i12/1374-1384.

Ridzuan, N. H., et al. (2022). Insect biodiversity of urban green spaces in Penang Island, Malaysia. *International Journal of Tropical Insect Science, 42*(1), 275–284. https://doi.org/10.1007/s42690-021-00543-2.

Robinson, W. H. (2005). *Urban Insects and Arachnids: A Handbook of Urban Entomology*. Cambridge University Press.

Rocha, E. A., & Fellowes, M. D. E. (2020). Urbanisation alters ecological interactions: Ant mutualists increase and specialist insect predators decrease on an urban gradient. *Scientific Reports, 10*(1), 6406. https://doi.org/10.1038/s41598-020-62422-z.

Roy, H., Beckmann, B., Comont, R., Hails, R., Harrington, R., Medlock, J., Purse, B., & Shortall, C. (2009). Nuisance insects and climate change. Available at http://nora.nerc.ac.uk/8332/ (accessed on 3 April 2025).

Samways, M. J., et al. (2020). Solutions for humanity on how to conserve insects. *Biological Conservation, 242*, 108427. https://doi.org/10.1016/j.biocon.2020.108427.

Semeraro, T., et al. (2021). Planning of urban green spaces: An ecological perspective on human benefits. *Land, 10*(2), 105. https://doi.org/10.3390/land10020105.

Soler, R., et al. (2022). Biodiversity islands at the world's southernmost city: Plant, bird, and insect conservation in urban forests and peatlands of Ushuaia, Argentina. In *Biodiversity Islands: Strategies for Conservation in Human-Dominated Environments* (pp. 419–437). Cham: Springer International Publishing.

Tamara, M. E. L., et al. (2021). Major insect groups show distinct responses to local and regional attributes of urban green spaces. *Landscape and Urban Planning, 216.*

Theodorou, P. (2022). The effects of urbanisation on ecological interactions. *Current Opinion in Insect Science, 52*, 100922. https://doi.org/10.1016/j.cois.2022.100922.

Vanderstock, A., et al. (2022). For the love of insects: Gardening grows positive emotions (biophilia) towards invertebrates. *Journal of Insect Conservation, 26*(5), 751–762. https://doi.org/10.1007/s10841-022-00419-x.

Vilenica, M., Vuataz, L., & Yanai, Z. (2022). Introduction to the special issue 'Aquatic insects: Biodiversity, ecology, and conservation challenges.' *Diversity, 14*(7), 573. https://doi.org/10.3390/d14070573.

Villarroya-Villalba, L., et al. (2021). Response of bats and nocturnal insects to urban green areas in Europe. *Basic and Applied Ecology, 51*, 59–70. https://doi.org/10.1016/j.baae.2021.01.006.

Watanabe, R., & Ohba, S. Y. (2022). Comparison of the community composition of aquatic insects between wetlands with and without the presence of *Procambarus clarkii*: A case study from Japanese wetlands. *Biological Invasions, 24*(4), 1033–1047. https://doi.org/10.1007/s10530-021-02700-7.

Wignall, V. R., et al. (2023). Food for flower-visiting insects: Appreciating common native wild flowering plants. *People and Nature, 5*(4), 1072–1081. https://doi.org/10.1002/pan3.10475.

Wilson, R. J., & Fox, R. (2021). Insect responses to global change offer signposts for biodiversity and conservation. *Ecological Entomology, 46*(4), 699–717. https://doi.org/10.1111/een.12970.

Wintergerst, J., et al. (2021). Partial mowing of urban lawns supports higher abundances and diversities of insects. *Journal of Insect Conservation, 25*(5, 6), 797–808. https://doi.org/10.1007/s10841-021-00331-w.

Wooster, E. I. F., et al. (2022). Urban green roofs promote metropolitan biodiversity: A comparative case study. *Building and Environment, 207*, 108458. https://doi.org/10.1016/j.buildenv.2021.108458.

Zverev, V., Kozlov, M. V., Forsman, A., & Zvereva, E. L. (2018). Ambient temperatures differently influence color morphs of the leaf beetle *Chrysomela lapponica*: Roles of thermal melanism and developmental plasticity. *Journal of Thermal Biology, 74*, 100–109.

CHAPTER 7

Insect-Plant Interactions in a Changing World: Navigating Climate and Environmental Shifts

MOHSIN IKRAM[1] and MANSI SAHU[2]

[1]*School of Agriculture, Uttaranchal University, Dehradun, Uttarakhand, India*

[2]*School of Applied and Life Sciences, Uttaranchal University, Dehradun, Uttarakhand, India*

*Corresponding author

ABSTRACT

Global warming is reshaping the ecological and geographical dynamics of plant-insect coevolution. This chapter examines how rising temperatures, elevated atmospheric CO_2 levels, land-use changes, and pollution directly and indirectly affect plant-insect interactions. Indirect effects, such as altered plant nutritional quality and microclimatic shifts at the leaf surface, can significantly influence insect herbivores and pollinators. Pollinators, which rely on mutualistic relationships with specific plants, are especially vulnerable to climate-induced disruptions, potentially leading to pollination deficits. Despite trophic links, species often respond differently to global change, complicating predictions. The chapter underscores the importance of understanding plant-mediated effects of climate change on insects, rather than evaluating each partner in isolation. It also addresses the consequences of phenological shifts, species redistribution, and biochemical changes for plant-insect interactions, all of which affect ecosystem stability, pollination, and food web dynamics.

Insect-Plant Interactions, Volume 1: Influence of Biotic and Abiotic Factors.
Jamin Ali, Adil Tonğa, Adnan Shakeel, and Nasheeman Ashraf (Eds.)

DOI: 10.1201/9781779644282-7

Emphasis is placed on the adaptive capacities of both plants and insects and the need for integrated approaches to preserve these critical relationships. By identifying sustainable practices and conservation strategies, this chapter advocates for proactive efforts to maintain ecosystem resilience and biodiversity in a rapidly changing environment.

7.1 INTRODUCTION

Over millions of years, the interaction between plants and insects has developed into a complex web of adaptation and mutual reliance. Approximately 450 million years ago, ancestral plants first started to colonize terrestrial settings. Soon after, insects became their main herbivores. This signaled the beginning of a coevolutionary story in which plants and insects continually shaped each other's characteristics. Terrestrial ecosystems are dominated by plants and insects in terms of biomass and species diversity. Insects and plants have a wide range of ecological relationships, and these linkages are crucial to the structure and operation of ecosystems (David et al., 2000). With an estimated 5.5 million species, insects most likely have more diversity than any other group of creatures (Stork et al., 2015). As feeders of plant cells and disease vectors, they can have a beneficial impact on plants, such as when they serve as pollinators, or a harmful impact. Almost half of all insect species are herbivorous, meaning they feed on the tissues of living plants. This chapter explains how herbivorous insects can negatively impact the health of crops and the defense systems plants put in place (Figure 7.1).

Roughly 31% of Earth's land area is covered by forests (FAO UNEP, 2020). With over 60,000 different kinds of trees, there is a wide variety of forest types that support over 80% of the planet's biodiversity and offer numerous vital ecosystem benefits. Despite their great variability in morphology, trees usually have lengthy lifespans, elongated stems, and secondary (woody) development. In addition to giving humans energy, food, and other industrial resources, healthy forests also provide global ES like biodiversity, carbon storage, and natural climate regulation (Ashra and Nair, 2022). Forests that are healthy have adapted to withstand a certain amount of stress brought on by diseases, pests, and weather. Global climate change is a significant threat to forest health both currently and in the future, and reducing its effects will be crucial to preserving robust, healthy forests for coming generations (Whitehill, 2023).

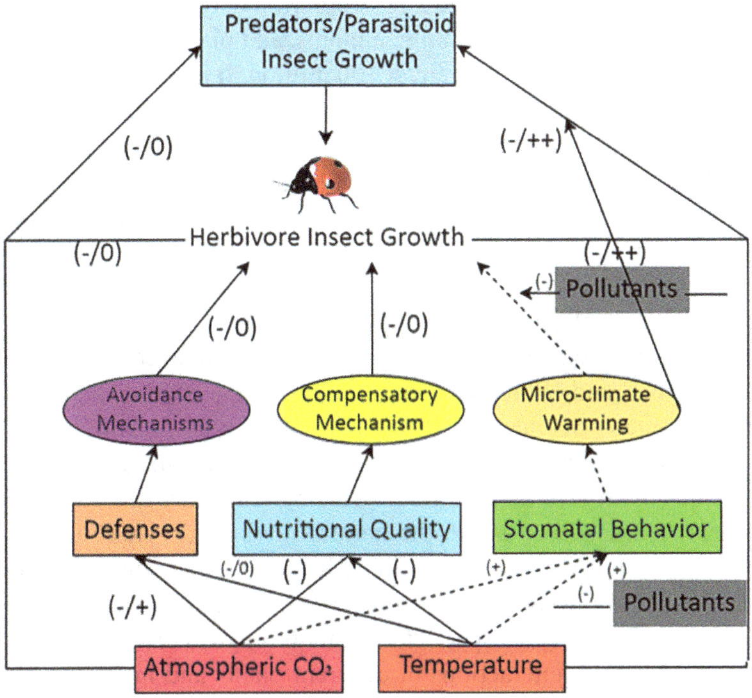

FIGURE 7.1 Plant-insect interaction shows positive and negative effects.

The interactions that species have with one another in natural ecosystems add another level of complexity to the challenges that organisms face. The capacity of individuals to adapt quickly enough to survive and maintain their fitness under erratic conditions is based on how quickly they can change their physiology, development, behavior, or morphology. This capacity is known as phenotypic plasticity, and it refers to the ability of a single gene to react to variation in environmental variables by generating alternative phenotype(s) (Coolen et al., 2022). Currently, it is widely accepted that almost all traits are changed in response to environmental variation. Essentially, this is the result of intricate, modified genotype-environment relationships that involve modifications in the expression of one or more genes, changed or novel environmental circumstances (Sharma et al., 2021).

Expanding and increasing biodiversity has benefits for all living things, including humans. Numerous domesticated animals, plants, and their derivatives are the source of many novel medications that have been developed to address problems like drought or salinized soils. The biodiversity of organisms is vast and perplexing. Approximately 1.7 million different species of organisms, such as fungi, animals, and plants, have been identified worldwide, out of a total of over 100 million. It is referred to as a wide biological annihilation that countless species of life have vanished from the face of the earth (it is estimated that half of all animals have gone extinct since 1970). Due to the extensive decline in wildlife, meticulous planning is necessary for the sustainable use of biodiversity, and the appropriate management of their use is both essential and unavoidable (Shah et al., 2023). Due to historical conditions, availability, abundance, and a wide range of biological and pharmacological effects, various ethnic groups have shown interest in medicinal herbs that grow according to Iran's climate (Groen and Whiteman, 2022). Nutritional, medicinal, cosmetic, health, and even ornamental and fuel needs have traditionally been satisfied by these resources (Pirttila, 2023). Iran has a remarkably rich and varied flora, with rare plant species growing there that are less frequent elsewhere in the globe, as are the natural goods they yield. In Persian, "mannas" or "angabins" are one of these lesser-known natural ingredients. Scholars from antiquity, like Ibn Sina and Al-Biruni, have studied them (Biniaz et al., 2023). One aspect of this rare biological diversity that arises from the ecological interactions of different plants and insects is Persian manna. Mannas, the plural of food, are sticky, sweet secretions that some insects create on particular plants as a means of eating or as a response from the plant to external temperature changes or mechanical stresses. The distribution of the host and breeding insects, the timing of production, the exploitation technique, and other variables have all been engaged in the production of manna, even if the precise methods and conditions are still unknown. Several ecological and biological elements have also been shown to have a part in the development and propagation of manna-producing insects (McNeil et al., 2022). Basic research on rangeland ecosystem management may be found in individual ecology studies (Kaur et al., 2022), and understanding the environmental elements that influence the emergence and growth of plants helps us understand how each species has adapted to its new surroundings.

7.2 THE COEVOLUTIONARY ARMS RACE

It is common to refer to the relationship among insects and plants as an "arms race," in which both parties develop strategies to defeat the other.

To stave off herbivory, plants evolved a variety of chemical defenses, including secondary metabolites and alkaloids. In order to take advantage of plant resources, insects developed physiological modifications that allowed them to detoxify or sequester these compounds. This cyclical process of evolution has caused remarkable specialization and diversification in both plants and insects (Irving et al., 2020). Mutual natural selection in more than one interacting species, or coevolution, couples the evolutionary histories of the interacting lineages in ecological interactions like plant-insect mutually beneficial relationships and antagonisms that can last for extended periods of time. Since volatile organic compounds (VOCs) constitute a component of the exchange of information between interacting lineages, the development of plant-insect interactions is probably linked to the evolution of VOCs. This mutualism is twofold: insects have the ability to modify floral fragrance production patterns (Friberg et al., 2019), and floral fragrance compositions significantly influence which insects are drawn to the plant. Therefore, it would be excellent to model both the development of plant VOCs and mechanisms to decipher the sequence and timing of evolving events in the development of relationships with insects (Akiyama et al., 2019). Novel approaches to modeling and comprehending the evolution of host collections in a clade of symbionts have been made possible by recent developments in the phylogenetic induction of coevolutionary histories. The process of host taxonomic acquisition and loss along a symbiotic clade's diversification is modeled by Braga et al. (2020). In its initial state, the genetic distance between the newly acquired host and the one that the symbiont has previously used can influence its rate of host gain. To incorporate plant-insect chemical communication, this approach can be easily modified by substituting a matrix that measures pairwise distances in VOC production for the phylogenetic relationship between hosts. Thus, one can approximate how much simpler it is for an insect to settle on a fresh host plant that emits VOCs comparable to those it already consumes, as opposed to hosts that emit extremely different VOCs (Manley et al., 2023).

7.2.1 MUTUALLY BENEFICIAL RELATIONS

Numerous plant and insect species have developed mutualistic connections that are essential to their existence, in addition to the antagonistic dynamics of herbivory-defenses. Pollination is among the most well-known instances.

Drawn to floral resources, insects unintentionally spread pollen from a single crop to another, aiding in reproduction. In addition to influencing the evolution of particular plant and insect species, this reciprocal dependence has made a substantial contribution to ecosystem stability and biodiversity (Gaugherty and Davies, 2021).

7.3 EFFECTS OF CHANGES IN THE CLIMATE ON PLANTS, INSECTS, AND THEIR RELATIONSHIPS

There are substantial effects of climate change on all aspects of agriculture. Changes in temperature, humidity, precipitation, and other meteorological factors affect how insect pests interact with plants. Because of their increased spread, increased rates of survival, and increased adaptation, insect pests have become more numerous and have the potential to cause more harm due to climate change. Changes in precipitation patterns, rising temperatures, disruptions in the gaseous state of the atmosphere, and other factors are the causes of the changes in population, movement, and insect pest behavior. Climate change affects many factors that determine how much a plant can grow. However, it would be more difficult for plants to thrive if there were more frequent high temperatures, a reduction in water availability, and modifications to the soil's composition. The connections between temperature and rising CO_2 affect plants and insects, which has significant effects on food security. Plant phenology accelerates due to warming, which causes incompatibilities between plant and insect pollinators. Similarly, altering the pace at which a plant develops in comparison to an insect's development can amplify or lessen the effects of an herbivore (Gvozdenac et al., 2022).

7.3.1 CLIMATE CHANGE'S EFFECTS ON INSECT PESTS

Changes in weather patterns, extreme weather events, and an overall increase in world average temperature have a significant impact on insect pests. Numerous insect pests' population dynamics would be impacted by these long-term and seasonal shifts. The main impact of varying climatic patterns on insect habitats and their ability to survive is observed. Important factors of climate change, such as increased CO_2 and temperature, as well as decreased soil humidity, impact the composition of insect pest populations, which in turn affects agricultural losses. By modifying the growth, survival, reproductive capacity, dispersal, and number of generations per season of insect pest populations, abiotic conditions directly affect the pace of spread and abundance of these populations. Insect pests are establishing more overwintering stages and generations as a result of the swift alterations in temperature (Dixit et al., 2023).

7.3.2 CLIMATE CHANGE'S EFFECTS ON BENEFICIAL INSECTS

The natural enemies of insect pests are impacted by climate change in many different ways. Higher temperatures, increased CO_2 levels, and decreased precipitation allow plants to provide a variety of nutritional options for diverse insect pests, which ultimately affects the fitness of predators and parasitoids that feed on insect pests (Cotrozzi et al., 2021). The primary factor causing variations in caterpillar parasitism is precipitation fluctuation, even in the presence of a diverse range of hosts and parasitoid species. Ladybird beetles (*Coccinella septempunctata*) are more effective at reducing aphid populations (*Sitobion avenae*) in hot summers than in temperate summers (Wood et al., 2022). Plants produce and release more floral nectar and volatile chemicals in response to temperature increases. The insects can fend off attacks from their natural adversaries thanks to these secretions. After surviving temperature extremes, natural enemies must adapt to climate change in order to reproduce. They must effectively locate hosts across a wide range of humidity and temperature conditions (Berasategui and Salem, 2020). The egg parasitoid *Trichogramma carverae* decreases fertility above 30?C and is incapable of identifying hosts at temperatures higher than 35?C (Chiu et al., 2019). In reaction to temperature, certain parasitoids evolve faster than their hosts, and when their hosts are absent, they frequently contribute to the extinction of a parasitoid population.

7.3.3 EFFECTS OF CHANGING THE CLIMATE ON INTERACTIONS BETWEEN POLLINATORS AND PLANTS

Because of the way abnormal temperatures impact plant and flower growth, there is a clear correlation between the decline of habitat, nutritional inadequacies, lack of diversity in diets, and climate change. Plants are currently flowering an entire month earlier than they did 45 years ago due to climate change, which forces flowers to open just over a day earlier annually. Lastly, early-flowering plants indicate that they have not been pollinated, which leaves them without food for butterflies and bees (Chorbadjian et al., 2019). In a shared habitat, there may be a decline in the temporal or spatial co-occurrence of interacting partners and mismatches in plant-pollinator interactions. These kinds of temporal conflicts between pollination insects and plants have drawn more attention. These mismatches may be caused by changes to the pollinator's phenology, which might occur sooner or later, or by adjustments to the plant's flowering time. Spatial disruption may also occur in the co-occurrence of pollinators and plants, which is necessary for interaction to take place. Global warming may cause a decrease or rise in the geographical overlap of interacting partners, contingent upon the species' plasticity, adaptability, and life cycle traits. Apart from incongruities in time or space, climate change can also impact the relationships between pollinators and plants that are controlled by morphological or physiological traits. In addition to crop morphology, the physical fit of the relationship can be influenced to gain access to plant resources because pollination success is dependent on morphological traits like tongue length or body size overall. For instance, it has been demonstrated that average temperature increases negatively impact body size in numerous species. Temperature increases will also impact the attractiveness of plants, the quality and quantity of plant resources, and the foraging behavior of pollinators (Anderson et al., 2020).

7.4 IMPORTANCE TO ENVIRONMENTAL SYSTEMS

Globally, interactions between insects and plants shape ecosystems' resilience, functionality, and overall structure such as:

1. **Biodiversity:** As a result of these interactions, organisms become more diverse and adapt to one another. Certain plants and insects have unique connections that greatly contribute to the terrestrial ecosystems' total biodiversity.
2. **Ecosystem Services:** Insect pollination is essential to both crop yields and the preservation of natural plant ecosystems. In addition, insect herbivory influences the dynamics of plant populations and the makeup of communities as an inherent regulatory mechanism.

3. **Energy Flow:** The exchange of energy between plants and insects is essential to ecosystems. Higher trophic levels mostly obtain their energy from insects that graze on plants, which affects the dynamics of the food web and the transmission of energy.

4. **Ecosystem Stability:** Ecosystems are resilient and stable because of the delicate balance that exists between plants and insects. Changes in these relationships can have a domino effect, affecting different trophic levels and changing the dynamics of the ecosystem (Aranjuelo et al., 2014).

7.5 A CHANGING WORLD WITH FLORAL REWARDS

An important moderator of plant-pollinator interactions is the amount and makeup of floral nectar. Visitors to flowers are rewarded with sugar and amino acids, but nectar can additionally include substances that shorten and/or stop bee visits, thus preventing pollination (Awmack et al., 1997). Secondary metabolites of nectar can affect pollinators negatively, neutrally, or positively (Banerjee et al., 2019). These effects may influence pollination across the population. Given that PSMs may be important for insect pollinator well-being and pollinator-pathogen interactions (Bellard et al., 2014), their functional roles in nectar are poorly understood and deserve more research. Furthermore, it is essentially unknown how global change drivers affect the secondary metabolites of nectar. Due to their frequent correlation with leaf secondary metabolites (Bultman et al., 2018), nectar secondary metabolites may reflect variations in the chemical defenses of the foliage under various GEC conditions. There has only been one study that we are aware of that looked at the impact of GEC on secondary metabolites in nectar. The reaction of floral attractiveness and defense features to leaf herbivory was examined by Halpern et al. (2023). Stress from drought in *Nicotiana quadrivalvis* discovered that nectar nicotine was unaffected by the presence of water. The general response of nectar to GEC variables is not well understood. Few studies have looked at secondary metabolites; most research thus far has concentrated on main metabolites. Temperature and water availability influence nectar quantity, sugar concentration, and sugar composition. However, these variables have not been extensively examined in relation to global change. Research on the effects of raising the temperature in open-top chambers on nectar quantity and sugar concentrations has produced contradictory results. While increased CO_2 can either increase or decrease nectar production, greenhouse drought treatments have the potential to lower nectar volume (Ferguson et al., 2019). The method by which different species respond to variations in nectar caused by GECs will probably differ. Lepidopteran fitness, for instance, may be impacted by changes in nectar amino acid content. However, the effects of nectar quantity and composition variations on other pollinators may vary depending on individual preferences and resource needs (Figure 7.2). Furthermore, different reactions in nectar can be elicited by interactions between GEC parameters, such as temperature, nitrogen, and CO_2, rather than by single elements. This can have implications for pollinator longevity and preference. Few studies have looked at how changes in nectar caused by GEC variables will affect pollinators, despite the widespread knowledge of nectar's significance as a pollinator resource (Frago et al., 2012).

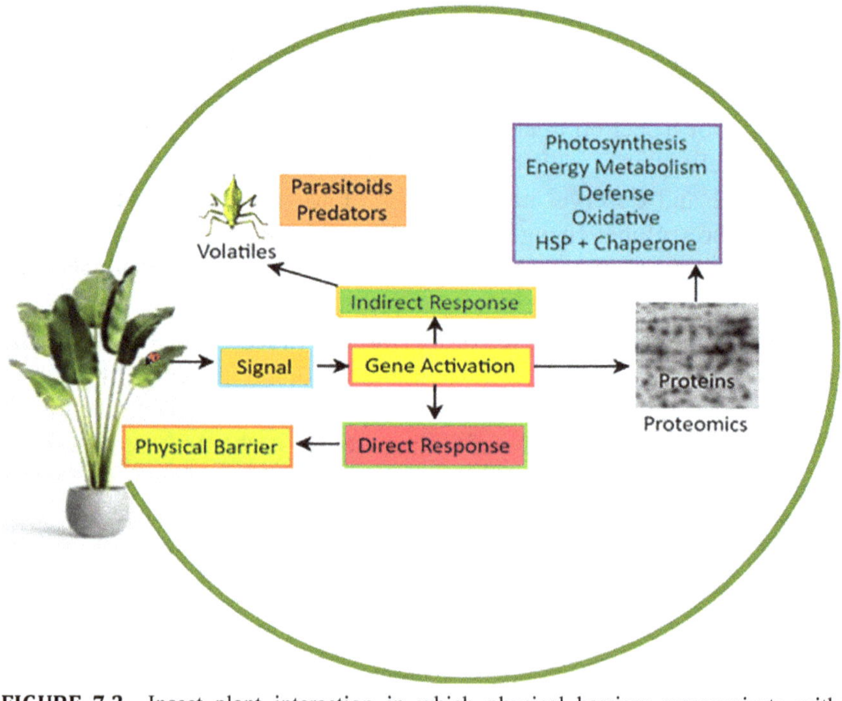

FIGURE 7.2 Insect plant interaction in which physical barriers communicate with parasitoids predators.

7.6 PLANT PERCEPTION AND RESPONSES TO INSECT ATTACK

Over the course of their almost 400-million-year coexistence, insects and terrestrial plants have changed in how they interact. There are certain partnerships that are mutually advantageous, such as defense aid and pollination. Furthermore, certain plants (including those in the genus Nepenthes, Drosera, and Dionaea) actually eat insects in order to survive. Carnivorous plants is another term for these. Over the course of development, plants have evolved a variety of defense mechanisms to keep out insects and other animals. The plant's ability to defend itself depends on these defense systems. Two well-known forms of plant defenses that are crucial for plant survival and reproduction in the face of predators are constitutive and induced defenses. Herbivores' invasive behavior is efficiently curbed by the various defense mechanisms found in plants. Plants find it difficult to recognize herbivore attacks because they can trigger their defensive mechanisms and cause them to react in ways that are necessary to survive in a harsh ecological environment. Enhancing crop resilience to biotic stress will be made easier for researchers if they comprehend the mechanisms behind this response (Gerard et al., 2013).

Plants' perception systems have evolved to detect damage during the duration of their long coexistence, and they contain unique receptors that allow them to distinguish between damage produced by random events and damage caused by feeding insects. The plant saves energy for development and reproduction when harm is assessed to be mechanical in nature, caused by natural external sources rather than activating its particular defensive mechanism. However, certain metabolites, primarily fatty acid-amino acid conjugates (FACs), along with others referred to as herbivore-associated stimulants (HAEs), bind to certain receptors on the cell surface when damage is produced by insect feeding (Hansen and Moran, 2014). These receptors start defense responses and phosphorylation cascades; they are referred to as pattern recognition receptors (PRRs). The understanding of mechanical injury transmission of signals has improved in recent years due to an increase in in-depth study on this signaling pathway. Cell wall breakage or alteration, as well as cell rupture, will initiate the defensive reaction. During this process, the first messengers to convey damage information may be glutamate, peptides, ATP, oligogalacturonides (OGs), high-mobility group (HMG), and so forth. Damage-associated molecular patterns (DAMPs) is the term for them. As of right now, at least 18 different species of DAMPs have been found and studied. DAMPs bind to particular receptors to initiate signal transduction that is then carried out by phosphorylation cascades and downstream CaBP reactions. Damage-related genes are regulated by this process. Additionally, these signals are further mediated by either long-range signals like glutamate, ROS, and others that carry damage signals to nearby tissues and organs, or short-range signals like Ca^{2+} waves and induced wound-activated surface potential (WASP) signals, and even throughout the entire organism. Certain plants also use VOCs to signal damage to other members of the population. This allows for the activation of the defense response throughout the entire organism, or perhaps the entire population (Jamieson et al., 2012). It is now generally accepted that the initial signals originate from damaged cells and their cell walls, in addition to the biological components of the attackers, such as glutamate, ATP, and HMG, which are derived from the damaged cells' cytoplasm (Ramachandran et al., 2019). Additionally, the destruction of cell walls by polygalacturonases (PG) results in the production of OGs, insect salivary protein molecules, and so on (Pontiggia et al., 2020). When these signaling molecules attach themselves to cell surface receptors, they have cellular consequences that are downstream (Kimura et al., 2020; Toyota et al., 2018). For instance, intracellular Ca^{2+} concentration rises, and CaBP controls the expression of genes relevant to damage when glutamate functions as the initial messenger. Simultaneously, the influx and outflow of ions modify the internal and exterior electrical potential of the cell, thereby relaying the signal to nearby cells. MAPK is triggered to control the expression of damage-related genes through receptors when cell wall damage breaks down under the influence of PG to generate OGs, which function as the initial messenger to relay damage signals (Wang et al., 2023). Furthermore, the expression of genes can be regulated by ATP, HMG, and other DAMPs analogously. These genes control the synthesis of JA, callose, secondary metabolites, and other substances that help heal wounds by enhancing the development of callus and strengthening the body's defenses. It is important to note that the interior of injured cells in plants should be closely examined because they may contain unique DAMPs. While the majority of studies on DAMPs have centered on the metabolites or fragments of fungi and insects, there are few investigations on the immune system reactions triggered by tiny molecules in plants. A unique strategy that can improve plant defense and pest and disease resistance is to use plant small molecules as triggers, offering a fresh approach to helping crops withstand biotic stress (Wang et al., 2023).

Glutamate and ATP, which are components of cells, separate in the periplasmic region following cell injury. Following their binding to adjacent cell membrane receptors like GLR and P2K1, they cause an inflow of Ca^{2+} and an outflow of Cl^{-}, which polarizes the cells and may produce a difference. Removing H^{+} via the AHA1 receptor channel causes depolarization. It has been suggested that by influencing GLR, H^{+} may also aid in depolarization. Concurrently, second messengers such as Ca^{2+} are delivered to peripheral cells via plasmodesmata, causing these cells to produce comparable reactions. WASPs are the result of this. Furthermore, Ca^{2+} controls the expression of pertinent genes via CaBP. Through the action of PG, cell wall destruction triggers the creation of OGs, which in turn controls the expression of associated genes via the MAPK pathway. This control affects defensive systems, particularly the formation of callose, ROS signals, and JA synthesis. HMG is another biological component that is regulated by the MAPK pathway, albeit the receptor is still unknown (Li et al., 2020).

It is thought that plants recognize predators by detecting differences in insect HAEs, which involve things like insect saliva, egg effusion, dung, and other materials. Plants are able to choose the best protection tactic thanks to this ability. Plants can differentiate between physical injury and herbivore predation by identifying PRRs and HAEs on their cell surface. This allows them to recognize the assailant and allocate resources optimally. However, there is currently little research on HAE, and the method by which plants recognize predators is still a topic of study (Rizwanuddin et al., 2023).

7.6.1 PLASTICITY OF PHENOTYPE IN PLANT DEFENSES

The main mechanism by which animals quickly adapt to their surroundings and deal with stress is phenotypic plasticity, a common phenomenon that is different from the drawn-out and sluggish process of trait development linked with genetic variation. The ability of an organism to express distinct phenotypes in response to its biological or abiotic surroundings is known as phenotypic plasticity. Induced defenses have been linked to the development of plasticity traits in plants. Just like genetic diversity, adaptations aren't always exclusive to the environment. A phenotypically flexible trait is one that enables an organism to alter its features in order to adapt to its surroundings. However, there are also non-adaptive or neutral phenotypes, which are far from an optimal characteristic (Platella et al., 2022).

7.7 THE INSECT AS A MICROBIAL HOST

Insects choose their hosts based on the ability of their endosymbionts to synthesize nutrients. Nevertheless, along the course of evolution, certain symbionts have lost the capacity to synthesize distinct chemicals in various hosts. For instance, it has been reported that genes for the synthesis of biotin in *Acerthosiphon pisum*, arginine biosynthesis in *Baizongia pistacea*, and tryptophan and riboflavin in *Cinara cedri* have been lost in *Buchnera aphidicola*. The insect may feel pressured to choose a host plant that can supply the lacking nutrient if its potential for biosynthesis is lost. Furthermore, certain mutualists or symbionts alter their insect host's behavior for their own adaptive benefit. One can witness such "Parasite Manipulation" in the "symbiont capsules" that the stink bug, *Megacopta punctatissima*, deposits on the underside of its egg masses. In order to acquire these symbionts and go quiescent, newly hatched nymphs exhibit a roaming behavior (Pineda et al., 2020).

7.7.1 SECRETIONS AND PRODUCTION OF SALIVA

The insect's foregut and midgut release saliva, an oral fluid and regurgitant that lubricates its mouthparts while it feeds. Fatty acid conjugates, proteins derived from plants and insects, insect enzymes, and elicitors make up the regurgitant portion of saliva. When food is chewed by insects, saliva may contain enzymes that break down the food before intestinal enzymes can process it. The main enzymes found in oral secretions are amylase and invertase, but they can also include other enzymes that aid in preventing infection during feeding, such as the enzymes glucose oxidase (GOX) and phospholipase C. Saliva solidifies to create a stylet sheath over the mouthparts of phytophagous Hemipterans, preventing leaks while eating plant sap. Additionally, oral secretions contain elicitor molecules that are responsible for starting induced plant responses, such as volicitin, inceptin, and N-linolenoyl-Gln (Pixley et al., 2019).

Additionally, salivary secretion, elicitor chemicals, proteins, and insect defense enzymes like GOX are all influenced by insect endosymbionts. Insects with and without endosymbionts show significant differences in salivary protein composition. Furthermore, microorganisms found in insect regurgitant may defend the insects during feeding by blocking crucial plant-protective enzymes such as polyphenol oxidase (PPO). These symbionts may affect not only the expression of insect genes but also the modification of the ensuing plant defense mechanism. Conversely, it has been demonstrated that specific endosymbiotic proteins found in insect saliva can activate plant defense mechanisms (Raffa et al., 2020).

7.8 THE SIGNIFICANCE OF MODIFIED PHENOLOGY FOR INSECTS AND PLANTS

Climate change affects both plant and insect phenology by delaying the timing of climatic events. Important life cycle events, including flowering, development, and migration, are influenced by changing patterns of precipitation and rising temperatures. A mismatch in the emergence of some bugs and the accessibility of plant hosts, for example, might result from earlier springs and longer growing seasons, which can have an impact on important interactions like herbivory and pollination. The survival and reproductive success of animals that depend on coordinated life cycle events are threatened by this mismatch (Ricciardi et al., 2021).

7.8.1 CHANGES IN ABUNDANCE AND DISTRIBUTION

The geographic ranges of plants and insects are influenced by variations in precipitation patterns and temperature. In response, a large number of species are changing their geographic range and migrating toward more hospitable environments. Nevertheless, not every species has the same rate of adaptation or migration, which could result in incompatibilities between interacting species. These changes in distribution may also have an impact on the biodiversity of the area, changing ecosystem functions and community structures (Saidi et al., 2021).

7.8.2 PHYSIOLOGICAL AND CHEMICAL REACTIONS

Climate change modifies the surrounding environment, which affects the physiology and chemistry of plants. Increased carbon dioxide concentrations can change the nutritional makeup of plants and the way secondary metabolites are produced, which can have an impact on the plants' ability to repel insects and their appeal to them. In response, insects may show alterations in their physiological responses and eating habits. For instance, changed plant chemistry may affect herbivorous insects' capacity to grow, develop, and survive by making their host plants less palatable (Blanchard and Bowers, 2020).

7.8.3 COPING AND ADAPTATION TECHNIQUES

Despite the obstacles posed by climate change, certain adaptive techniques are shared by plants and insects. Certain species have phenotypic plasticity, which enables them to modify their physiology, behaviors, or life cycles in accordance with shifting environmental factors. Furthermore, dispersing species may seek new habitats or change their range to accommodate more hospitable climates. However, species differ in their capacity for adaptation, which could result in disparate effects on other taxa and ecosystems (Davis et al., 2022).

7.8.4 IMPACTS ON PLANT REPRODUCTIVE SUCCESS AND POLLINATION DYNAMICS

One important ES that insects help with, pollination, is vulnerable to disturbances brought on by climate change. Ineffective pollination can be caused by alterations in flowering timings, distribution patterns, or the mismatch between pollinators and flowering plants. Decreased pollination efficiency can impact entire plant groups as well as individual species by lowering plant reproductive success. The stability of ecosystems that depend on pollination services, native plant development, and agricultural productivity are all at risk in this scenario (Harvey et al., 2020).

One of the primary causes of the reported pollination crisis is frequently identified as urbanization. Since 1950, the global population has been rapidly increasing, and by 2050, there will be an equal number of people living in urban areas as there are people on the earth today (United Nations Report, 2014). The effectiveness of pollination can also be lowered by other variables, such as air pollution, one of the most pervasive environmental consequences caused by humans, which decreases plant attractiveness and inhibits interactions between plants and pollinators. The floral volatiles that honey bees use to identify flowers can be quickly destroyed by diesel exhaust. Additionally, it has been demonstrated that bees trained with contaminated volatiles took longer to recognize an attraction and did not reliably recall a polluted stimulus over the years (Hu et al., 2021).

Interactions between pollinators and plants are crucial for the welfare of both humans and biodiversity. Numerous species of plants and animals, primarily insects but also certain birds, lizards, and mammals, are mutually dependent on one another for food and pollination, and how they interact may have an impact on the persistence of populations (Secretariat, 2021).

7.9 TROPHIC INTERACTIONS AND BIODIVERSITY: CONSEQUENCES

Trophic linkages in ecosystems are based on interactions between plants and insects. Changes to these relationships may have a domino effect on trophic dynamics, affecting the patterns of food webs and the relationships between different species. For example, decreased accessibility to specific plant resources could affect the higher trophic levels, which include insect-eating birds or predators that are impacted by herbivorous insects. These disturbances may result in modifications to community structures, alterations in species composition, and even a decline in the total biodiversity of ecosystems (Kopaczyk et al., 2020).

7.9.1 EXAMINATION OF CASCADE IMPACTS ON ECOSYSTEM FUNCTIONS

Disrupted insect-plant interactions have a domino impact on ES and processes. For example, alterations in plant-insect interactions leading to changed herbivory rates might impact seed distribution, vegetation dynamics, and nutrient cycling. Reduced pollination efficiency can also have an impact on fruit and seed production, which in turn affects other creatures' access to food sources (Figure 7.3). These cascade effects have the power to modify how an ecosystem functions, including resilience to environmental changes, production, and nutrient cycling (Lehmann et al., 2020).

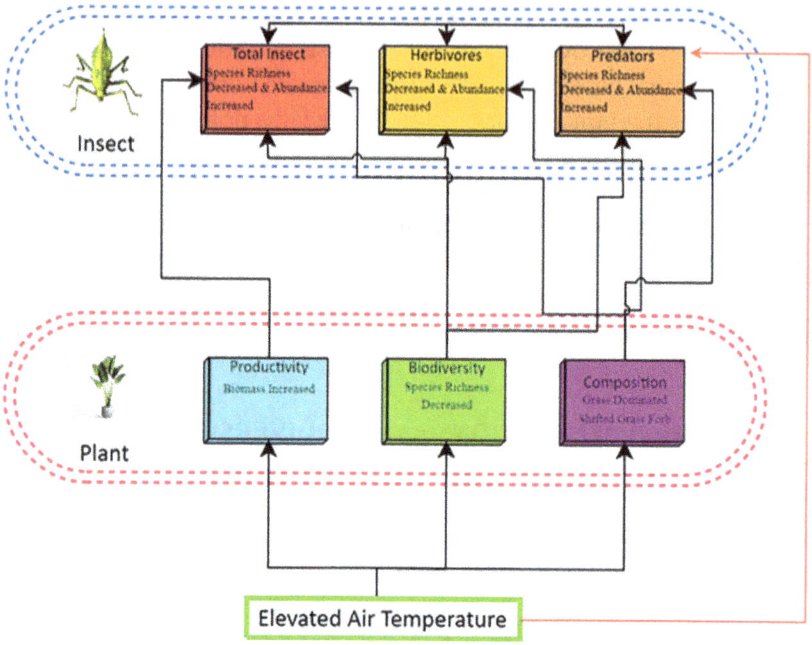

FIGURE 7.3 Changes in both plant and insect in changing world.

7.9.2 COMPREHENDING FEEDBACK LOOPS IS CRITICAL FOR CONSERVATION

It is essential to comprehend feedback loops that arise from interrupted interactions for practical conservation tactics. Ecosystem impacts can be amplified by either beneficial or adverse feedback loops that are created when important interactions are disrupted. For example, a decrease in pollinator numbers may result in fewer plants being reproduced, which may impact the availability of food for other organisms and hence have an adverse effect on the environment. Designing conservation initiatives with the goal of restoring and maintaining healthy ecosystem functioning requires an understanding of and ability to manage these feedback loops (Lüdecke et al., 2021).

7.9.3 PRINCIPLES OF CONSERVATION

Sustaining Biodiversity and Ecological Resilience: Given the interdependence of species within ecosystems, conservation efforts ought to place a high priority on maintaining biodiversity. In order to preserve ecological balance, this entails creating protected areas, preserving important ecosystems, and preserving habitats that sustain a variety of insect and plant populations. The conservation and restoration of habitats are essential for preserving the beneficial interactions between insects and plants. Restoring damaged habitats and establishing pathways for species migration can help make ecosystems more resilient. Insect populations are essential for pollinating and other ecological functions, and they can be preserved by creating a variety of habitats that support different plant species (Piao et al., 2019).

7.9.4 ECO-FRIENDLY METHODS AND REDUCTION

1. **Function of Sustainable Agriculture:** Maintaining insect-plant interactions is largely dependent on sustainable farming methods. Diverse environments that support beneficial insects are fostered by the application of agroecological techniques that enhance the environment, such as polyculture, farming, and integrated pest management (IPM). The equilibrium between plants and insects in agricultural settings can be preserved by avoiding the overuse of pesticides and promoting natural pest management systems.
2. **Methods for Prevention to Reduce Adverse Effects:** Proactive steps are needed to reduce the adverse effects of changes in the environment on insect-plant interactions. This includes preventing habitat fragmentation, managing invasive plants and animals that interfere with native interactions, and lowering other stressors like pollution and habitat loss. Furthermore, raising public awareness and educating the public about the significance of these interactions can help conservation initiatives succeed.

7.9.5 INTEGRATED METHODS

The key to developing resilient insect-plant interactions is to combine conservation strategies with sustainable management practices:

1. **Ecosystem-Based Management:** By applying management techniques that take into account the health and resilience of the

ecosystem as a whole, as opposed to just concentrating on individual species, ecosystem health is promoted.

2. **Collaborative Conservation:** Effective preservation and utilization depend on cooperation between scientists, decision-makers, land management, and local communities. Involving stakeholders in decision-making procedures encourages a sense of collective accountability for maintaining these connections (Cárcamo et al., 2024).

7.10 PLANT–MICROBIAL AND INSECT INTERACTIONS

Plants connect two distinct ecosystems—one above ground and one below—making them complex systems in and of themselves. Much research has been done on the interactions between plants and various pathogens, focusing on interactions between plants and a single bacterium or a particular type of bug, either in a greenhouse or in a lab setting. Still, the circumstances are not utopian. In addition to leaf-chewing insects, bacterial and soil-dwelling herbivores, and a variety of fungal pathogens both above and below ground, plants must contend with multiple hazards, both biotic and abiotic, simultaneously in harsh environmental conditions like extreme heat, salt, or heavy metal stresses, and so forth. This section discusses how different bacteria affect the relationships between plants and insects. In this part, we address the plant-insect interactions affected by different microbes. These interactions are significantly influenced by abiotic variables, yet this field's research continues to be in its early stages (Zhao et al., 2024).

7.10.1 IMPACT OF BACTERIA ON PLANT-INSECT RELATIONSHIPS

Microbiomes in the rhizosphere are made up of both harmful and benign microorganisms. The onslaught of herbivores and other pathogens causes the plant to modify its rhizosphere microbiome to defend itself against both above-ground and below-ground adversaries. Thus, the chemical composition of the roots varies based on mutualistic or antagonistic interactions, increasing the complexity of the interactions between the insect and the host. Nodule herbivory has grown as a result of the mutualistic relationship that nitrogen-fixing bacteria have evolved with leguminous plants, providing them with nitrogen (N) as a substitute for carbon (C). Root nodules provide food for insects in the groups Coleoptera and Diptera, meeting their nutritional needs (Johnson and Rasmann). For instance, *S. lineatus* and *S. hispidulus* are typically regarded as obligatory nodule feeders since, in their early developmental stages, they feed on nodules; yet, later on, they start to feed on other portions of the root system. It is unknown if plants may benefit from this kind of relationship. However, given that *Trifolium repens* plants are known to attract nodule herbivores by raising the level of formononetin, an iso-flavonoid in their roots that serves as a chemical signal for the insects and aids in their location, it is possible that there is a positive or negative connection that exists between the two. Still, additional research is needed to comprehend the behavior of this interaction between MPI (Riaz et al., 2024).

The symbiotic relationships of microorganisms below the ground have also been shown to increase the susceptibility of the plant to above-ground herbivory; however, the exact function of this interaction is yet unknown. For instance, the Colorado potato beetles impede the plant's JA-associated defense by using the symbiotic bacteria present in the plant, which can increase the plant's susceptibility to herbivores that nibble on it. The Pseudomonas bacteria release a protein called flagellin, which also causes JA-SA cross-talk and prevents the plant from defending itself against beetles, assisting students in demonstrating better results. The phloem-feeding whitefly, *Bemisia tabaci*, exhibits resistance to the root-based bacterial disease *Ralstonia solanacearum* in a manner akin to that of the above-ground herbivore. However, by boosting the number of gram-positive bacteria, it also initiates a favorable alteration in the rhizosphere microbiome's composition.

Bacillus thuringiensis is the best example to comprehend how plant microorganisms affect insect herbivory (Bt). These bacteria are typically found close to the soil's surface and in the phyllosphere. These bacteria are consumed by insects like *Diabrotica virgifera* when they feed on plants and plant parts. The Cry proteins, which are released by bacteria, function as an insecticide by breaking down the gut membrane of insects and killing them. It is unclear what part each organism plays in adjusting the defense and ensuring their survival because there has been little research on the plant-bacteria-herbivore defense. Thus, additional research in this area is needed to gain more insight into the various roles that bacteria play in plant-insect interaction networks (Abbasi et al., 2024).

7.10.2 THE EFFECT OF FUNGI ON PLANT-INSECT RELATIONS

The relationships that fungi have with plants vary greatly, from mutualistic relationships to harmful relationships. Instead of forming a harmful or symbiotic connection with the non-pathogenic fungus, plants adopt a variety of defense mechanisms to protect themselves against the pathogenic fungus. However, in forests and agricultural fields, these kinds of plants are attacked simultaneously or consecutively by both fungi and insects. Together, these intricate interactions mold a plant's final defensive characteristics against all of its adversaries. As an illustration, the plants *Arachis hypogaea* and *Theobroma cacao* generate the pathogenic fungus-killing compounds AhPR-10 and TcPR-10, respectively (Elmquist et al., 2024).

7.11 CONCLUSION

The complex interactions between plants and insects provide the foundation of ecosystems and act as a pivot in the larger scheme of life. These significant interactions are facing previously unheard-of difficulties due to changes in the climate and environmental upheavals, which are upsetting phenological synchronization, changing distribution patterns, and altering the chemistry of plants that are essential for different ecological partnerships. Despite these obstacles, there are signs of hope due to the adaptability and resiliency displayed by plants and insects. The potential for resiliency within these relationships is highlighted by their capacity to adjust behaviorally, exhibit phenotypic plasticity, and change in response to changing conditions. To protect these fragile interactions, conservation activities that prioritize habitat restoration, biodiversity preservation, and environmentally friendly methods are essential. Recognizing how altered interactions affect trophic connections, ecological stability, and pollination dynamics in far-reaching ways emphasizes how urgent it is to take proactive conservation efforts. The complex web of life that depends on these basic interactions for the resilience and stability of our ecosystems must be preserved. To do this, joint efforts and ongoing multidisciplinary study are essential.

ACKNOWLEDGEMENT

This research is supported by Division of Research and Innovation, Uttaranchal University, Dehradun, India under the seed money grant reference no. UU/DRI/SM/2023-24/004.

KEYWORDS

- **ecosystem**
- **plant defense**
- **plants-insect interaction**
- **pollination**
- **sustaining ecosystem**
- **trophic interactions**

REFERENCES

Abbasi, A., et al. (2024). Climate change and biodiversity of beneficial and social insects. In *Climate Change and Insect Biodiversity* (pp. 160–178). CRC Press.

Akiyama, K., Alberdi, A., Alef, W., Asada, K., Azulay, R., Baczko, A. K., Ball, D., Baloković, M., Barrett, J., Bintley, D., & Blackburn, L. (2019). First M87 event horizon telescope results. V. Physical origin of the asymmetric ring. *The Astrophysical Journal Letters, 875*(1), L5.

Anderson, R., Bayer, P. E., & Edwards, D. (2020). Climate change and the need for agricultural adaptation. *Current Opinion in Plant Biology, 56*, 197–202. https://doi.org/10.1016/j.pbi.2019.12.006.

Aranjuelo, I., Arrese-Igor, C., & Molero, G. (2014). Nodule performance within a changing environmental context. *Journal of Plant Physiology, 171*(12), 1076–1090. https://doi.org/10.1016/j.jplph.2014.04.002.

Ashra, H., & Nair, S. (2022). Review: Trait plasticity during plant-insect interactions: From molecular mechanisms to impact on community dynamics. *Plant Science: An International Journal of Experimental Plant Biology, 317*(111188), 111188. https://doi.org/10.1016/j.plantsci.2022.111188.

Awmack, C. S., Woodcock, C. M., & Harrington, R. (1997). Climate change may increase vulnerability of aphids to natural enemies. *Ecological Entomology, 22*, 366–368. https://doi.org/10.1046/j.1365-2311.1997.00069.x.

Banerjee, S., et al. (2019). Agricultural intensification reduces microbial network complexity and the abundance of keystone taxa in roots. *The ISME Journal, 13*(7), 1722–1736. https://doi.org/10.1038/s41396-019-0383-2.

Bellard, C., et al. (2014). Vulnerability of biodiversity hotspots to global change. *Global Ecology and Biogeography: A Journal of Macroecology, 23*(12), 1376–1386. https://doi.org/10.1111/geb.12228.

Berasategui, A., & Salem, H. (2020). Microbial determinants of folivory in insects. In *Cellular Dialogues in the Holobiont* (pp. 217–232). CRC Press.

Biniaz, Y., et al. (2023). Pest and disease management under changing climate. In *Strategizing Agricultural Management for Climate Change Mitigation and Adaptation* (pp. 149–165). Cham: Springer International Publishing.

Blanchard, M., & Bowers, M. D. (2020). Critical phenological events affect chemical defense of plant tissues: Iridoid glycosides in a woody shrub. *Journal of Chemical Ecology, 46*(2), 206–216. https://doi.org/10.1007/s10886-019-01135-8.

Braga, L. P. P., Spor, A., Kot, W., et al. (2020). Impact of phages on soil bacterial communities and nitrogen availability under different assembly scenarios. *Microbiome, 8*, 52. https://doi.org/10.1186/s40168-020-00822-z.

Bultman, T. L., et al. (2018). Complex interactions among sheep, insects, grass, and fungi in a simple New Zealand grazing system. *Journal of Chemical Ecology, 44*(10), 957–964. https://doi.org/10.1007/s10886-018-0993-6.

Cárcamo, H., et al. (2024). Revising economic injury levels for *Lygus* spp. in canola: The value of historical yield and insect data to improve decision making. *Crop Protection (Guildford, Surrey), 176*, 106467. https://doi.org/10.1016/j.cropro.2023.106467.

Chiu, C. C., Keeling, C. I., & Bohlmann, J. (2019). The cytochrome P450 CYP6DE1 catalyzes the conversion of α-pinene into the mountain pine beetle aggregation pheromone trans-verbenol. *Scientific Reports, 9*(1), 1477. https://doi.org/10.1038/s41598-018-38047-8.

Chorbadjian, R. A., Phelan, P. L., & Herms, D. A. (2019). Tight insect-host phenological synchrony constrains the life-history strategy of European pine sawfly. *Agricultural and Forest Entomology, 21*(1), 15–27. https://doi.org/10.1111/afe.12299.

Coolen, S., Magda, R. der-Molen, & Welte, C. U. (2022). The secret life of insect-associated microbes and how they shape insect-plant interactions. *FEMS Microbiology Ecology, 98*(9). https://doi.org/10.1093/femsec/fiac083.

Cotrozzi, L., et al. (2021). In the tripartite combination ozone-poplar-*Chrysomela populi*, the pollutant alters the plant-insect interaction via primary metabolites of foliage. *Environmental Research, 201*(111581), 111581. https://doi.org/10.1016/j.envres.2021.111581.

Davis, E. C., Sohngen, B., & Lewis, D. J. (2022). The effect of carbon fertilization on naturally regenerated and planted US forests. *Nature Communications, 13*(1), 5490. https://doi.org/10.1038/s41467-022-33196-x.

Dixit, S., et al. (2023). Plant responses to concurrent abiotic and biotic stress: Unravelling physiological and morphological mechanisms. *Plant Physiology Reports, 49*, 6–17.

Elmquist, D. C., et al. (2024). Soil arthropod communities collected from agricultural soils influence wheat growth and modify phytohormone responses to aboveground herbivory in a microcosm experiment. *Applied Soil Ecology, 194*, 105197.

Ferguson, C. M., et al. (2019). Quantifying the economic cost of invertebrate pests to New Zealand's pastoral industry. *New Zealand Journal of Agricultural Research, 62*(3), 255–315. https://doi.org/10.1080/00288233.2018.1478860.

Frago, E., Dicke, M., & Godfray, H. C. J. (2012). Insect symbionts as hidden players in insect–plant interactions. *Trends in Ecology & Evolution, 27*(12), 705–711. https://doi.org/10.1016/j.tree.2012.08.013.

Friberg, M., Schwind, C., Guimarães, P. R., Raguso, R. A., & Thompson, J. N. (2019). Extreme diversification of floral volatiles within and among species of *Lithophragma* (Saxifragaceae). *PNAS, 116*(10), 4406–4415.

Gerard, P. J., et al. (2013). Potential effects of climate change on biological control systems: Case studies from New Zealand. *BioControl, 58*(2), 149–162. https://doi.org/10.1007/s10526-012-9480-0.

Gougherty, A. V., & Davies, T. J. (2021). Towards a phylogenetic ecology of plant pests and pathogens. *Philosophical Transactions of the Royal Society B: Biological Sciences, 376*(1837), 20200359. https://doi.org/10.1098/rstb.2020.0359.

Groen, S. C., & Whiteman, N. K. (2022). Ecology and evolution of secondary compound detoxification systems in caterpillars. In *Caterpillars in the Middle: Tritrophic Interactions in a Changing World* (pp. 115–163). Cham: Springer International Publishing.

Gvozdenac, S., et al. (2022). Impact of climate change on integrated pest management strategies. In *Climate Change and Agriculture* (pp. 311–372).

Halpen et al. (2023). *Science*. American Association for the Advancement of Science (AAAS).

Hansen, A. K., & Moran, N. A. (2014). The impact of microbial symbionts on host plant utilization by herbivorous insects. *Molecular Ecology, 23*(6), 1473–1496. https://doi.org/10.1111/mec.12421.

Harvey, J. A., et al. (2020). Climate change-mediated temperature extremes and insects: From outbreaks to breakdowns. *Global Change Biology, 26*(12), 6685–6701. https://doi.org/10.1111/gcb.15377.

Hooper, D. U., Bignell, D. E., Brown, V. K., Brussard, L., Dangerfield, J. M., Wall, D. H., Wardle, D. A., Coleman, D. C., Giller, K. E., Lavelle, P., Van Der Putten, W. H., De Ruiter, P. C., Rusek, J., Silver, W. L., Tiedje, J. M., & Wolters, V. (2000). Interactions between aboveground and belowground biodiversity in terrestrial ecosystems: Patterns, mechanisms, and feedbacks. *BioScience, 50*(12), 1049–1061. https://doi.org/10.1641/0006-3568(2000)050[1049:IBAABB]2.0.CO;2.

Hu, G., et al. (2021). Environmental drivers of annual population fluctuations in a trans-Saharan insect migrant. *Proceedings of the National Academy of Sciences of the United States of America, 118*(26). https://doi.org/10.1073/pnas.2102762118.

Irving, M. R., et al. (2023). Temperature alters the toxicological impacts of plant terpenoids on the polyphagous model herbivore *Vanessa cardui*. *Journal of Chemical Ecology*. https://doi.org/10.1007/s10886-023-01449-8.

Jamieson, M. A., et al. (2012). Consequences of climate warming and altered precipitation patterns for plant-insect and multitrophic interactions. *Plant Physiology, 160*(4), 1719–1727. https://doi.org/10.1104/pp.112.206524.

Kaur, I., et al. (2022). Surface warfare: Plant structural defenses challenge caterpillar feeding. In *Fascinating Life Sciences* (pp. 65–92). Cham: Springer International Publishing.

Kopaczyk, J. M., Warguła, J., & Jelonek, T. (2020). The variability of terpenes in conifers under developmental and environmental stimuli. *Environmental and Experimental Botany, 180*, 104197. https://doi.org/10.1016/j.envexpbot.2020.104197

Lehmann, P., et al. (2020). Complex responses of global insect pests to climate warming. *Frontiers in Ecology and the Environment, 18*(3), 141–150. https://doi.org/10.1002/fee.2160.

Li, S., et al. (2020). Research advance in forage diseases, insect pests, and rodents in China. *Journal of Biological Control, 36*, 9–16. https://doi.org/10.16409/j.cnki.2095-039x.2020.01.002.

Lüdecke, D., et al. (2021). Performance: An R package for assessment, comparison, and testing of statistical models. *Journal of Open-Source Software, 6*(60), 3139. https://doi.org/10.21105/joss.03139.

Manley, R., et al. (2023). Conservation measures or hotspots of disease transmission? Agri-environment schemes can reduce disease prevalence in pollinator communities. *Philosophical Transactions of the Royal Society B: Biological Sciences, 378*(1873), 20220004. https://doi.org/10.1098/rstb.2022.0004.

McNeill, M. R., et al. (2022). Sustainable management of *Medicago sativa* for future climates: Insect pests, endophytes, and multitrophic interactions in a complex environment. *Frontiers in Agronomy, 4*. https://doi.org/10.3389/fagro.2022.825087.

Palatella, M., Guillaume, S. M., Linterman, M. A., & Huehn, J. (2022). The dark side of Tregs during aging. *Frontiers in Immunology, 13*, 940705.

Piao, S., et al. (2019). Plant phenology and global climate change: Current progresses and challenges. *Global Change Biology, 25*(6), 1922–1940. https://doi.org/10.1111/gcb.14619.

Pineda, A., et al. (2020). Conditioning the soil microbiome through plant–soil feedbacks suppresses an aboveground insect pest. *The New Phytologist, 226*(2), 595–608. https://doi.org/10.1111/nph.16385.

Pirttilä, A. M., et al. (2023). Exchange of microbiomes in plant-insect herbivore interactions. *mBio, 14*(2), e0321022. https://doi.org/10.1128/mbio.03210-22.

Pixley, K. V., et al. (2019). Genome editing, gene drives, and synthetic biology: Will they contribute to disease-resistant crops, and who will benefit? *Annual Review of Phytopathology, 57*(1), 165–188. https://doi.org/10.1146/annurev-phyto-080417-045954.

Raffa, K. F., Bonello, P., & Orrock, J. L. (2020). Why do entomologists and plant pathologists approach trophic relationships so differently? Identifying biological distinctions to foster synthesis. *The New Phytologist, 225*(2), 609–620. https://doi.org/10.1111/nph.16181.

Riaz, A., et al. (2024). Climate change and household insect pests. In *Climate Change and Insect Biodiversity* (pp. 196–207). CRC Press.

Ricciardi, A., et al. (2021). Four priority areas to advance invasion science in the face of rapid environmental change. *Environmental Reviews, 29*(2), 119–141. https://doi.org/10.1139/er-2020-0088.

Rizwanuddin, S., Kumar, V., Singh, P., Naik, B., Mishra, S., Chauhan, M., Saris, P. E. J., Verma, A., & Kumar, V. (2023). Insight into phytase-producing microorganisms for phytate solubilization and soil sustainability. *Frontiers in Microbiology, 14*, 1127249.

Saidi, S., et al. (2021). Improvement of *Medicago sativa* crops productivity by the co-inoculation of *Sinorhizobium meliloti–Actinobacteria* under salt stress. *Current Microbiology, 78*(4), 1344–1357. https://doi.org/10.1007/s00284-021-02394-z.

Secretariat, I. (2021). Scientific review of the impact of climate change on plant pests: A global challenge to prevent and mitigate plant-pest risks in agriculture, forestry, and ecosystems. *Forestry and Ecosystems.*

Shah, F. M., et al. (2023). Crop protection under climate change: The effect on tri-trophic relations concerning pest control. In *Climate Change Impacts on Agriculture* (pp. 329–354). Cham: Springer International Publishing.

Sharma, G., Malthankar, P. A., & Mathur, V. (2021). Insect–plant interactions: A multilayered relationship. *Annals of the Entomological Society of America, 114*(1), 1–16. https://doi.org/10.1093/aesa/saaa032.

Wang, X., Kang, J., Wang, H., Wang, S., Tang, B., & Lu, J. (2023). Phenotypic plasticity plays an essential role in the confrontation between plants and herbivorous insects. *CABI Agriculture and Bioscience, 4*(1), 58.

Whitehill, J. G., Bohlmann, J., & Krokene, P. (2023). Forest insect-plant interactions. In *Forest Entomology and Pathology* (pp. 169–204). Cham: Springer International Publishing.

Woods, H. A., Legault, G., Kingsolver, J. G., Pincebourde, S., Shah, A. A., & Larkin, B. G. (2022). Climate-driven thermal opportunities and risks for leaf miners in aspen canopies. *Ecological Monographs, 92*(4), e1544.

Zhao, R., et al. (2024). Aboveground antagonists mitigate belowground plant-antagonist interactions but not affect plant-mutualist interactions. *European Journal of Soil Biology, 120,* 103577.

CHAPTER 8

Potential Effects of Climate Change on Plant-Insect Cascades in Agroecosystems: Implications for Food Security

ALI RAJABPOUR* and FATEMEH YARAHMADI

Department of Plant Protection, Agricultural Sciences and Natural Resources University of Khuzestan, Mollasani, Khuzestan, Iran

*Corresponding author

ABSTRACT

Each year, pests destroy around 40% of global agricultural crops, causing an estimated $220 billion in losses, according to the FAO. Climate change is expected to exacerbate this issue, with projected temperature increases of 1–2.5°C potentially leading to a 10–25% rise in crop losses. Warming temperatures can enhance pest survival, reproduction, and dispersal, thereby intensifying their impact. Simultaneously, climate-induced physiological changes in crops, such as altered carbon-nitrogen ratios and protein content—can increase plant vulnerability and influence pest feeding behavior. These shifts threaten global food security, especially as food production must rise by 70% by 2050 to meet growing demand. Currently, over 70,000 pest and disease species already affect global plant health, and climate change, globalization, and habitat disruption continue to drive the emergence and spread of new pest species. This chapter examines the complex interactions among plants, arthropod herbivores, and natural enemies under changing environmental conditions. It also highlights the growing frequency of pest outbreaks and their implications

Insect-Plant Interactions, Volume 1: Influence of Biotic and Abiotic Factors.
Jamin Ali, Adil Tonğa, Adnan Shakeel, and Nasheeman Ashraf (Eds.)

DOI: 10.1201/9781779644282-8

for crop productivity and food availability. Lastly, the chapter discusses the need for informed policies and strategic decision-making to mitigate the risks posed by climate change-driven pest dynamics and to support sustainable agricultural systems.

8.1 CLIMATE CHANGES: CHALLENGE FOR AGRICULTURE AND FOOD SECURITY

The burning of coal, oil, and natural gas, as well as activities like deforestation and changes in land cover, are rapidly boosting the levels of greenhouse gases (GHG) in the atmosphere. This rise in gases is causing a warming trend of around 0.1?C–0.2?C per decade, which is likely the main reason for the reduction in snow cover, melting Arctic Sea glaciers, more frequent heavy rainfall, rising sea levels, and shifts in plant and animal habitats. Currently, the global average temperature is approximately 0.8?C higher than it was before the industrialization period, and existing GHG levels are expected to lead to an additional warming of 0.5?C–1?C as they stabilize (Lenton et al., 2012). The warming effect continues to be more prominent in mid and high latitudes compared to low latitudes, over land rather than oceans, and during the night rather than the day. As emissions keep rising, both the rate of warming and the commitment to future warming are increasing by about 0.2?C per decade, with warnings that this rate could rise further without emission control measures. These changes are likely to have severe impacts on critical societal and environmental systems. Current estimates suggest that to avoid the worst effects of climate change, it will be necessary to limit the increase in global average surface temperature to no more than 2?C–2.5?C above its level in 1750 (15?C), although not all consequences can be prevented (Lenton et al., 2012; MacCracken, 2018). There is strong certainty that anthropogenic GHG emissions have escalated consistently over each decade, with persistent growth seen across various types of gases. It is also highly certain that global human-caused GHG emissions were at their peak during the period of 2009–2018 compared to any previous decade, with emissions continuing to rise in the most recent decade. Notably, no sectors worldwide have demonstrated lasting decreases in GHG emissions.

While several countries have managed to lower their GHG emissions in the past decade, these reductions are relatively modest and overshadowed by significantly larger emissions increases in countries like China, India, and Indonesia (Minx et al., 2022). Several years have elapsed since the Paris Agreement was ratified, calling upon nations to define their individual roles in combating climate change. The Agreement lacks a universal metric for tracking advancements and instead depends on regular assessments to guide efforts in enhancing ambitions. Despite these mechanisms, it appears that policies aimed at halting the gradual increase in GHG emissions have not yielded entirely successful results (Nascimento et al., 2022). This suggests the need for a closer examination of the strategies in place to address this ongoing challenge.

The FAO of the United Nations has announced that approximately 13% of the population in developing nations is currently facing undernourishment (Food and Agriculture Organization of the United Nations Statistical Database, 2020). The challenge of feeding the global population is expected to grow more severe in the coming years, as projections indicate a future world population of 9.2 billion by 2050, leading to a surge in food demand ranging between 59% and 102%. Given these projections, it appears imperative to enhance agricultural production by approximately 60%–70% to sustainably feed the global population by 2050. Consequently, there is a pressing need to essentially double food production by 2050 to address the escalating demand (Janni et al., 2024; Pawlak and Kołodziejczak, 2020). Therefore, it is crucial to change the agricultural system to enhance production efficiency, addressing the rising global demands and averting potential food security risks faced by diverse nations and populations. Global warming can lead to significant influences on many weather and abiotic conditions in different ecosystems, including agroecosystems (Figure 8.1), affecting agricultural production in the world.

About 40% of agricultural crops have been globally destroyed by various pests every year. The economic value of the losses is about $220 billion, according to a new survey by the FAO (Rajabpour and Yarahmadi, 2024). Across the globe, over 10,000 species of insects and mites inflict damage upon agricultural plants. The impact of pest-related issues fluctuates with advancements in agricultural technology and alterations in farming methods. These pests target various parts of their host plants, potentially leading to economic losses in yields (Dhaliwal et al., 2010). Therefore, given the threats to food security posed by the growth of the global population and climate change, it is crucial to implement

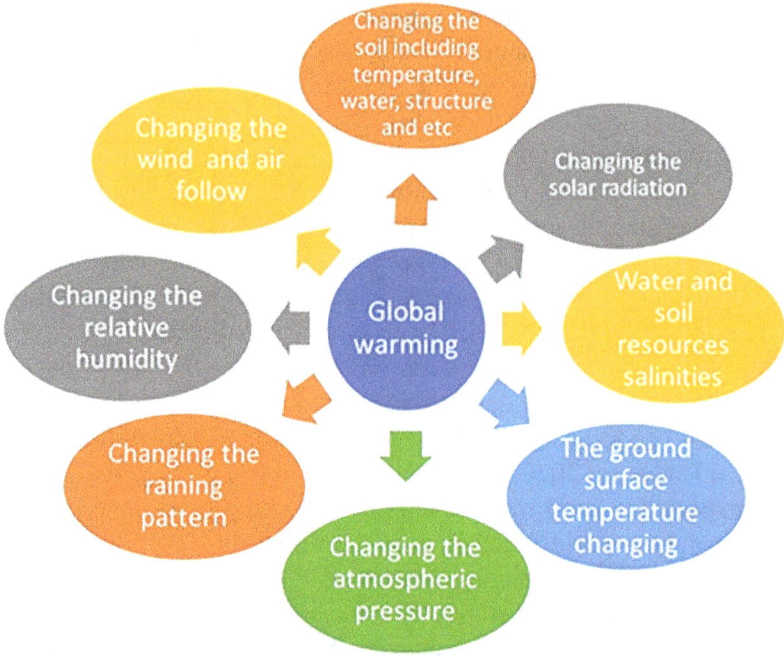

FIGURE 8.1 Possible scenarios for the effect of global warming on ambient abiotic conditions of an agroecosystem.

effective policies for sustainable pest management to address the challenges that await. Understanding the impact of climate change-induced global temperature rise on agricultural systems is essential to mitigate resulting complications and develop appropriate strategies for ensuring future food security. Agriculture operates within a unique ecosystem involving interconnected plants and animals, underscoring the importance of comprehending how climate changes affect all levels of the agricultural ecosystem to address their significant impacts on agroecosystem organisms. The possible effects of climate change on different trophic levels that may lead to more pest outbreaks are shown in Figure 8.2.

8.2 THE EFFECTS OF CLIMATE CHANGES ON AGRICULTURAL PLANTS

The current comprehension of the enduring consequences of global warming on wild plant species can be succinctly encapsulated as follows:

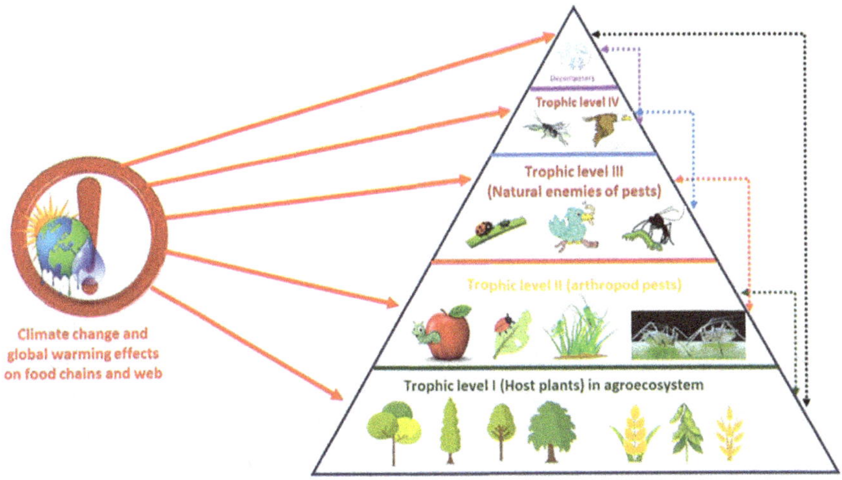

FIGURE 8.2 The schematic diagram illustrates the effects of climate change on trophic cascades within an agroecosystem.

The anticipation is that global warming will diminish the adaptability of numerous wild plant species within their native habitats. These species might opt to relocate to cooler environments and/or adjust to the shifting climate through natural selection. Migration entails the dispersion of plants via seeds, spores, or vegetative material across vast distances, leading to rivalry with indigenous species in the new surroundings. This rivalry frequently culminates in the suppression or potential eradication of one of the contending species, rather than coexistence. Natural selection typically progresses slowly, relying on chance mutations. Nevertheless, for many species, the pace of natural selection is predicted to be slower than the speed of global warming, rendering it inadequate for swift adaptation. Consequently, a substantial portion of the global plant life is anticipated to confront extinction within the current century (Lippmann et al., 2019). Therefore, wild plants, such as weeds, can alter the interactions among organisms coexisting with crop plants in the agroecosystem. For example, the introduction or elimination of a wild plant species in a new agroecosystem can directly and indirectly influence the abundance of currently inhabiting arthropod pests. Many wild plant species are considered alternative hosts of polyphagous herbivores and provide suitable habitats for animals during hibernation or aestivation. Moreover, allelochemicals from wild plants affect pest host selection and may trigger induced resistance in crop plants. Wild plant species in agroecosystems have a significant impact on the natural enemies as they often provide shelter and food sources. Therefore, any changes in the flora of coexisting wild plant species can lead to complex and potentially unpredictable consequences. These consequences can result in a serious pest outbreak or the emergence of new pest species in an agroecosystem.

Agricultural crops are also profoundly affected by global warming. Studies show that not only the envisioned future temperature escalation but also historical rises in average ambient temperature have significantly reduced global yields of key staple crops such as wheat, maize, and rice by about 5% per °C. In contrast to wild species, crops have been acclimated to human-cultivated agricultural environments through domestication and plant breeding. Conventional plant breeding exploits the observable characteristics resulting from naturally occurring genetic variances, which can be amplified through physical or chemical mutagenesis. Unlike wild species, the extensive crossing and selection by plant breeders expedite the adaptation process considerably. This provides crops a greater likelihood to promptly adjust to climate change scenarios, provided that the appropriate genes for selection are judiciously chosen and functional genetic variances are present in the crop's gene pool or can be generated through contemporary technologies such as genome editing (Mall et al., 2017). Hence, while wild plants depend on natural selection for adaptation, the advancement of climate-resilient crops through plant breeding is an achievable objective that demands a profound understanding of the repercussions of global warming on both wild species and crops (Lippmann et al., 2019).

The general impact of temperature on plant growth has been understood for an extended period. Virtually every characteristic in plants has an optimal ambient temperature at which it can achieve its maximum potential. Deviations from this ideal temperature commonly result in a marked decline in trait performance, with temperature response patterns following a bell-shaped curve until severe temperature stress leads to irreversible harm. Given the disquieting prospects of global warming, there is an escalating necessity for a comprehensive understanding of the consequences and fundamental principles of both moderately heightened ambient temperatures and the growing occurrence of heatwaves on plant growth and development. Moderately elevated ambient temperatures denote temperature increases that permit seed-to-seed development under prolonged exposure (Lippmann et al., 2019).

Agriculture and corresponding plants deeply depend on water resources. Research on Global Climate Models indicates substantial shifts in regional and global average precipitation and air temperature, which are expected to bring about corresponding effects on water resources (Kurylyk and MacQuarrie, 2013). According to the IPCC report from 2008, climate change in the coming century is projected to influence rainfall patterns, river flows, and sea levels worldwide. Various studies suggest that agricultural yields are likely to face significant challenges in the next century due to rapid and unprecedented changes in the climate system (Jarvis et al., 2010; Misra, 2014; Thornton et al., 2011). Therefore, water stress will be added to temperature stress by increasing global warming in agriculture. The responses of plant species to water stress vary depending on various factors such as the plant species, the severity and duration of the water stress, and the environmental conditions. A common response to the stress is a change in plant metabolism. Water stress can trigger changes in plant metabolism, including altered hormone levels and enzyme activities (Janni et al., 2024; Ximénez-Embún et al., 2017). Furthermore, under water stress conditions, plants may produce stress proteins that help protect cellular structures and maintain plant function. This response can significantly influence the host plant's defense against herbivorous arthropods. When a plant senses water stress conditions, it initiates a series of tolerance mechanisms. Initially, it closes its stomata; however, prolonged stress may halt its growth and trigger a metabolic reset. To prevent dehydration, cells undergo osmotic adjustments by increasing levels of free sugars and amino acids, particularly proline, known as a stress index (Janni et al., 2024; Ximénez-Embún et al., 2017). These metabolic alterations in the plant lead to an enhancement in nutritional value for herbivores, potentially boosting their performance, for example, increasing their reproduction, survival, and generation number (White, 2009; Ximénez-Embún et al., 2017).

The climate change can affect the nitrogen (N): carbon (C) ratio in the host plant tissues. For instance, in studies involving two *Eucalyptus* species, elevated CO_2 levels and temperature notably amplified foliar C:N ratios, particularly when ambient temperatures were maintained (Gherlenda et al., 2015; Murray et al., 2013). Conversely, in other research, nitrogen content exhibited a marked decrease under the combined influence of elevated CO_2 and temperature, with elevated temperatures in isolation showing minimal impact (Gherlenda et al., 2015; Murray et al., 2013). An analysis of 42 studies by Zvereva and Kozlov (2006) validated that the joint impacts of elevated CO_2 and temperature on plant physiology and chemistry often differed from the effects observed when these factors were studied independently. Consequently, multifactorial experiments have the potential to unveil distinct biological reactions to climate change that might not be evident in studies focusing on individual factors (Robinson et al., 2012; Zvereva and Kozlov, 2006). Plants employ a diverse array of secondary metabolites as part of their defense mechanisms against herbivores. Nitrogen-based secondary metabolites such as alkaloids, cyanogenic glycosides, and glucosinolates (GSL) function as chemical deterrents against herbivores, possessing toxicity or unpalatability that dissuades consumption of the plant. Similarly, carbon-based secondary metabolites like terpenoids and phenolics also play a vital role in plant defense against herbivores. Terpenoids can exhibit repellent or toxic properties towards herbivores, while phenolics can act as antifeedants or disrupt the digestive processes of herbivores. In essence, the synthesis of N and C-based secondary metabolites in plants represents an intricate defense strategy aimed at repelling herbivores and safeguarding the plant from harm or consumption (Price, 1997; Price et al., 2011). Both carbon-based and nitrogen-based secondary metabolites play a crucial role in the defense of host plants and influence the biology, behavior, and feeding of arthropod pests (Price, 1997; Rajabpour et al., 2018a). As a result, environmental changes attributed to global warming can impact the chemical defensive metabolites of host plants and their resistance to herbivores, including arthropod pests.

In response to rising temperatures in the environment, plants undergo significant changes in their physical and morphological characteristics, including elongated stems, upward leaf movement, and the development of small, narrow leaves, collectively referred to as thermomorphogenesis. This process is closely linked to the global phenomenon of climate change, posing a significant ecological concern due to its impact on plant growth and crop yield (Classen et al., 2015; Lee et al., 2020). The biology, reproduction, settlement, and behavior of insects, including herbivorous pests and their natural enemies, deeply depend on the host plant's physical and morphological characteristics, as plants serve as their food source, shelter, and foraging environments. For instance, the morphology and abundance of trichomes on the host plant are influenced by environmental temperature (Azadi et al., 2013; Mohammadi et al., 2015a, b). Trichome thickness reached its maximum at the coldest and driest location and was minimal at the wettest site. The presence of thicker trichomes led to a noticeable rise in leaf temperature. Densely distributed trichomes on leaves can hinder gas exchange by creating higher resistance to gas diffusion, yet they can also indirectly promote gas exchange by elevating leaf temperature, which in turn raises resistance to heat diffusion. This dual effect may aid in the plant's adaptation to dry or cold conditions (Azadi et al., 2018; Mohammadi et al., 2015a, b). Moreover, higher leaf trichome density could potentially reduce daily water-use efficiency across various elevations (Amada et al., 2023). Numerous studies indicate that plant trichomes significantly affect herbivory and the populations, behavior, and biology of their natural enemies (Azadi et al., 2018; Mohammadi et al., 2015a, b; Salehi et al., 2016; Shahbi and Rajabpour, 2017).

Furthermore, the timing of key developmental stages in plants, such as flowering, is intricately connected to shifts in climate patterns. This relationship implies that alterations in climate can affect the timing of crucial events in the plant life cycle. Recent research has demonstrated that fluctuations in temperature can influence the flowering schedule of various plant species. For instance, flowering patterns in British plants have undergone notable changes, resulting in annual plants blooming earlier than perennial ones. This accelerated flowering provides an advantage to plants needing early pollination to attract pollinators efficiently. However, such changes may also lead to mismatches between different plant species and their interactions with insects, including herbivorous, parasitic, predatory, and pollinator insects (Lee et al., 2020). The alignment of plant phenology with herbivorous arthropods or their natural predators plays a crucial role in determining their population dynamics and survival, thus impacting the occurrence of pest outbreaks. Modifying this synchronization can lead to phenomena such as pest resurgences or secondary pest outbreaks, particularly in the context of climate change within an agroecosystem.

As another aspect of global warming effects on agricultural plants, the repercussions of climate change on the synchronization of plant species could have a profound impact on ecosystem biodiversity. The increased likelihood of population bottlenecks due to frequent extreme weather events could significantly reduce genetic diversity within populations, thereby diminishing the occurrence of genetic mutations and lowering species richness, ultimately elevating the risk of species extinction (Cobben et al., 2011; Lee et al., 2020). By applying the Red List criteria of the International Union for Conservation of Nature and Natural Resources to the forecasts, it is evident that numerous plant species in Europe may face significant threats in the future. The analysis indicates that over 50% of the species examined could be at risk or endangered by the year 2080. The projected decline in species and the shift in species composition were shown to be closely linked to changes in only two climate variables related to temperature and moisture. The study suggests that the likelihood of extinction for European plant species could be substantial, even under moderate climate change scenarios, highlighting the considerable risk despite variations between models (Thuiller et al., 2005). Within an agroecosystem, biodiversity encompasses the variety of plants and crops present in fields, orchards, greenhouses, and so forth, along with the organisms associated with them, such as herbivores, carnivores, and decomposers. These elements collectively play a crucial role in influencing the sustainability of the ecosystem. The interaction between cultivated plants and surrounding vegetation, like weeds, can have mutual effects on ecosystem stability and possible pest outbreaks (Rajabpour and Yarahmadi, 2024) (Figure 8.3). For example, the shade provided by trees on a farm might favor the growth of only sun-tolerant crops. The diverse array of plants and their flowers can bolster populations of pollinators, parasitoids, and generalist predators by offering nectar and shelter resources. Research has shown that supplying pollen resources can significantly boost the reproductive success and longevity of *Orius* sp. (Hemiptera: Anthocoridae), which are vital generalist predators of numerous insect and mite pests (Rajabpour and Yarahmadi, 2024; Safaei et al., 2016). Enhancing plant biodiversity through practices like intercropping or multiculture can improve host plant resilience by enhancing soil nutrients, such as legumes enriching cereal fields with *Rhizobium* bacteria, or boosting plant resistance through induced resistance mechanisms. Moreover, a variety of host plants can benefit beneficial organisms by mitigating the negative impacts of certain agricultural activities, like offering shelter and alternative food sources during pesticide applications or crop harvesting in fields or orchards (Rajabpour and Yarahmadi, 2024). It is documented that the biodiversity of the family Anthocoridae (Hemiptera), as important natural enemies of many agricultural pests, is significantly influenced by climate changes (Davari et al., 2015; Sadegh Nejad et al., 2019; Sepahvand et al., 2019). In many cases, insect pests locate their host plants through plant-derived volatile compounds emitted, known as synomones. Increasing the diversity of plant species in fields can disrupt pest attacks by masking the volatile synomones with a mix of compounds from different plants (that can potentially be used as companion plants), thus confusing the pests and reducing their efficacy (Pedigo, 2002).

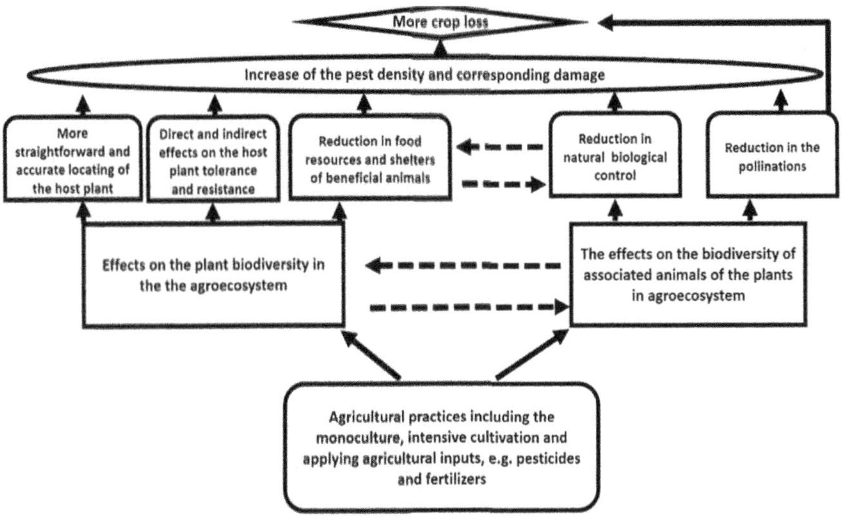

FIGURE 8.3 Potential impacts of agroecological interventions on the diversity of organisms and crop damage resulting from pest infestations in an agroecosystem.

Global warming also affects the reproductive strategies of plants, prompting investigations into the molecular and biochemical mechanisms involved in plant growth under changing climatic conditions. Plants are known to be more responsive to temperature variations compared to animals, as they are immobile and cannot seek out warmer or cooler environments. A common response observed in plants to temperature increases is the tendency to grow taller and slimmer in an effort to regulate body temperature (Huey, 2002). This results in taller stems and narrower leaves spaced further apart from each other, making the plants more susceptible to instability (Park and Park, 2019). This vulnerability becomes evident during activities like grain harvesting, where plants with weaker stalks are prone to bending in adverse weather conditions, leading to reduced biomass production. Additionally, temperature shocks can lead to a decrease in essential substances, such as proteins stored in grain kernels (Lee et al., 2020).

The changing climate directly and indirectly impacts interactions between soil microbes, between microbes and plants, and the potential consequences of shifts in these interactions on ecosystem composition and function (Jansson and Hofmockel, 2020; Pugnaire et al., 2019). As soil microorganisms play a key role in nutrient cycling, nutrient provision to plants, fostering neighbor coexistence, and regulating plant populations, alterations in soil microbe-plant interactions could profoundly influence the composition of plant communities and the functioning of ecosystems (Classen et al., 2015). The soil microbial community significantly influences the nutrition of host plants and their tolerance and induced resistance against herbivorous arthropod pests.

8.3 EFFECTS OF CLIMATE CHANGES ON HERBIVORE ARTHROPODS IN AGROECOSYSTEMS

The global warming and corresponding climate change can affect different aspects of insect and other arthropod agricultural pests (Figure 8.4). Climate and weather conditions, particularly ambient temperature, play a pivotal role in determining the pace and success of invertebrate development, encompassing both embryonic and post-embryonic phases. The application of mathematical models to elucidate how climate impacts the growth and development of insect populations garners significant interest among applied entomologists and managers involved in plant pest control. Initially, efforts to elucidate the correlation between environmental temperature and growth rate were spearheaded by botanists. However, the models devised to assess the influence of environmental temperature on plants can also be repurposed for poikilothermic creatures such as arthropods. Arthropods, including insect pests, are poikilothermic organisms, and their physiological processes, including those governing growth and development, are heavily influenced by ambient temperature. Typically, optimal growth and development occur within specific temperature ranges for pests, with these ranges varying across arthropod species. Generally, insect developmental rates are higher in warmer years compared to colder ones. Notably, research has demonstrated that insect enzyme systems, including those crucial for growth and reproduction, are significantly impacted by environmental temperature (Rajabpour and Yarahmadi, 2024). However, raising the environmental temperature above the growth threshold is anticipated to accelerate the developmental rate of arthropods significantly. At a specific temperature called the optimum temperature, the rate of development is expected to peak. However, beyond this optimal temperature, the developmental rate diminishes, eventually reaching zero at a temperature known as the lethal temperature (Damos and Savopoulou-Soultani, 2012). Usually, there are similar trends for the survival and reproduction of the animals with increasing environmental temperature (Figure 8.5).

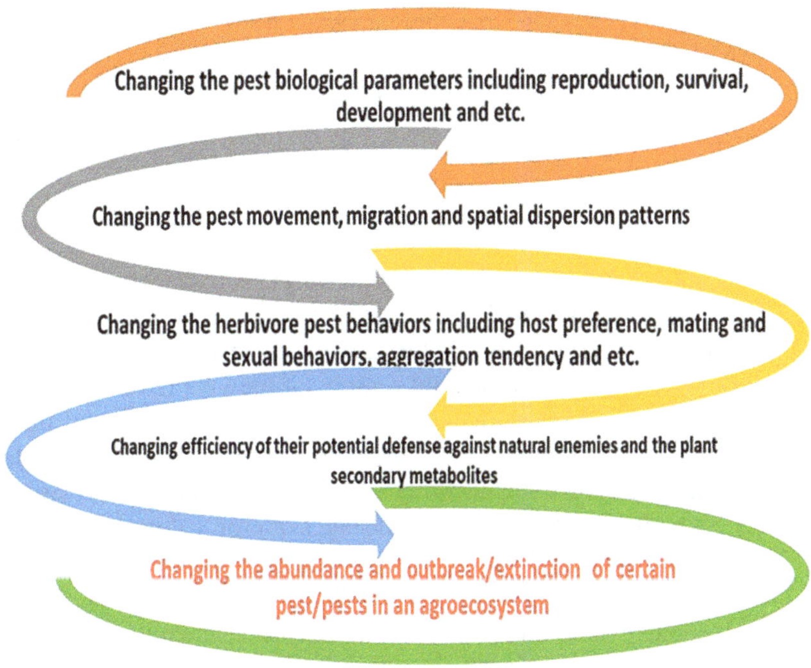

FIGURE 8.4 The potential effects of climate changes on different aspects of the arthropod's pests of an agricultural pest, leading to their outbreaks or extinctions.

Beyond the impacts of temperature fluctuations on the growth, development, biology, and survival of herbivorous arthropods, the alterations brought about by global warming could lead to shifts in population densities, sizes, genetic makeup, levels of host plant utilization, and local as well as geographical distribution associated with colonization and extinction. These effects are expected to be most pronounced in above-ground

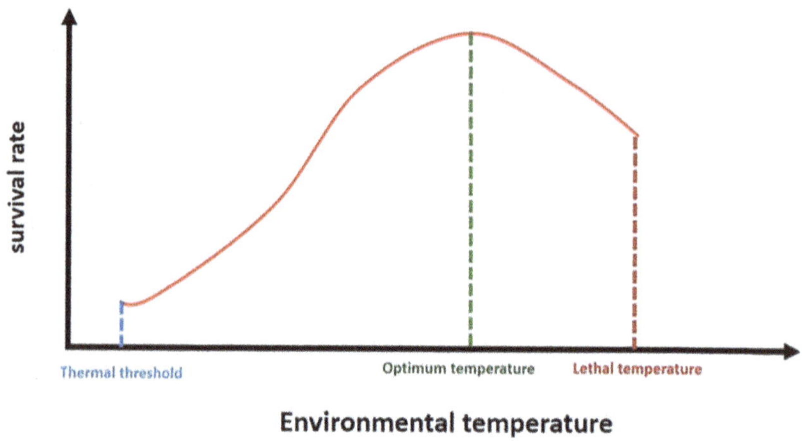

FIGURE 8.5 Trends of the insect survival rate by changes in the environmental temperature.

herbivores, which are subject to the complete spectrum of micro- and macroclimatic conditions, whereas soil-dwelling species encounter thermal conditions that are moderated by the dense soil surroundings (Bale et al., 2002). Numerous herbivorous arthropods enter diapause, halting their growth and reproduction in the cold seasons of the year (overwintering). However, due to global warming, the growth and reproduction of insect populations may persist throughout the cold seasons. This uninterrupted population growth could significantly elevate the likelihood of early and intense pest outbreaks during the initial stages of crop cultivation.

Temperature can influence the feeding and host preference of herbivorous arthropods, suggesting that temperature-driven impacts on herbivory could affect plant community composition and nutrient cycling. However, the effects of rising temperatures on herbivore preferences and performances across different plant species, which play a crucial role in determining their overall impact, have been minimally explored in current research. For instance, the host plant preferences and performances of the generalist herbivore *Popillia japonica* Newman (Coleoptera: Scarabaeidae) across nine plant species indicated that, on average, consumption rates and herbivore performances increased at higher temperatures. Nevertheless, there was significant variation among plant species, with some exhibiting increased consumption and performance at elevated temperatures while others showed a decrease. These patterns seemed to be influenced by the nutritional quality of the plants, as beetles tended to feed more on high-nitrogen plants as temperatures increased, indicating a stronger nitrogen limitation. Furthermore, in addition to changes in feeding rates, the feeding preferences of *P. japonica* shifted with varying temperatures, primarily due to the varying repellent effects of plant chemical extracts at different temperature levels. Notably, the alterations in the efficacy of plant chemical extracts caused by temperature led *P. japonica* to narrow its dietary choices at higher temperatures. The findings suggest that herbivores' feeding behavior will be affected by rising temperatures through changes in the importance of plant nutritional and chemical traits, indicating that climate change will alter the dynamics and intensity of interactions between plants and insects (Lemoine et al., 2014). For instance, the rise in damage caused by bugs infesting rice and fruit, coupled with simultaneous outbreaks and the expansion of their geographical range towards the poles, are phenomena that could potentially be linked to global warming in Japan. Projections suggest that a 1?C increase could lead to a 15% decrease in winter mortality for adult *Nezara viridula* L. and *Halyomorpha halys* Stål (Hemiptera: Pentatomidae) (Kiritani, 2006). Furthermore, climate change is believed to be a factor contributing to the recent decrease in the population of *Plutella xylostella* L. (Lepidoptera: Plutellidae), along with the rise in populations of *Helicoverpa armigera* Hübner and *Trichoplusia ni* Hübner (Lepidoptera: Noctuidae).

The increasing presence of alien species originating from tropical regions, alongside the escalation in pesticide usage, is a trend that may continue. It is imperative to intercept alien pests through plant quarantine measures, followed by Integrated Pest Management strategies (IPMS) to address these challenges effectively. The climate plays a vital role in many behavioral aspects of insects, for example, their dispersal, movement, migration, sexual behavior, and so forth, directly or indirectly by changing their environments and resources. The phenomenon of global warming has profound and enduring impacts on various meteorological elements, including wind patterns, precipitation levels, relative humidity, as well as salinity and the conditions of water and soil. These changes in environmental factors can significantly influence various behaviors of insects and exacerbate issues related to plant pests. The interconnected nature of these variables within an ecosystem leads to a complex web of interactions among abiotic components, living organisms, and the complicated interactions among these biotic and abiotic components in an agroecosystem.

For instance, variations in salinity levels and water quality can affect the availability of resources for insect populations, leading to shifts in their behavior and distribution. Furthermore, changes in soil conditions, such as temperature and moisture levels, can also impact many behaviors of insects, including their distributions, oviposition, and so forth. By understanding the intricate relationships between these environmental factors and insect behaviors, researchers and policymakers can better predict and mitigate the undesired ecological consequences of global warming on pest management and ecosystem dynamics.

The act of movement arises from the complex interplay of various internal factors, encompassing internal motivation (the reason for movement), physical ability to move (the mechanics of movement), navigational ability (the direction of movement), and external influences such as wind, temperature, and light. The actual execution of movement results from the integration of processes occurring at different levels, including life history, physiological functions, morphology, and behavior. External factors have the potential to impact multiple internal factors, creating a web of direct and indirect interactions through feedback and feedforward mechanisms, collectively defining the movement process. Movement is accompanied by potential hidden costs, such as decreased reproductive output, diminished physical condition, compromised immune response, missed opportunities for mating and foraging, and the time invested in exploring and mapping new territories. The significance of wind extends beyond its role as a facilitator of planktonic movement, playing a critical part in enabling long-distance dispersal and likely influencing various forms of migration. An illustrative depiction featuring the monarch butterfly *Danaus plexippus* L. (Lepidoptera: Nymphalidae) showcases key attributes that enable its remarkable migrations across North America, with the mean annual temperatures indicated. Notably, while the southward migration is completed within a single generation, the northward journey may span three generations to accomplish.

Various meteorological factors, with a specific focus on temperature and wind patterns, play a crucial role in influencing the dispersal of arthropods. Moreover, weather systems, including the formation and movement of air masses, the presence of cold and warm fronts, and the airflow associated with geographical features like mountains, valleys, and bodies of water, also impact arthropod dispersal. Recent advancements have led to the development of models that can simulate how these weather systems affect the airborne dispersal of insects. Furthermore, some research has highlighted the significance of arthropod behavior during favorable weather conditions for dispersal. It is evident that both winged and wingless arthropods exhibit specific responses to various stimuli that promote dispersal, enhancing their potential for migration. The ongoing efforts in understanding the intricate relationship between insect movement and mesoscale weather conditions hold great promise for effective pest management strategies (McManus, 1988). Temperature stands out as the primary influencer of aerial dispersal for both winged and wingless invertebrates, surpassing all other physical factors. Insects face a critical temperature threshold below which they are incapable of taking flight, and a lower threshold where flying becomes impossible, thereby constraining their ability to ascend to higher elevations actively. The daily fluctuations in temperature can significantly impact the takeoff timing, as the threshold may be influenced by a period of preconditioning. Several studies have delved into the impact of temperature on the flight behavior of aphids. The threshold temperature required for takeoff varies not only across different species but also among individuals within the same species. For instance, in the case of *Rhopalosiphum padi* L. (Hemiptera: Aphididae), the spring emigrants exhibit a higher threshold temperature (16?C–17?C) compared to the alate exules (13?C–14?C), who, in turn, have a higher threshold than the fall migrants (9?C–10?C). Even in conditions where the air temperature surpasses the threshold, additional triggers such as variations in light intensity or wind may be crucial for initiating takeoff. Factors like physiological condition, population density, and the quality of the host plant can further delay the takeoff process. This discussion sheds light on the intricate interplay between temperature and other influencing factors, underscoring the recurring concerns raised by Wellington regarding entomologists' historical focus on evaluating isolated physical factors and the inherent limitations associated with this approach. In Japan, over the period from 1966–1987, more than 50 butterfly species exhibited northward expansions in their ranges, with 10 previously migratory species establishing themselves on the Nansei Islands (Kiritani, 2006). In a long-term study in China, the effects of increasing surface temperatures over a three-decade period on the long-distance migration behavior of the economically significant cutworm species, *Agrotis ipsilon* Hufn. (Lepidoptera: Noctuidae), were investigated. Through extensive monitoring in China, it is observed that high-altitude migrant populations of *A. ipsilon* exhibit two distinct seasonal peaks, occurring during both spring and summer. A gradual shift in the origins of migrant populations, with ground-level abundance of *A. ipsilon* reflecting the levels of high-altitude migrants, was identified. Over a span of 27 years, it is demonstrated that the population peaks for overwintering and the first generation of adult *A. ipsilon* occurred 17 and 8 days earlier, respectively. Furthermore, the northern boundary of the potential overwintering area for *A. ipsilon* moved northward by 58–232 km, while its migration range expanded by 708 km. The research demonstrates how the progressive rise in surface temperatures impacts the long-distance migration patterns of this common agricultural pest, leading to changes in population dynamics and geographical distribution in eastern Asia. The influence of global warming on such ecological processes in temperate agroecosystems has the potential to exacerbate the economic consequences of migratory pests on a larger scale (Zeng et al., 2020).

While the spatial distribution of pests is recognized as a species-specific feature of insects in ecology (Taylor, 1984), it can be influenced by various biotic and abiotic factors within the ecosystem. For instance, plant characteristics and the presence of other plant species in an agricultural setting can alter spatial distribution patterns of an insect pest (Dinarvand et al., 2020; Kafeshani et al., 2018; Shahbi and Rajabpour, 2017). Consequently, the anticipated effects of global warming on plants are likely to impact pest distribution. Furthermore, temperature variations across different parts of a plant canopy may exceed the pest's tolerance limit, leading to changes in pest establishment and spatial distribution parameters. It is worth noting that beyond its theoretical significance in insect ecology, spatial distribution plays a pivotal role in pest population sampling and monitoring initiatives.

Temperature plays a pivotal role in shaping the behaviors observed during insect and arachnid mating interactions and reproduction processes. It influences various aspects such as precopulatory activity levels, competition dynamics among rivals, communication with potential mates, and the overall costs and benefits associated with mating. The temperature conditions during reproductive activities and early life experiences significantly impact mating-related behaviors. To better anticipate the impacts of global warming on insect and arachnid mating behavior, we recommend integrating a function-valued approach to assess thermal sensitivity comprehensively. Temperature affects precopulatory activity, mating likelihood, intra- and intersexual interactions, and the overall fitness consequences of engaging in mating behaviors. Both environmental temperatures during reproduction and developmental temperatures can influence a broad array of behavioral traits used by insects and arachnids to attract, compete for, and select mates. By measuring changes in mating-related behaviors as continuous functions of temperature, researchers can accurately pinpoint peaks in mating activity, differentiate thermal sensitivity in mating behaviors, and predict outcomes of interactions across environmental variations. Furthermore, the diverse thermal environments at small spatial scales can lead to differences in body temperatures among interacting individuals, resulting in variations in thermal sensitivity due to developmental adaptability. Understanding how fine-scale thermal heterogeneity shapes responses to temperature variations can offer valuable insights into the ecological and evolutionary processes of the Anthropocene. Given the significance of reproduction for population sustainability, the thermal sensitivity of mating behaviors is crucial for how insects and arachnids have adapted to climatic changes historically. A detailed understanding of the thermal sensitivity of mating behaviors and the information conveyed through inter- and intra-sexual signals is vital for comprehending how temperature influences sexual selection, reproduction, and persistence in a warming world (Leith et al., 2021).

8.4 EFFECTS OF CLIMATE CHANGES ON NATURAL ENEMIES OF AGRICULTURAL PESTS

Pest arthropods have a wide range of natural enemies that play a crucial role in regulating their populations in natural control programs or integrated pest management (IPM). These natural enemies encompass various vertebrates, invertebrates, and pathogenic microorganisms such as fungi, bacteria, viruses, and protozoa. Among these, predatory and parasitoid insects are particularly significant in controlling the population of plant pests. Global warming and climate change significantly impact the activity and population of natural enemies, which serve as vital biological control agents for insects, including pest insects. Climate changes, along with alterations in the abiotic environment, host plants, and herbivorous insects as primary sources of habitat and food, have led to notable changes in the activity, population, and effectiveness of natural enemies (Figure 8.6). These changes should be considered in IPM policies and when forecasting challenges related to pest management in the future. The effects of temperature, relative humidity, precipitation, light, and other climate change-induced factors on the natural enemies of arthropod pests, akin to those affecting host plants and herbivorous arthropods, can influence patterns of survival, reproduction, movement, and various behaviors (Banihashemi et al., 2017; Salehi et al., 2016). These effects may have direct or indirect significance, stemming from changes in plant quantities or variations in prey/host herbivorous arthropods.

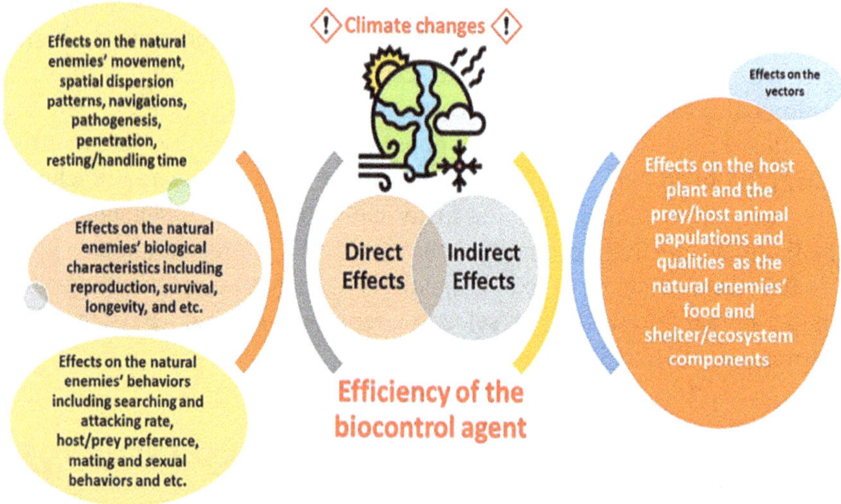

FIGURE 8.6 Possible effects of climate changes on different aspects of natural enemies and their corresponding biological control efficient in an agroecosystem.

The impact of climate change on microbiological agents of arthropod pests can be substantial. Changes in temperature, humidity, and precipitation patterns directly affect the survival, reproduction, and effectiveness of pathogenic microorganisms. These organisms require specific environmental conditions, such as optimal ambient temperature and relative humidity. The life and distribution of microbiological agents, acting as parasites, depend entirely on their arthropod hosts' life and dispersion. As previously discussed, climate change significantly influences arthropod biology, distribution, and behavior. Warmer temperatures and altered precipitation levels can create more favorable conditions for some pathogens to thrive while adversely affecting others. Moreover, unsuitable environmental conditions can rapidly disinfect these agents, with global warming potentially exacerbating this issue. This can reduce the efficiency of microbiological control, leading to a rise in pest population densities and outbreaks.

KEYWORDS

- **agroecosystem**
- **climate change**
- **global warming**
- **outbreaks**
- **pests**
- **sustainable agriculture**
- **wingless invertebrates**

REFERENCES

Amada, G., Yoshiko, K., Kitayama, K., & Onoda, Y. (2023). Roles of lower-side leaf trichomes in diffusion resistance and gas-exchange characteristics across environmental gradients in *Metrosideros polymorpha*. *Tree Physiology, 43*(8), 1400–1415. https://doi.org/10.1093/treephys/tpad053.

Azadi, F., Rajabpour, A., Lotfi Jalal-Abadi, A., & Mahjoob, M. (2018). Resistance of tomato cultivars to *Tuta absoluta* (Lepidoptera: Gelechiidae) under field condition. *Journal of Crop Protection, 7*(1), 87–92.

Bale, J. S., Masters, G. J., Hodkinson, I. D., Awmack, C., Bezemer, T. M., Brown, V. K., Butterfield, J., Buse, A., Coulson, J. C., Farrar, J., Good, J. E. G., Harrington, R., Hartley, S., Jones, T. H., Lindroth, R. L., Press, M. C., Symrnioudis, I., Watt, A. D., & Whittaker, J. B. (2002). Herbivory in global climate change research: Direct effects of rising temperature on insect herbivores. *Global Change Biology, 8*(1), 1–16. https://doi.org/10.1046/j.1365-2486.2002.00451.x.

Banihashemi, A. S., Seraj, A. A., Yarahmadi, F., & Rajabpour, A. (2017). Effect of host plants on predation, prey preference and switching behavior of *Orius albidipennis* on *Bemisia tabaci* and *Tetranychus turkestani*. *International Journal of Tropical Insect Science, 37*(3), 176–182. https://doi.org/10.1017/S1742758416000229.

Classen, A. T., Sundqvist, M. K., Henning, J. A., Newman, G. S., Moore, J. A. M., Cregger, M. A., Moorhead, L. C., & Patterson, C. M. (2015). Direct and indirect effects of climate change on soil microbial and soil microbial-plant interactions: What lies ahead? *Ecosphere, 6*(8), 1–21. https://doi.org/10.1890/ES15-00217.1.

Cobben, M. M. P., Verboom, J., Opdam, P. F. M., Hoekstra, R. F., Jochem, R., Arens, P., & Smulders, M. J. M. (2011). Projected climate change causes loss and redistribution of genetic diversity in a model metapopulation of a medium-good disperser. *Ecography, 34*(6), 920–932. https://doi.org/10.1111/j.1600-0587.2011.06713.x.

Damos, P., & Savopoulou-Soultani, M. (2012). Temperature-driven models for insect development and vital thermal requirements. *Psyche: A Journal of Entomology, 2012*, 1–13. https://doi.org/10.1155/2012/123405.

Davari, H., Seraj, A. A., & Rajabpour, A. (2015). Biodiversity of genus *Orius* (Hemiptera: Anthocoridae) in various climate regions and seasons of Kohgiloyeh and Boyerahmad Province and evaluation of agro-ecosystem effects on their biodiversity. *Journal of Entomological Society of Iran, 35*(3), 1–14.

Dhaliwal, G. S., Jindal, V., & Dhawan, A. K. (2010). Insect pest problems and crop losses: Changing trends. *Indian Journal of Ecology, 37*(1), 1–7.

Dinarvand, N., Rajabpour, A., Zandi Sohani, N., & Farkhari, M. (2020). Effect of weedy culture on population densities, spatial distributions and sampling procedures of *Spodoptera exigua* and *Sesamia cretica* (Lep., Noctuidae) in corn fields. *Bulletin of Entomological Research, 110*(1), 84–95. https://doi.org/10.1017/S0007485319000312.

Food and Agriculture Organization of the United Nations Statistical Database. (2020). Annual population. Retrieved from: http://www.fao.org/faostat/en/#data/OA (accessed on 3 April 2025).

Gherlenda, A. N., Haigh, A. M., Moore, B. D., Johnson, S. N., & Riegler, M. (2015). Responses of leaf beetle larvae to elevated [CO_2] and temperature depend on *Eucalyptus* species. *Oecologia, 177*(2), 607–617. https://doi.org/10.1007/s00442-014-3182-5.

Huey, R. B. (2002). Plants versus animals: Do they deal with stress in different ways? *Integrative and Comparative Biology, 42*(3), 415–423. https://doi.org/10.1093/icb/42.3.415.

Hummel, I., Pantin, F., Sulpice, R., Piques, M., Rolland, G., Dauzat, M., Christophe, A., Pervent, M., Bouteillé, M., Stitt, M., Gibon, Y., & Muller, B. (2010). *Arabidopsis* plants acclimate to water deficit at low cost through changes of carbon usage: An integrated perspective using growth, metabolite, enzyme, and gene expression analysis. *Plant Physiology, 154*(1), 357–372. https://doi.org/10.1104/pp.110.157008.

Janni, M., Maestri, E., Gullì, M., Marmiroli, M., & Marmiroli, N. (2024). Plant responses to climate change, how global warming may impact on food security: A critical review. *Frontiers in Plant Science, 14*, 1297569. https://doi.org/10.3389/fpls.2023.1297569.

Jansson, J. K., & Hofmockel, K. S. (2020). Soil microbiomes and climate change. *Nature Reviews Microbiology, 18*(1), 35–46. https://doi.org/10.1038/s41579-019-0265-7.

Jarvis, A., Ramirez, J., Anderson, B., Leibing, C., & Aggarwal, P. (2010). Scenarios of climate change within the context of agriculture. In *Climate Change and Crop Production* (pp. 9–39). CABI International.

Kafeshani, F. A., Rajabpour, A., Aghajanzadeh, S., Gholamian, E., & Farkhari, M. (2018). Spatial distribution and sampling plans with fixed level of precision for citrus aphids (Hom., Aphididae) on two orange species. *Journal of Economic Entomology, 111*(2), 931–941. https://doi.org/10.1093/jee/tox380.

Kiritani, K. (2006). Predicting impacts of global warming on population dynamics and distribution of arthropods in Japan. *Population Ecology, 48*(1), 5–12. https://doi.org/10.1007/s10144-005-0225-0.

Kurylyk, B. L., & MacQuarrie, K. T. B. (2013). The uncertainty associated with estimating future groundwater recharge: A summary of recent research and an example from a small unconfined aquifer in a northern humid-continental climate. *Journal of Hydrology, 492*, 244–253. https://doi.org/10.1016/j.jhydrol.2013.03.043.

Lee, J. H., Kim, J. Y., Kim, J. I., Park, Y. J., & Park, C. M. (2020). Plant thermomorphogenic adaptation to global warming. *Journal of Plant Biology, 63*(1), 1–9. https://doi.org/10.1007/s12374-020-09232-y.

Leith, N. T., Macchiano, A., Moore, M. P., & Fowler-Finn, K. D. (2021). Temperature impacts all behavioral interactions during insect and arachnid reproduction. *Current Opinion in Insect Science, 45*, 106–114. https://doi.org/10.1016/j.cois.2021.03.005.

Lemoine, N. P., Burkepile, D. E., & Parker, J. D. (2014). Variable effects of temperature on insect herbivory. *PeerJournal, 2*, e376. https://doi.org/10.7717/peerj.376.

Lenton, T. M., Livina, V. N., Dakos, V., Van Nes, E. H., & Scheffer, M. (2012). Early warning of climate tipping points from critical slowing down: Comparing methods to improve robustness. *Philosophical Transactions of the Royal Society A, 370*(1962), 1185–1204. https://doi.org/10.1098/rsta.2011.0304.

Lippmann, R., Babben, S., Menger, A., Delker, C., & Quint, M. (2019). Development of wild and cultivated plants under global warming conditions. *Current Biology, 29*(24), R1326–R1338. https://doi.org/10.1016/j.cub.2019.10.016.

MacCracken, M. C. (2008). Prospects for future climate change and the reasons for early action. *Journal of the Air & Waste Management Association, 58*(6), 735–786. https://doi.org/10.3155/1047-3289.58.6.735.

Mall, R. K., Gupta, A., & Sonkar, G. (2017). Effect of climate change on agricultural crops. In S. K. Dubey, A. Pandey, & R. S. Sangwan (Eds.), *Current Developments in Biotechnology and Bioengineering* (pp. 23–46). Elsevier. https://doi.org/10.1016/B978-0-444-63661-4.00002-5.

McManus, M. L. (1988). Weather, behavior and insect dispersal. *Memoirs of the Entomological Society of Canada, 120*(S146), 71–94. https://doi.org/10.4039/entm120146071-1.

Minx, J. C., Lamb, W. F., Andrew, R. M., Canadell, J. G., Crippa, M., Döbbeling, N., Forster, P., Guizzardi, D., Olivier, J., Pongratz, J., Reisinger, A., Rigby, M., Peters, G., Saunois, M., Smith, S. J., Solazzo, E., & Tian, H. (2022). A comprehensive and synthetic dataset for global, regional and national greenhouse gas emissions by sector 1970–2018 with an extension to 2019. https://doi.org/10.5281/ZENODO.6483002.

Misra, A. K. (2014). Climate change and challenges of water and food security. *International Journal of Sustainable Built Environment, 3*(1), 153–165. https://doi.org/10.1016/j.ijsbe.2014.04.006.

Mohammadi, S., Seraj, A. A., & Rajabpour, A. (2015a). Evaluation of six cucumber greenhouse cultivars for resistance to *Tetranychus turkestani* (Acari: Tetranychidae). *Journal of Crop Protection, 4*(4), 545–556.

Mohammadi, S., Seraj, A. A., & Rajabpour, A. (2015b). Effects of six greenhouse cucumber cultivars on reproductive performance and life expectancy of *Tetranychus turkestani* (Acari: Tetranychidae). *Acarologia, 55*(2), 231–242. https://doi.org/10.1051/acarologia/20152164.

Murray, T. J., Ellsworth, D. S., Tissue, D. T., & Riegler, M. (2013). Interactive direct and plant-mediated effects of elevated atmospheric [CO_2] and temperature on a *Eucalypt*-feeding insect herbivore. *Global Change Biology, 19*(5), 1407–1416. https://doi.org/10.1111/gcb.12142.

Nascimento, L., Kuramochi, T., & Höhne, N. (2022). The G20 emission projections to 2030 improved since the Paris Agreement, but only slightly. *Mitigation and Adaptation Strategies for Global Change, 27*(6), 39. https://doi.org/10.1007/s11027-022-10018-5.

Park, Y. J., & Park, C. M. (2019). Physicochemical modeling of the phytochrome-mediated photothermal sensing. *Scientific Reports, 9*(1), 10485. https://doi.org/10.1038/s41598-019-47019-5.

Pawlak, K., & Kołodziejczak, M. (2020). The role of agriculture in ensuring food security in developing countries: Considerations in the context of the problem of sustainable food production. *Sustainability, 12*(13), 5488. https://doi.org/10.3390/su12135488.

Pedigo, L. P. (2002). *Entomology and Pest Management*. Prentice Hall.

Price, P. W. (1997). *Insect Ecology*. John Wiley & Sons.

Price, P. W., Denno, R. F., Eubanks, M. D., Finke, D. L., & Kaplan, I. (2011). *Insect Ecology: Behavior, Populations and Communities*. Cambridge University Press. https://doi.org/10.1017/CBO9780511975387.

Pugnaire, F. I., Morillo, J. A., Peñuelas, J., Reich, P. B., Bardgett, R. D., Gaxiola, A., Wardle, D. A., & Van Der Putten, W. H. (2019). Climate change effects on plant-soil feedbacks and consequences for biodiversity and functioning of terrestrial ecosystems. *Science Advances, 5*(11), eaaz1834. https://doi.org/10.1126/sciadv.aaz1834.

Rajabpour, A., & Yarahmadi, F. (2024). *Decision System in Agricultural Pest Management*. Springer Nature Singapore.

Rajabpour, A., Abdali-Mashhadi, A. R., & Ghorbani, M. R. (2018a). Acaricidal and repellent properties of some plant extracts against poultry red mite, *Dermanyssus gallinae* (Mesostigmata: Dermanyssidae). *Persian Journal of Acarology, 7*(1). https://doi.org/10.22073/pja.v7i1.34098.

Rajabpour, A., Seraj, A. A., & Mehrnia, A. (2018b). Developing an artificial diet for rearing *Orius albidipennis* Reuter (Het., Anthocoridae). *Journal of Plant Protection Research, 58*(3), 276–281. https://doi.org/10.24425/jppr.2018.124635.

Robinson, E. A., Ryan, G. D., & Newman, J. A. (2012). A meta-analytical review of the effects of elevated CO_2 on plant–arthropod interactions highlight the importance of interacting environmental and biological variables. *New Phytologist, 194*(2), 321–336. https://doi.org/10.1111/j.1469-8137.2012.04074.x.

Sadegh Nejad, A., Rajabpour, A., & Cultler, C. (2019). Effect of climate and ecosystem manipulation on biodiversity of Anthocoridae in Hormozgan Province (Iran). *Plant Protection (Scientific Journal of Agriculture, 42*(2), 37–49. https://doi.org/10.22055/ppr.2019.14745.

Safaei, N., Rajabpour, A., & Seraj, A. A. (2016). Evaluation of various diets and oviposition substrates for rearing *Orius albidipennis* Reuter. *Journal of Entomological Society of Iran, 35*(4), 29–37.

Salehi, Z., Yarahmadi, F., Rasekh, A., & Sohani, N. Z. (2016). Functional responses of *Orius albidipennis* Reuter (Hemiptera, Anthocoridae) to *Tuta absoluta* Meyrick

(Lepidoptera, Gelechiidae) on two tomato cultivars with different leaf morphological characteristics. *Entomologia Generalis, 36*(2), 127–136. https://doi.org/10.1127/entomologia/2016/0339.

Sepahvand, Z., Seraj, A. A., Rajabpour, A., & Kocheili, F. (2019). Biodiversity of minute pirate bugs (Hemiptera: Anthocoridae) in agricultural and natural habitats of Lorestan Province, West Iran. *Iranian Journal of Applied Biosciences, 15*(1). https://doi.org/10.22067/ijab.v15i1.77875.

Shahbi, M., & Rajabpour, A. (2017). A fixed-precision sequential sampling plan for the potato tuberworm moth, *Phthorimaea operculella* Zeller (Lepidoptera: Gelechidae), on potato cultivars. *Neotropical Entomology, 46*(4), 388–395. https://doi.org/10.1007/s13744-017-0487-1.

Taylor, L. R. (1984). Assessing and interpreting the spatial distributions of insect populations. *Annual Review of Entomology, 29*(1), 321–357. https://doi.org/10.1146/annurev.en.29.010184.001541.

Terblanche, J. S., Clusella-Trullas, S., & Lehmann, P. (2024). How climate change might impact insect movement via physiological mechanisms. *One Earth, 7*(4), 608–622. https://doi.org/10.1016/j.oneear.2024.03.008.

Thornton, P. K., Jones, P. G., Ericksen, P. J., & Challinor, A. J. (2011). Agriculture and food systems in sub-Saharan Africa in a 4?C+ world. *Philosophical Transactions of the Royal Society A, 369*(1934), 117–136. https://doi.org/10.1098/rsta.2010.0246.

Thuiller, W., Lavorel, S., Araújo, M. B., Sykes, M. T., & Prentice, I. C. (2005). Climate change threats to plant diversity in Europe. *Proceedings of the National Academy of Sciences, 102*(23), 8245–8250. https://doi.org/10.1073/pnas.0409902102.

White, T. C. R. (2009). Plant vigor versus plant stress: A false dichotomy. *Oikos, 118*(6), 807–808. https://doi.org/10.1111/j.1600-0706.2009.17495.x.

Ximénez-Embún, M. G., Glas, J. J., Ortego, F., Alba, J. M., Castañera, P., & Kant, M. R. (2017). Drought stress promotes the colonization success of an herbivorous mite that manipulates plant defenses. *Experimental and Applied Acarology, 73*(3, 4), 297–315. https://doi.org/10.1007/s10493-017-0200-4.

Zeng, J., Liu, Y., Zhang, H., Liu, J., Jiang, Y., Wyckhuys, K. A. G., & Wu, K. (2020). Global warming modifies long-distance migration of an agricultural insect pest. *Journal of Pest Science, 93*(2), 569–581. https://doi.org/10.1007/s10340-019-01187-5.

Zvereva, E. L., & Kozlov, M. V. (2006). Consequences of simultaneous elevation of carbon dioxide and temperature for plant–herbivore interactions: A metaanalysis. *Global Change Biology, 12*(1), 27–41. https://doi.org/10.1111/j.1365-2486.2005.01086.x.

CHAPTER 9

Induced Plant Defenses Against Insect Herbivores: The Chemical Communication Underlying Induced Defenses Against Invasive Insect Pests

GHOLAMREZA ABDI,[1,*] SUDIPTA SAMANTA,[2] AMISHA RANI,[3] MD. SABIR AHMED MONDOL,[4] and ACHYUTA BASAK[5]

[1]*Department of Biotechnology, Persian Gulf Research Institute, Persian Gulf University, Bushehr, Iran*

[2]*Department of Genetics and Plant Breeding, Indian Agricultural Research Institute, Kolkata Hub, Barrackpore, Kolkata, West Bengal, India*

[3]*Department of Forest Products, Dr. Yashwant Singh Parmar University of Horticulture and Forestry, Nauni, Solan, Himachal Pradesh, India*

[4]*Department of Agricultural Biochemistry, Bidhan Chandra Krishi Viswavidyalaya, Mohanpur, Nadia, West Bengal, India*

[5]*Department of Genetics and Plant Breeding, Uttar Banga Krishi Viswavidyalaya, Pundibari, Cooch Behar, West Bengal, India*

*Corresponding author

ABSTRACT

Insects are major herbivores, exerting significant pressure on plants, which have evolved diverse defense mechanisms despite lacking mobility and an animal-like immune system. While some plant-insect interactions, like

Insect-Plant Interactions, Volume 1: Influence of Biotic and Abiotic Factors.
Jamin Ali, Adil Tonğa, Adnan Shakeel, and Nasheeman Ashraf (Eds.)

DOI: 10.1201/9781779644282-9

pollination, are mutualistic, most are antagonistic, with insects feeding on plant tissues and plants responding with morphological barriers and chemical defenses. Herbivorous insects employ various strategies to extract nutrients, but plants counter by producing toxins, protease inhibitors, and volatiles that attract insect predators. Early theories focused on constitutive defenses; compounds produced during normal growth that deter herbivores. However, plants also deploy induced defenses, rapidly modifying their transcriptomes, proteomes, and metabolomes in response to herbivore cues such as compounds in insect saliva or oviposition fluids. These changes involve complex biochemical pathways, including reactive oxygen species (ROS) generation and scavenging enzymes, defense-related phytohormones, and the synthesis of secondary metabolites like terpenes, volatile organic compounds (VOCs), and green leaf volatiles (GLVs). This chapter explores the intricate ways plants perceive and respond to insect herbivory, emphasizing the dynamic and sophisticated nature of these defense systems. Understanding these mechanisms provides valuable insights for developing sustainable pest management strategies and enhancing plant resilience in agricultural ecosystems.

9.1 INTRODUCTION

The connections between flora and herbivores are essential ecological relationships in nature (Johnson, 2011). As main producers, nearly all plants inherently evade consumption by herbivores (Zhao et al., 2009). Therefore, these interactions will impact the nutrient cycles and energy transfers within food chains (Coley and Barone, 1996). According to reports, these plant-eating animals devour more than 15% of the total amount of plant matter generated each year in temperate and tropical habitats. Therefore, herbivory is the primary means by which energy is transferred into food chains (Agrawal, 2011; Johnson, 2011). Over 75% of animals are naturally herbivores, and they have a crucial impact on the structure and function of ecosystems (Cebrian, 2004; Huntly, 1991). Herbivores exert a substantial impact on the plant distribution as well as the quantity of plants through their consumption (Houlahan and Findlay, 2004; Xiong et al., 2008). Furthermore, they exert considerable evolutionary pressure on the plant population by elevating its mortality rate and reducing its biomass (Coley et al., 1985). Instead, the eating habits of plant insects will fluctuate based on their environmental factors, including the kind of ecosystem, the quantity of vegetation, and the level of light intensity. The interaction between flora and herbivores alters the arrangement and dynamics of plant populations as well as influences the variety and composition of communities, along with the broader environment, via the food web and nutrient cycles (Jiang et al., 2005). Plants deploy many defense strategies when confronted with herbivores, which are crucial for studying herbivore-plant interactions and ultimately facilitate the co-evolution of plant species and herbivorous organisms. Therefore, comprehending the development and environmental interactions of plant defenses is essentially the same as comprehending the emergence and purpose of existing ecosystems (Johnson, 2011). Moreover, plant defense research includes several subdisciplines and incorporates multiple levels of study, including genetics, genomics, community ecology, physiology, chemistry, and global patterns of herbivory and plant defense mechanisms (Johnson, 2011). Investigating how plants protect themselves from herbivores is crucial, as the impact of herbivory results in considerable financial setbacks in agriculture, horticulture, and forestry, amounting to billions of dollars (Oerke and Dehne, 2004). A brief synopsis of the herbivore classes is given in this chapter. So, we then looked at the major groups of protection that plants have against herbivores. Finally, we evaluated the progress that has been made in understanding how plants and herbivores interact. Then, in order to provide a strong theoretical framework for our upcoming studies, we discussed possible avenues for future study in this area.

9.2 HERBIVORY

In addition to taking resources from other consumers, herbivores also supply resources to other consumers, which influences trophic relationships. Ecosystem processes, including production, decomposition, and nutrient cycling, are influenced by herbivore characteristics such as resource constraints, feeding strategies, nutrient intake versus predation risk, and elemental chemistry balance. A strategy for investigating herbivore-mediated ecological consequences is presented in this chapter. There has been discussion over the possible advantages of herbivory for plants. Certain scholars contend that herbivores can improve plant lifetime, seed output, or net primary productivity by removing or harming plant tissues.

This might result in an improvement in plant fitness and the evolution of herbivore-plant mutualisms. However, increased plant biomass from tissue removal has primarily been observed under controlled conditions and in cultivated crops. There is no compelling evidence that herbivory consistently benefits grazed plants, although it may reduce competition or remove senescent tissue (Belsky, 1986).

In terrestrial ecosystems, herbivorous insects play a vital role in shaping plant populations and communities. High-resolution, quantitative worldwide forecasts of insect herbivory patterns and determinants are still difficult to achieve despite this. Strong geographical heterogeneity was found when 9,682 recordings of insect herbivory severity across natural populations were analyzed. Among herbaceous plants, herbivory did not significantly vary with latitude; however, for woody plants, it did increase with latitude. Herbivory on herbaceous plants was significantly predicted by soil's cation-exchange capacity (CEC), but herbivory on woody plants was mostly governed by the climate. Future climate projections indicate that while herbivore pressure on woody plants (e.g., northern coniferous forests) would rise globally, herbivore impact on herbaceous plants is expected to become more confined. This evaluation helps forecast plant community vulnerabilities and ecosystem functions in the Anthropocene by offering quantitative insights into how environmental factors affect insect herbivory (Liu et al., 2024). The harm that herbivores do to plants, as well as their density-dependent survival and growth, are strongly influenced by both plant resilience and predators. The impacts of plant defenses and natural predators on herbivores and herbivory are well reported; nonetheless, the extent and distribution of plant damage, along with the interactions and density-dependent effects on herbivores, remain little understood. Understanding these relationships is essential to understanding the mutual influences between plant and herbivore populations. Herbivore population dynamics and plant damage are influenced by different factors, such as plant traits, the presence of predators, and herbivore density. Research on tritrophic relationships emphasizes the importance of herbivore predators as well as plants. Examining the interactions between these variables is crucial to comprehending the control mechanisms of herbivore populations and damage in both wild and controlled systems. The consumptive (predation) and non-consumptive (fear and threat perception) impacts of predators on herbivore populations modify their vital rates and damage levels. Plant resilience, predation pressures, herbivore density, and their interactions can all have an impact on the harm that herbivores inflict. In situations where herbivory is reliant on density, herbivores can help one another by getting past plant defenses like trichomes or stiff leaves (Paniagua et al., 2024). Reductions in the nutritional value of their food sources can especially affect herbivores. The presence of CO_2 pollution exacerbates the dilution of nutrients in plant tissues, making it harder for herbivores to survive by lowering vital elements like zinc, salt, and nitrogen. Increased CO_2 absorption by plants can result in "green deserts," or regions where herbivores can find poor-quality food. It is suggested that this nutrient dispersion is the primary reason for the worldwide decline in herbivore abundance, with notable consequences anticipated in nutrient-poor environments such as open waterways and tropical forests. Regardless of habitat size or pesticide use, fertilization can offset this dilution impact and support other management approaches (Kaspari and Welti, 2024). The key to understanding plant-herbivore interactions is knowing how herbivore damage is distributed across leaves and plants. Damage patterns that are not uniform have historically been linked to differences in plant quality or herbivore activity. Stochastic processes, however, could potentially be involved in the structure of these relationships. The field has mostly concentrated on finding trait-based mechanisms that govern herbivore damage patterns; nevertheless, because herbivory is inherently stochastic, it can be challenging to separate important variables from random fluctuations and to create predictions (Pan and Wetzel, 2024). Foundational ecosystems are being impacted by climate change, and individual species and their interactions are predicted to be affected by warming. Tropical herbivores moving poleward in maritime environments have the potential to change important ecosystems and food web structures. The phenomena referred to as "tropicalization" have been noted in kelp forests as diminishing; nevertheless, its consequences on other systems remain little comprehended. We studied the effects of increasing herbivory (simulated by leaf clipping) on turtle grass, a fundamental maritime plant, using a network of locations spanning 23° of latitude. The findings demonstrated that latitudinal variations in temperature and light caused turtle grass close to its northern range limit to exhibit decreased resistance and productivity. Because they received more sunlight, low-latitude meadows exhibited greater resistance to herbivory. There is little concrete proof that agriculture has an effect on herbivory, despite the assumption that insect herbivory is stronger in agroecosystems than in wild systems. It is crucial to contrast farmed crops with their wild equivalents in their original locations in order to comprehend these consequences. In Mexico, a region known for its agricultural domestication in Mesoamerica, we examined the herbivory trends in four crops: maize, beans, squash, and husk tomatoes. The aim of this research was to investigate potential differences between domesticated crops and their wild equivalents in terms of overall leaf consumption, along with the kinds and frequency of harm inflicted by leaf-eating insects (Ruiz-Arocho et al., 2024).

9.2.1 CATEGORIES OF HERBIVORES

9.2.1.1 BASED ON CRITERIA OF ZOOLOGICAL CLASSIFICATION

Herbivores can be classified into invertebrates and vertebrates based on zoological categorization standards. The majority of these animals are typically herbivorous mammals, mostly ungulates. This characteristic is widely acknowledged as a crucial element in preserving the biodiversity of grasslands (Allan and Crawley, 2011). In contrast, the latter is mostly made up of Mollusca (usually Gastropoda, such as snails, slugs, and other similar creatures) and Arthropoda (which includes herbivorous insects and crustaceans). At first, several writers documented significant connections between mammals and plants. In contrast, there was no focus on the contribution of invertebrate herbivores to the formation of plant communities and the dynamics of populations. Multiple studies have shown that invertebrate herbivores significantly impact the process of secondary succession in plant communities (Brown and Gange, 1989, 1992; Edwards et al., 1987; Gibson et al., 1987; Hanley et al., 1995). Many animals and mollusks eat plant seeds, but insects do a lot of damage to grown plants in the field. A lot of studies have been done on how plants and insects from different types of forests connect with each other. Because they have different mouthparts, insects hurt leaves in many ways, such as by chewing, sucking, skeletonizing, insect galling, rolling, and mining (Allen and Coufal, 1984). The majority of species on our planet consist of plants and insects, and their interactions have important implications for ecological and evolutionary processes (Coley, 1983; Hartley and Jones, 2003; Kursar and Coley, 2003), as well as for the recycling of nutrients and transfer of energy within ecosystems (Coley, 1983; Herrera and Pellmyr, 2002). Currently, there is insufficient study on the interactions between mollusks and plants, unlike the extensive studies on the links between insects and plants. Nevertheless, the majority of studies on mollusk herbivory have demonstrated that mollusks, despite consuming minimal biomass, do contribute to the heightened mortality of seedlings of less dominant herbaceous species (Allan and Crawley, 2011; Buschmann et al., 2005). Mollusks typically prefer to consume seedlings rather than mature plants, resulting in a significant impact on individual plants that is not proportional to the amount of biomass taken (Duthoit, 1964; Fenner et al., 1999). For instance, the consumption of a single bite of the hypocotyl by a mollusk can result in the death of an entire seedling, but a comparable bite on an old or mature leaf would have very little impact on the plant's survival (Dirzo and Harper, 1980). Thus, mollusks significantly influence the composition of the herb layer in a community (Crawley, 1989; Hulme, 1994), particularly when it comes to the establishment of seedlings, which is a critical stage in a species' life cycle (Hulme, 1994, 1996; Scheidel and Bruelheide, 2005). Lodge (1991) noted that aquatic herbivores have a negligible impact on the growth and development of aquatic plants. Subsequent research has shown that aquatic herbivores significantly impact the biomass and species mix of aquatic plants (Jeppensen et al., 1998; Søndergaard et al., 1996; van Donk, 1998). Snails and crustaceans, such as crayfish, are extensively spread in aquatic environments and are commonly found as herbivores. These organisms belong to the Gastropoda and Crustacea classes, respectively. The majority have been employed as versatile herbivorous organisms in bioassay studies to clarify the relationship between aquatic herbivores and aquatic plants (Cronin et al., 2002; Morrison and Hay, 2011; Parker and Hay, 2005; Xiong et al., 2008).

9.2.1.2 BASED ON HERBIVORE'S PREFERENCE FOR PLANT SPECIES

Herbivores may be categorized into two main groups based on their preferred plant types: generalists and specialists, the latter including oligophagous and monophagous herbivores. Generally, herbivorous animals and mollusks are categorized as generalist consumers, whereas phytophagous insects are recognized as specialist consumers (Morrison and Hay, 2011). As an illustration, crayfish (a kind of crustacean) are frequently employed as versatile plant-eating animals in scientific research (*see* Figure 9.1). Generalist herbivorous animals have the ability to consume a diverse range of plants. They do not have specific eating preferences that would force them to avoid particular plant types. Specialized herbivores mostly depend on a single food supply due to the presence of plant defenses and natural enemies (Coley and Barone, 1996). Considering the distinct reliance and impact on plants exhibited by generalists and specialists (Abrahamson, 1989), two perspectives emerge: Some scientists suggest that generalist consumers exert a more significant influence on plant fitness and community composition (Agrawal et al., 2005; Morrison and Hay, 2011; Parker et al., 2006; Xiong et al., 2008). In contrast, other researchers argue that skilled eaters are more detrimental to plants due to their higher abilities in food finding, food use, food placement, and detoxification. Furthermore, they possess a reduced likelihood of encountering natural adversaries (Zhang, 2008). The variations in experimental results may be ascribed to the varying study focus on either generalists or specialists, in conjunction with the evolutionary stage of the plants. Temperate forest ecosystems predominantly feature generalist herbivores, whereas tropical forest ecosystems are primarily inhabited by specialized herbivores. Basset (1994) posited that this phenomenon may be ascribed to the heightened palatability of foliage in temperate woods. The endurance of fully grown or undeveloped foliage in tropical forests was double that of temperate forests. However, the concentrations of nitrogen and water content in mature leaves in tropical forests were much lower than those in temperate forests (Coley et al., 1991) (Table 9.1).

9.3 REACTIVE OXYGEN SPECIES (ROS), A SIGNALING MOLECULE FOR REGULATING PLANT DEFENCE SYSTEMS

When plants encounter certain stressors, they rapidly produce a significant amount of ROS. For example, it only takes 13 minutes for the intracellular concentration of hydrogen peroxide (H_2O_2) to reach 1M when a plant is exposed to certain specific stressed conditions (Jacks and Davidonis, 1979). Hydrogen peroxide is generated when *Spodoptera littoralis* damages the leaves of the Lima bean (*Phaseolus lunatus*), and it is expected that levels will increase as long as the attacks persist (Maffei et al., 2006). ROS are undesirable byproducts of normal cellular metabolism seen in plants and other aerobic organisms (Gechev et al., 2006; Halliwell, 2006). Molecular oxygen (O_2) serves as a terminal electron acceptor in plants.

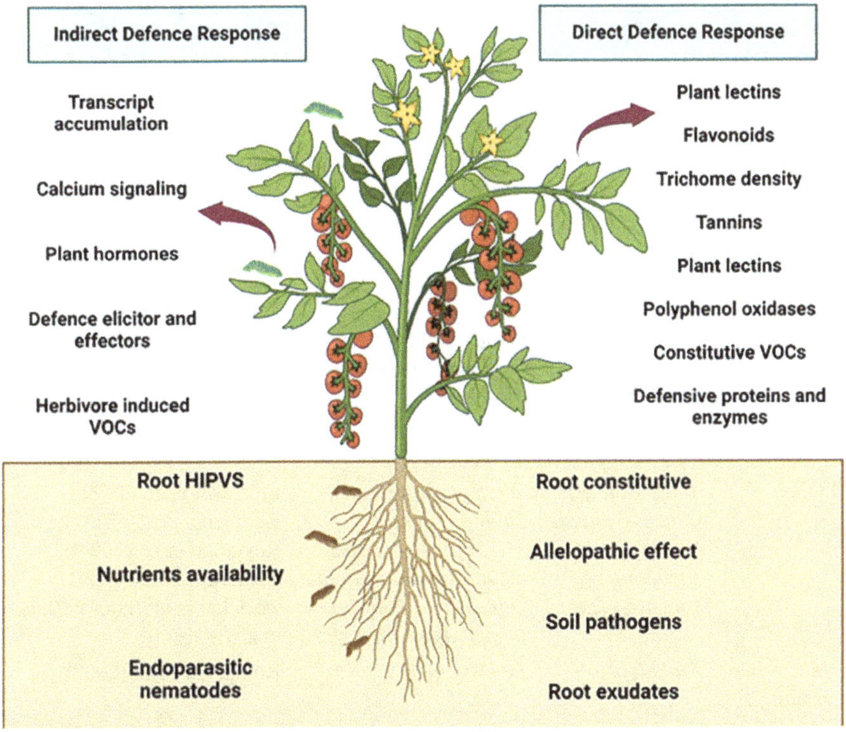

FIGURE 9.1 Diagrammatic representations of the systems governing plant defense against insect herbivores.

ROS are generated when there is a reduction in molecular oxygen (O_2). Halliwell (2006) emphasized this process. Superoxide (O_2^-) and hydroperoxide (HO_2) radicals are formed during the first stage of this reduction (Figure 9.2). After that, H_2O_2 (hydrogen peroxide) is produced, which can translocate from its subcellular synthesis locations to adjacent cells or compartments (Bienert et al., 2006). O_3^+ and H_2O_2 have a significant oxidizing capacity, which can be dangerous for the surrounding cellular environment. H_2O_2 is remarkable because it may pass through biological membranes, start the oxidation of lipids, and oxidize the thiol groups of enzymes to render them inactive (Halliwell, 2006). The primary organelles responsible for the production of ROS are glyoxysomes, peroxisomes, and chloroplasts. ROS are generated in peroxisomes during photorespiration, in glyoxysomes during fatty acid oxidation, and during photosynthetic

TABLE 9.1 List of Putative Defense Proteins in Plants Corresponding to Insect Species

Putative Defense Protein	Plant Species	Insect Species	References
SOD	*Medicago sativa*	*Aphis medicaginis*	Huang et al. (2007)
PPOs	Buffalo grass; Tomato	*Manduca sexta; Blissus oxiduus; Spodoptera frugiperda Helicoverpa armigera*	Bhonwong et al. (2009); Chen et al. (2005); Heng-Moss et al. (2004)
Peroxidases	*Alnus glutinosa Arabidopsis;* Buffalo grass; Corn; *Medicago sativa;* Poplar Rice	*Agelastica alni; Aphis medicaginis; Bemisia tabaci* (whitefly); *Blissus occiduus; Lymantria dispar; Spodoptera frugiperda; Spodoptera littoralis*	Barbehenn et al. (2009); Chen et al. (2009); Gulsen et al. (2010); Heng-Moss et al. (2004); Huang et al. (2007); Kempema et al. (2007); Stout et al. (2009); Tscharntke et al. (2001)
LOXs	*Alnus glutinosa; Cucumis sativus; Nicotiana attenuata;* Tomato; Wheat	*Spodoptera littoralis; Sitobion avenae; Myzus persicae; Myzus nicotianae Macrosiphium euphorbiae; Bemisia tabaci; Agelastica alni*	Fidantsef et al. (1999); Kempema et al. (2007); Reymond et al. (2004); Tscharntke et al. (2001); Voelckel et al. (2004); Zhao et al. (2009)
Hevein-like protein	*Arabidopsis*	*Bemisia tabaci*	Kempema et al. (2007)
Chitinases	*Sorghum bicolor*	*Schizaphis graminum*	Voelckel and Baldwin (2004)
Catalase	Buffalograss	*Blissus oxiduus*	Heng-Moss et al. (2004)

electron transport in chloroplasts (Asada, 2006). Another source of O_3^- and H_2O_2, but at lower concentrations than in chloroplasts, is mitochondrial respiration. According to Robson and Vanlerberghe (2002), mitochondrial ROS are crucial for regulating various biological processes, including stress adaptation and programmed cell death (PCD). To regulate ROS levels, plants have evolved many defense systems, including enzymes and small antioxidant molecules (Van et al., 2001). They have created a sophisticated network of detoxifying and ROS-producing enzymes that regulate the amount of ROS produced in different cell types, organs, time periods, and developmental stages in accordance with biological requirements. The evolutionary adaptation of ROS has enabled them to function as signaling molecules that regulate cell growth and apoptosis. This ability significantly impacts a plant's growth and development, its adaptability to environmental stress, and its responses to diseases (Dat et al., 2000; Gapper and Dolan, 2006; Gechev and Hille, 2005). Upon effective identification of a pathogen, one of the initial cellular responses is the generation of ROS (Torres et al., 2006). This process happens in two phases: a non-specific initial phase that occurs minutes after pathogen detection, and a secondary phase that occurs one to three hours later and is dependent on the host's recognition of incompatible pathogens (Dat et al., 2000). The function of ROS in pathogen defense is ambiguous, specifically regarding their involvement as signaling molecules or as direct destructive agents against pathogens and host cells via the hypersensitive response (HR) (Torres et al., 2006). After the discovery of pathogens, many enzymes, such as NADPH oxidases and cell wall peroxidases, are believed to play a role in the production of ROS in the apoplast (Torres et al., 2006). While the apoplast is the site of the major oxidative burst, ROS produced in other cellular compartments may also contribute to pathogen defense (Asada, 2006; Rhoads et al., 2006).

9.3.1 ENZYMES INVOLVED IN REACTIVE OXYGEN SPECIES (ROS) PRODUCTION AND SCAVENGING

9.3.1.1 NAD(P)H OXIDASE

Recent studies indicate that NADPH oxidase in plant plasma membranes is crucial for the generation of ROS under stressful conditions (Takemoto et al., 2006; Tanaka et al., 2006). Superoxide anion (O_2^-) is formed when electrons from cytoplasmic NAD(P)H are transferred to molecular oxygen (O_2) via NADPH oxidase. The dismutation of O_2^- into hydrogen peroxide (H_2O_2) occurs next. Plant functions such as growth, development, defensive responses against pathogens or elicitors (Tanaka et al., 2006; Takemoto et al., 2006), and responses to abiotic challenges including cold, UVB radiation, ozone, and salt stress are all dependent on NADPH oxidase (Sagi and Fluhr, 2006).

9.3.1.2 SUPEROXIDE DISMUTASE (SOD)

The essential enzymes known as superoxide dismutase (SODs) scavenge superoxide radicals (O_2^-) by transforming them into hydrogen peroxide

(H_2O_2). Mn-SOD, Cu/Zn-SOD, and Fe-SOD are the three primary varieties of SODs, which are metalloproteins classified based on their metal cofactors. Prokaryotes and eukaryotes' mitochondrial matrices contain large amounts of Mn-SOD. Prokaryotes and certain plant groups' chloroplasts have Fe-SOD (Bannister et al., 1987; Bowler et al., 1992).

9.3.1.3 CATALASE

Catalases are tetrameric heme-containing enzymes that facilitate the decomposition of hydrogen peroxide (H_2O_2) into water and oxygen, protecting cells from oxidative damage (Sanchez-Casas and Klessig, 1994). According to Havir and McHale (1989), catalases are mainly linked to the elimination of H_2O_2 from microbodies like glyoxysomes and peroxisomes. In these environments, they may also oxidize substrates like formaldehyde, ethanol, formate, or elemental mercury. While catalases are important for scavenging H_2O_2, stress circumstances that hinder translation or promote degradation can cause a rapid reduction in their levels (Dat et al., 2000). The inhibition of catalase activity, which is crucial for defense mechanisms in tobacco plants, has been suggested to occur due to salicylic acid (SA) during pathogenesis.

9.3.1.4 PEROXIDASE

Peroxidases (PODs) are a class of enzymes that catalyze oxidative processes by using different peroxides (ROOH) as electron acceptors (Passardi et al., 2007). Plants have PODs in two primary classes: class I and class III. Class III PODs are secreted into the extracellular matrix and surrounding environment, while class I PODs are localized intracellularly. Ferriprotoporphyrin IX acts as a prosthetic group and consists of 10 α-helices that are common to both classes. Class III PODs have three supplementary α-helices, conserved amino acids, and four disulfide bonds (Passardi et al., 2004). PODs diminish H_2O_2 levels by absorbing electrons from diverse donor molecules, including phenolic compounds and lignin precursors. PODs are bifunctional enzymes that have the ability to both oxidize substrates and create ROS. They contribute to several physiological and developmental processes, encompassing germination to senescence.

9.3.1.5 ASCORBATE PEROXIDASE (APX)

Ascorbate peroxidase (APX) is essential for the removal of ROS and the protection of cells from oxidative damage in higher plants, algae, and other animals (Teixeira et al., 2006). APX is found as isoenzymes in the cytoplasm, mitochondria, and peroxisomes, among other cellular compartments. Responses to environmental conditions like drought and salinity stress have been shown to increase the activity of particular APX isoforms; this suggests that these isoenzymes have a functional specialization in the elimination of H_2O_2 (Sharma and Dubey, 2005; Teixeira et al., 2006).

9.3.1.6 GLUTATHIONE REDUCTASE (GR)

Glutathione, crucial for sustaining cellular redox balance, participates in both enzymatic and non-enzymatic activities and occurs in two forms: reduced glutathione (GSH) and oxidized glutathione (GSSG) (Valencia et al., 2001). GSH oxidizes more quickly in an oxidative stress environment. According to Stephensen et al. (2002), GR is a crucial enzyme that is responsible for producing GSH from GSSG. Its activity is thought to be a valuable biomarker for redox cycling chemicals.

9.4 THE ROLE OF PLANT HORMONES IN PROMOTING RESISTANCE TO STRESS

Plants defend against herbivores through diverse signal transduction pathways, underpinned by a complex network of phytohormones crucial for controlling growth, development, and defense mechanisms (Verhage et al., 2010). Several phytohormones have been linked to communication between plants, both inside a single plant and between different plants, when they are harmed by herbivores. The initiation of the majority of plant defensive responses to insects is activated by jasmonic acid (JA), salicylic acid (SA), and ethylene (ET) signal transduction pathways (Gill et al., 2010; Shivaji et al., 2010). Specific groups of genes associated with defense mechanisms are activated by these pathways when faced with injury or the feeding of insects. These hormones may have individual, synergistic, or antagonistic actions depending on the aggressor, with JA serving as the principal hormone linked to plant defense against herbivores, notwithstanding the participation of other plant hormones. JA is crucial in triggering both direct and indirect defense mechanisms in plants (Shivaji et al., 2010; Usha and Jyothsna, 2010; War et al., 2011a). In response to damage or herbivory, JA begins to accumulate in plant tissues; it is produced from linolenic acid via the octadecanoid pathway. Certain lipoxygenases (LOXs) in herbivorous insects facilitate the oxidation of linoleic and linolenic acids at carbon 9 or carbon 13, respectively, yielding (9S)- or (13S)-hydroperoxy-octadecadi(tri)enoic acids. These compounds are subsequently converted into 12-oxophytodienoic acid (12-OPDA) by allene oxide synthase and allene oxide cyclase, before being transported to the peroxisome, where OPDA is reduced by OPDA reductase 3 (OPR3), culminating in the biosynthesis of JA. The production of ROS during the oxidative burst transforms linolenic acid into phytoprostanes, which in turn initiate signal transduction pathways (Rayapuram and Baldwin, 2007). Jasmonates elicit a wide range of defensive reactions, such as the activation of antioxidative enzymes, creation of protease inhibitors (PIs), emission of VOCs, synthesis of alkaloids, construction of trichomes, and secretion of extrafloral nectar (EFN) (Dickens, 2006; Mao et al., 2007; Pauwels et al., 2009). A considerable array of genes that govern defense mechanisms against herbivores is regulated by JA (Shivaji et al., 2010). Jasmonates increase the level of indole glucosinolate (GSL), an important protective compound. OPDA not only helps produce JA, but also activates defense pathways on its own. OPDA signaling modulates the expression of Coronatin-Insensitive 1 (COI1) in a COI1-dependent and COI1-independent manner (Ribot et al., 2008). Furthermore, it influences the concentrations of calcium within cells and the state of cellular redox (Walter, 2007). Jasmonates, particularly the JA-amino acid conjugate jasmonoyl-isoleucine, are acknowledged for their interaction with the COI1 component of the E3 ubiquitin ligase complex known as SCFCOI1 (Skip/Cullin/F-box–COI1). This interaction promotes the binding of COI1 to JAZ (jasmonate ZIM-domain) proteins, resulting in their degradation and the subsequent alleviation of inhibition on the expression of jasmonic acid-inducible genes (Sheard et al., 2010). JA's influence encompasses the modulation of both the transcript and activity of calcium-dependent protein kinases (CDPK) in potato plants (Ulloa, 2002). CDPKs, present in plants, encompass a varied collection of serine/threonine kinases, with Arabidopsis alone containing 34 members. These kinases play a pivotal role in plant defense against a range of biotic and abiotic stressors by aiding in signal transduction (Ludwig et al., 2004).

JA not only has a significant function in directly combating insect pests by triggering the production of defensive chemicals, but it also has a well-established role in indirectly enhancing resistance. For instance, the EFN generated by JA serves as an alternative food source for the insect pests. JA also stimulates the production of defensive enzymes, such as POD (Shivaji et al., 2010; Usha and Jyothsna, 2010; War et al., 2011a) and PPO (Usha and Jyothsna, 2010; War et al., 2011a; Zhao et al., 2009). SA plays a crucial role as a phytohormone in controlling plant defense mechanisms. It is derived from benzoic acid and is known as SA (War et al., 2011b). It is an essential internal plant growth regulator that elicits numerous metabolic and physiological responses in plants, which contribute to their defensive mechanisms as well as influence their growth and development (Rivas et al., 2011). NPR1, a regulatory protein, affects the reaction to systemic acquired resistance (SAR) (Pieterse and Van Loon, 2004). SA accumulation results in the activation of NPR1 via redox pathways. Subsequently, it is transported to the nucleus and influences gene expression through transcription factors, rather than attaching directly to DNA. SAR provides stronger protection against piercing and sucking insect pests compared to chewing bugs. The SA signaling molecule contributes to local defense and the initiation of systemic resistance. The SA route has been suggested as a means of triggering the production of ROS in plants, which can help plants develop resistance to insect pests. For example, in tomato plants, this mechanism has been seen to provide protection against *Helicoverpa armigera* (Peng et al., 2004). The presence of H_2O_2, caused by SA in plants, serves as a defense mechanism against different insect pests. H_2O_2 adversely impacts the digestive systems of insects, resulting in stunted growth and development (Maffei et al., 2007; Peng et al., 2004), whereas SA triggers the emission of plant volatiles that lure indigenous predators of insect pests. For example, when Lima bean and tomato plants are infested by spider mites, they emit volatile compounds that draw in the natural predators of these pests (de Boer et al., 2004). According to Maffei et al. (2007), it has been established that SA and JA exhibit antagonistic interactions, with salicylic acid suppressing the activity of JA and the reverse also occurring. Methyl salicylate (MeSA) functions as a volatile signaling molecule that activates induced defenses in plants, including the emission of herbivore-induced plant volatiles (HIPVs). In natural environments, MeSA draws in various predatory arthropods. ET is another crucial phytohormone that actively contributes to the defensive systems of plants against diverse insects (Van et al., 2006). ET signaling is crucial for plant defense against herbivores and pathogens, producing both direct and indirect effects (Van et al., 2006); however, there are few data concerning its role in indirect defense through the release of HIPVs (Yujie et al., 2006). The ET signaling pathway can either promote or suppress the development of plant defensive responses against pathogens and herbivorous insects, in combination with JA (Kunkel and Brooks, 2002). It has been observed that ET and JA collaborate in the expression of PIs in tomato (Kunkel and Brooks, 2002). The presence of *Agelastica alni* caused *Alnus glutinosa* L. leaves to produce ET and release a variety of volatile compounds, including mono-, sesqui-, and homoterpenes (Tscharntke et al., 2001). The study by Horiuchi et al. (2003) found that 1-aminocyclopropane-1-carboxylic acid, which is an ET precursor, can increase the release of volatile compounds from detached leaves treated with JA. ET furthermore stimulated the release of volatile compounds triggered by volicitin, JA, or (Z)-3-hexen-ol in maize (Ruther and Kleier, 2005).

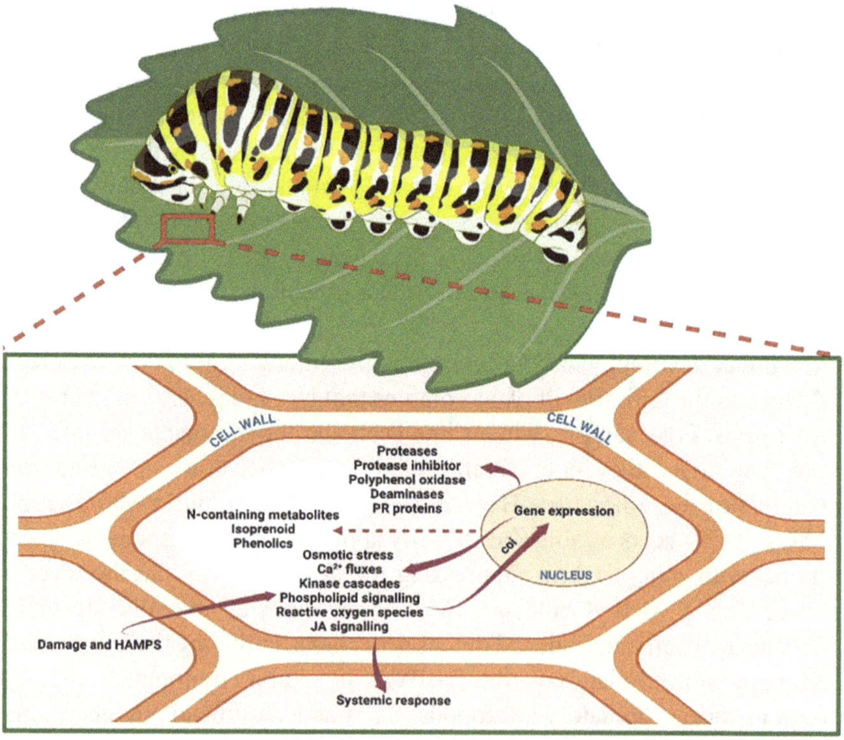

FIGURE 9.2 Plant defense response by the insect attack.

9.5 VOLATILE ORGANIC COMPOUNDS (VOCS)

Many plants emit VOCs in response to herbivore attacks. As a kind of defense, these VOCs can either draw in predatory arthropods or repel herbivores (Heil and Bueno, 2007). These chemicals are released in large quantities throughout the kingdom of plants. Herbivore-infested plants release complex scent mixtures that frequently include more than 100 distinct components (Arimura et al., 2005). At least three distinct biosynthetic pathways account for the variety of volatiles emitted by injured plants: the fatty acid (or octadecanoid) pathway, which yields jasmone and C6 GLVs; the shikimic acid (or tryptophan) pathway, which produces methyl salicylate and indole; and the isoprenoid-derived pathway, which produces terpenes (Agrawal et al., 1999). Usually, the primary volatile substances emitted by plants after being damaged by herbivores consist of terpenes, indole derivatives, and GLVs. While volatile terpenoids are released many hours after herbivore damage, specific volatiles are released considerably faster, such as the C6-aldehyde (3Z)-hex-3-enal. After disturbing Arabidopsis leaf tissue, this chemical was found 20 seconds later (Matsui et al., 2000).

9.5.1 TERPENES

Terpenes, a type of secondary metabolite often found in plants, exhibit diverse forms and are crucial for interactions with plants, insects, and diseases (Dudareva et al., 2004; Paschold et al., 2006). The compounds are typically produced in glandular trichomes or secretory structures and are usually emanated from plant epidermal cells and discharged into the rhizosphere or atmosphere (Dudareva et al., 2004). Isopentenyl diphosphate (IDP) serves as the precursor in the terpenoid biosynthesis pathway (Figure 9.3), and IDP-isomerase catalyzes its conversion to dimethylallyl diphosphate (DMADP). Prenyltransferases subsequently catalyze the head-to-tail condensation of IDP and DMADP to yield geranyl diphosphate (GDP). Farnesyl diphosphate (FDP) is produced by the condensation of GDP with IDP, while geranylgeranyl diphosphate (GGDP) is produced by a subsequent reaction. All monoterpenes (C10) are derived from GDP, sesquiterpenes (C15) are produced by FDP, and diterpenes (C20) are produced by GGDP (Bartram et al., 2006). Sesquiterpene and sterol precursors are often obtained via the cytosolic mevalonate route (MVA), whereas monoterpene precursors are obtained from the plastidial methylerythritol pathway (MEP). Evidence suggests that these pathways may not be fully independent, since there may be some cross-talk between them (Dudareva et al., 2004). After herbivore damage, highly volatile monoterpenes like linalool and β-ocimene are freshly produced. Significant amounts of sesquiterpenes, which are composed of three isoprene units and are frequently present in floral scents (Chen et al., 2003), are also produced from leaves injured by herbivores. According to Holopainen (2004), β-caryophyllene and α-farnesene are common inducible sesquiterpenes released in response to herbivore damage. (E)-nerolidol, a sesquiterpene alcohol produced from FDP, a prevalent precursor of sesquiterpenes, is the origin of DMNT in plants subjected to herbivore damage (Degenhardt and Gershenzon, 2000).

9.5.1.1 TERPENE SYNTHASE FAMILIES

There are seven subfamilies within the plant terpene synthase family, and each subfamily has at least 40% sequence similarity among its members. According to Cheng et al. (2007), evolutionary connections have a greater influence on the sequence similarity of terpene synthases (TPS) than do their catalytic products.

9.5.2 GREEN LEAF VOLATILES (GLVS)

Green Leaf Volatiles (GLVs) are C6 compounds that constitute over 50% of emissions from injured vegetation. They are synthesized by the LOX pathway. Chemically speaking, GLVs are mostly monounsaturated or saturated alcohols, esters, and aldehydes with a range of configurational isomers and distinct sensory characteristics (Ruther, 2000). GLVs could potentially originate from intact leaves of plants that have been consumed by herbivores, hinting at the activation of the LOX pathway in intact leaves. Upon physical damage, GLVs are frequently emitted from injured plant organelles within 1–2 seconds (Holopainen, 2004). TPSs are proteins that range in length from 550–850 amino acids. Monoterpene synthases consist of around 600–650 amino acids, which include a transit peptide of 50–70 amino acids at the N-terminal for targeting to the plastids. Diterpene synthases possess an additional conserved amino acid, resulting in an internal element that is about 210 amino acids longer (Bohlmann et al., 1998; Martin et al., 2004).

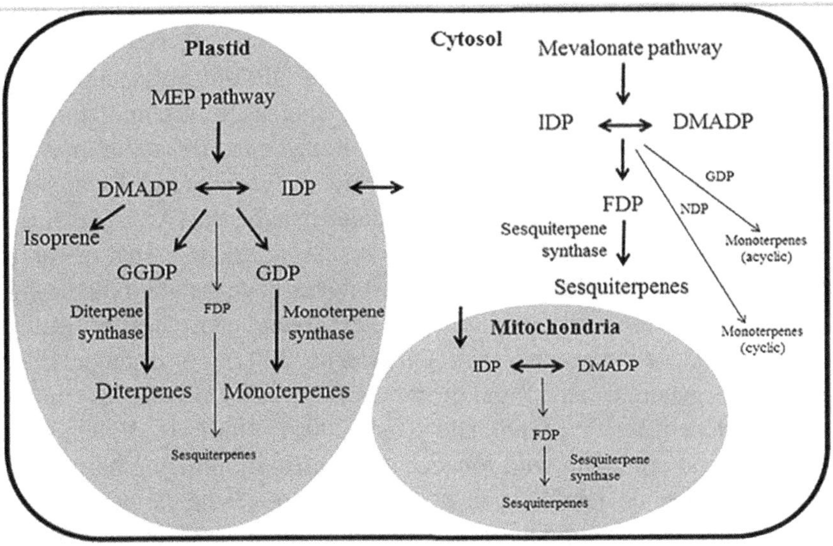

FIGURE 9.3 The biosynthetic pathway of terpenoids.

Source: Reprinted from Pazouki and Niinemets (2016).

9.6 PLANT DEFENSE AND SIGNALING PATHWAYS

Plants and herbivores engage in several activities, such as recognizing chemical and physical signals from herbivores, initiating signal transduction pathways, and activating defense-related genes. When saliva or regurgitant comes into close contact with plant tissues, it predigests plant material in chewing insects such as caterpillars (Bede et al., 2006). The strong elicitor Volicitin from the regurgitant of *Spodoptera exigua* induces modifications in the production of volatile plant compounds (Alborn, 1997). Glucose oxidase (GOX), which produces H_2O_2 through glucose oxidation, is present in the saliva of *Lepidopteran larvae* (Bede et al., 2006). Increased SA buildup from elevated H_2O_2 levels may render plants more vulnerable to insect herbivory (Cipollini et al., 2004). The underlying mechanism for this includes the suppression of JA production or disruption of JA-dependent gene expression, mediated by SA (Bede et al., 2006). The induction of volatiles depends on JA, a phytohormone produced from octadecanoid (Boland et al., 1998). Though not quite the same, the volatile profiles released following JA treatment are similar to those caused by natural herbivory. Herbivore-induced volatiles are not produced when JA biosynthesis is inhibited (Thaler, 1999). In addition to interacting with SA signaling pathways, JA promotes the isoprenoid and shikimic acid pathways. For instance, increased OPDA levels can promote the emission of certain volatiles like TMTT, whereas SA accumulation can lower JA levels and the biosynthesis of JA-linked volatiles (Arimura et al., 2005). ET, stimulated by wounding, herbivore feeding, and JA therapy, similarly enhances the expression of defensive genes (Arimura et al., 2005). Plant Reactions Induced by Volatiles According to recent research, VOCs generated by plants harmed by herbivores have the ability to trigger defense gene expression (Arimura et al., 2000) and cause nearby healthy plants to produce their own VOCs (Dicke et al., 2003). For instance, Lima bean plants generate extra-floral nectar upon exposure to volatiles emitted by injured conspecifics (Heil and Kost, 2006). Similarly, VOCs from sagebrush can stimulate wild tobacco's defenses, minimizing harm from herbivores later on (Baldwin et al., 2006). These insights significantly enhance the understanding of multitrophic interactions in ecosystems and the development of innovative plant defense mechanisms against herbivores.

9.7 TRANSGENERATIONALLY GENERATED RESISTANCE TO HERBIVORES

Both biotic and abiotic challenges have been found to elicit resistance in parent plants and their progeny (Agrawal, 2001), with maternally derived resistance, referred to as transgenerational immunity, proven to safeguard the offspring of plants exposed to herbivory by insect pests. This resistance results in the generation of resilient seeds and seedlings (Agrawal, 2001). However, there is a scarcity of studies regarding the transgenerational defensive mechanisms of plants in reaction to insect herbivory. The wild radish plants, *Raphanus raphanistrum*, wounded by *Pieris rapae* or treated with JA, display offspring with heightened levels of induced resistance to this insect (Agrawal, 2002). Subjecting *Arabidopsis* plants to environmental stressors, including heat, cold, and submerged conditions, resulted in an increased incidence of homologous recombination and augmented genome methylation. This consequently engendered resilience to stress in the progeny. Maternal plants experiencing mild herbivore damage can produce seeds with enhanced vigor and seedlings that display tolerance to insect pests (Agrawal, 2002). Further investigation is required to understand the genetic and molecular processes implicated in these signaling connections. Furthermore, it is essential for research investigating the interplay between plants and insects to account for both genetic influences and the epigenetic regulation of plant defensive strategies and insect responses. This is essential due to substantial evidence that supports the presence of movable siRNA signals and the passing down of gene expression changes induced by DNA methylation. Extensive research is required to thoroughly investigate this topic in order to utilize it for pest control through the manipulation of maternal ecology. Transgenerational induced resistance might provide insights into the complex mechanisms behind plants' capacity to tolerate herbivore harm.

9.8 CONCLUSION

In conclusion, the intricate relationship between plants and herbivorous insects, which has evolved over millions of years, highlights the dynamic and complex nature of plant defense mechanisms. Plants, despite their inability to move and lack of an immune system, have established multiple strategies to safeguard against insect herbivory. These strategies include both direct and indirect defenses, such as the production of toxins, defensive proteins, and VOCs that not only deter herbivores but also attract their predators. The biochemical interactions, particularly those involving phytohormones like JA and SA, play an important role in modulating plant defenses. The ability of plants to alter their gene expression, metabolite production, and protein activities in response to herbivore attacks underscores their adaptive capacity. Moreover, the phenomenon of plant-to-plant communication through VOCs reveals a sophisticated level of interspecies signaling that contributes to plant survival in ecosystems. The findings not only enhance our comprehension of the connections between plants and insects but also offer potential possibilities for developing advanced agricultural methods to enhance crops' ability to withstand pests.

KEYWORDS

- **farnesyl diphosphate**
- **geranylgeranyl diphosphate**
- **herbivores**
- **insect-plant interaction**
- **jasmonic acid**
- **plant defense**
- **systemic acquired resistance**
- **volatile organic compounds**

REFERENCES

Abrahamson, W. G. (1989). *Plant-Animal Interactions*. McGraw-Hill Inc.

Agrawal, A. A. (2001). Transgenerational consequences of plant responses to herbivory: An adaptive maternal effect? *American Naturalist, 157*, 555–569. https://doi.org/10.1086/319932.

Agrawal, A. A. (2002). Herbivory and maternal effects: Mechanisms and consequences of transgenerational induced plant resistance. *Ecology, 83*, 3408–3415. https://doi.org/10.1890/0012-9658(2002)083[3408]2.0.CO;2.

Agrawal, A. A. (2011). Current trends in the evolutionary ecology of plant defence. *Functional Ecology, 25*(2), 421–433.

Agrawal, A. A., Kotanen, P. M., Mitchell, C. E., Power, A. G., Godsoe, W., & Klironomos, J. (2005). Enemy release? An experiment with congeneric plant pairs and diverse above- and belowground enemies. *Ecology, 86*(11), 2979–2989.

Agrawal, A. A., Tuzun, S., & Bent, E. (1999). *Induced Plant Defenses Against Pathogens and Herbivores: Biochemistry, Ecology, and Agriculture* (Vol. ix, p. 390).

Alborn, H. T., Turlings, T. C. J., Jones, T., Stenhagen, G., Loughrin, J. H., & Tumlinson, J. H. (1997). An elicitor of plant volatiles from beet armyworm oral secretion. *Science, 276*(5314), 945–949.

Allan, E., & Crawley, M. J. (2011). Contrasting effects of insect and molluscan herbivores on plant diversity in a long-term field experiment. *Ecology Letters, 14*(12), 1246–1253.

Allen, D. C., & Coufal, J. E. (1984). *Introduction to Forest Entomology: A Training Manual for Forest Technicians*. Syracuse University Press.

Arimura, G. I., Kost, C., & Boland, W. (2005). Herbivore-induced, indirect plant defences. *Biochimica et Biophysica Acta (BBA)–Molecular and Cell Biology of Lipids, 1734*(2), 91–111.

Arimura, G. I., Ozawa, R., Shimoda, T., Nishioka, T., Boland, W., & Takabayashi, J. (2000). Herbivory-induced volatiles elicit defence genes in lima bean leaves. *Nature, 406*(6795), 512–515.

Asada, K. (2006). Production and scavenging of reactive oxygen species in chloroplasts and their functions. *Plant Physiology, 141*(2), 391–396.

Baldwin, I. T., Halitschke, R., Paschold, A., Von Dahl, C. C., & Preston, C. A. (2006). Volatile signaling in plant-plant interactions: "Talking trees" in the genomics era. *Science, 311*(5762), 812–815.

Bannister, J. V., Bannister, W. H., & Rotilio, G. (1987). Aspects of the structure, function, and applications of superoxide dismutase. *Critical Reviews in Biochemistry, 22*(2), 111–180.

Barbehenn, R. V., Jaros, A., Lee, G., Mozola, C., Weir, Q., & Salminen, J. P. (2009). Hydrolyzable tannins as "quantitative defenses": Limited impact against *Lymantria dispar* caterpillars on hybrid poplar. *Journal of Insect Physiology, 55*, 297–304. https://doi.org/10.1016/j.jinsphys.2008.12.001.

Bartram, S., Jux, A., Gleixner, G., & Boland, W. (2006). Dynamic pathway allocation in early terpenoid biosynthesis of stress-induced lima bean leaves. *Phytochemistry, 67*(15), 1661–1672.

Basset, Y. (1994). Palatability of tree foliage to chewing insects: A comparison between a temperate and a tropical site. *Acta Oecologica, 15*, 181–191.

Bede, J. C., Musser, R. O., Felton, G. W., & Korth, K. L. (2006). Caterpillar herbivory and salivary enzymes decrease transcript levels of *Medicago truncatula* genes encoding early enzymes in terpenoid biosynthesis. *Plant Molecular Biology, 60*, 519–531.

Belsky, A. J. (1986). Does herbivory benefit plants? A review of the evidence. *The American Naturalist, 127*(6), 870–892.

Bhonwong, A., Stout, M. J., Attajarusit, J., & Tantasawat, P. (2009). Defensive role of tomato polyphenol oxidases against cotton bollworm (*Helicoverpa armigera*) and beet armyworm (*Spodoptera exigua*). *Journal of Chemical Ecology, 35*, 28–38. https://doi.org/10.1007/s10886-008-9571-7.

Bienert, G. P., Schjoerring, J. K., & Jahn, T. P. (2006). Membrane transport of hydrogen peroxide. *Biochimica et Biophysica Acta (BBA)–Biomembranes, 1758*(8), 994–1003.

Bohlmann, J., Meyer-Gauen, G., & Croteau, R. (1998). Plant terpenoid synthases: Molecular biology and phylogenetic analysis. *Proceedings of the National Academy of Sciences, 95*(8), 4126–4133.

Boland, W., Hopke, J., & Piel, J. (1998). Induction of plant volatile biosynthesis by jasmonates. In *Natural Product Analysis* (pp. 255–269). Vieweg.

Brown, V. K., & Gange, A. C. (1989). Differential effects of above- and below-ground insect herbivory during early plant succession. *Oikos, 54*(1), 67–76.

Brown, V. K., & Gange, A. C. (1992). Secondary plant succession: How is it modified by insect herbivory? *Vegetatio, 101*(1), 3–13.

Buschmann, H., Keller, M., Porret, N., Dietz, H., & Edwards, P. J. (2005). The effect of slug grazing on vegetation development and plant species diversity in an experimental grassland. *Functional Ecology, 19*(2), 291–298.

Cebrian, J. (2004). Role of first-order consumers in ecosystem carbon flow. *Ecology Letters, 7*(3), 232–240.

Chen, F., Tholl, D., D'Auria, J. C., Farooq, A., Pichersky, E., & Gershenzon, J. (2003). Biosynthesis and emission of terpenoid volatiles from *Arabidopsis* flowers. *The Plant Cell, 15*(2), 481–494.

Chen, H., Wilkerson, C. G., Kuchar, J. A., Phinney, B. S., & Howe, G. A. (2005). Jasmonate-inducible plant enzymes degrade essential amino acids in the herbivore midgut. *Proceedings of the National Academy of Sciences of the United States of America, 102*, 19237–19242. https://doi.org/10.1073/pnas.0509026102.

Chen, Y., Ni, X., & Buntin, G. D. (2009). Physiological, nutritional, and biochemical bases of corn resistance to foliage-feeding fall armyworm. *Journal of Chemical Ecology, 35*, 297–306. https://doi.org/10.1007/s10886-009-9600-1.

Cheng, A. X., Lou, Y. G., Mao, Y. B., Lu, S., Wang, L. J., & Chen, X. Y. (2007). Plant terpenoids: Biosynthesis and ecological functions. *Journal of Integrative Plant Biology, 49*(2), 179–186.

Cipollini, D., Enright, S., Traw, M. B., & Bergelson, J. (2004). Salicylic acid inhibits jasmonic acid-induced resistance of *Arabidopsis thaliana* to *Spodoptera exigua*. *Molecular Ecology, 13*(6), 1643–1653.

Coley, P. D. (1983). Herbivory and defensive characteristics of tree species in a lowland tropical forest. *Ecological Monographs, 53*(2), 209–233.

Coley, P. D., & Aide, T. M. (1991). Comparison of herbivory and plant defenses in temperate and tropical broad-leaved forests. In P. W. Price, T. M. Lewinsohn, G. W. Fernandes, & W. W. Benson (Eds.), *Plant-Animal Interactions: Evolutionary Ecology in Tropical and Temperate Regions* (pp. 25–49). Wiley & Sons.

Coley, P. D., & Barone, J. A. (1996). Herbivory and plant defenses in tropical forests. *Annual Review of Ecology and Systematics, 27*(1), 305–335.

Coley, P. D., Bryant III, J. P., & Chapin, F. S. (1985). Resource availability and plant antiherbivore defense. *Science, 230*(4728), 895–899.

Crawley, M. J. (1989). Insect herbivores and plant population dynamics. *Annual Review of Entomology, 34*(1), 531–564.

Cronin, G., Lodge, D. M., Hay, M. E., Miller, M., Hill, M. A., Horvath, T., et al. (2002). Crayfish feeding preferences for freshwater macrophytes: The influence of plant structure and chemistry. *Journal of Crustacean Biology, 22*(4), 708–718.

Dat, J., Vandenabeele, S., Vranova, E. V. M. M., Van Montagu, M., Inzé, D., & Van Breusegem, F. (2000). Dual action of the active oxygen species during plant stress responses. *Cellular and Molecular Life Sciences (CMLS), 57*, 779–795.

de Boer, J. G., Posthumus, M. A., & Dicke, M. (2004). Identification of volatiles that are used in discrimination between plants infested with prey or nonprey herbivores by a predatory mite. *Journal of Chemical Ecology, 30*, 2215–2230. https://doi.org/10.1023/B.0000048784.79031.5e.

Degenhardt, J., & Gershenzon, J. (2000). Demonstration and characterization of (E)-nerolidol synthase from maize: A herbivore-inducible terpene synthase participating in (3E)-4,8-dimethyl-1,3,7-nonatriene biosynthesis. *Planta, 210*, 815–822.

Dicke, M., Agrawal, A. A., & Bruin, J. (2003). Plants talk, but are they deaf? *Trends in Plant Science, 8*(9), 403–405.

Dickens, J. C. (2006). Plant volatiles moderate response to aggregation pheromone in Colorado potato beetle. *Journal of Applied Entomology, 130*, 26–31. https://doi.org/10.1111/j.1439-0418.2005.01014.x.

Dirzo, R., & Harper, J. L. (1980). Experimental studies on slug-plant interactions. II. The effect of grazing by slugs on high density monocultures of *Capsella bursa-pastoris* and *Poa annua*. *Journal of Ecology, 68*(3), 999–1011.

Dudareva, N., Pichersky, E., & Gershenzon, J. (2004). Biochemistry of plant volatiles. *Plant Physiology, 135*(4), 1893–1902.

Duthoit, C. M. G. (1964). Slugs and food preferences. *Plant Pathology, 13*(2), 73–78.

Edwards, P. J., & Gillman, M. P. (1987). Herbivores and plant succession. In A. J. Gray, M. J. Crawley, & P. J. Edwards (Eds.), *Colonization, Succession and Stability* (pp. 295–314). Blackwell Scientific Publications.

Fenner, M., Hanley, M. E., & Lawrence, R. (1999). Comparison of seedling and adult palatability in annual and perennial plants. *Functional Ecology, 13*(4), 546–551.

Fidantsef, A. L., Stout, M. J., Thaler, J. S., Duffey, S. S., & Bostock, R. M. (1999). Signal interactions in pathogen and insect attack: Expression of lipoxygenase, proteinase inhibitor II, and pathogenesis-related protein P4 in the tomato, *Lycopersicon esculentum*. *Physiological and Molecular Plant Pathology, 54*, 97–114.

Gapper, C., & Dolan, L. (2006). Control of plant development by reactive oxygen species. *Plant Physiology, 141*(2), 341–345.

Gechev, T. S., & Hille, J. (2005). Hydrogen peroxide as a signal controlling plant programmed cell death. *The Journal of Cell Biology, 168*(1), 17.

Gechev, T. S., Van Breusegem, F., Stone, J. M., Denev, I., & Laloi, C. (2006). Reactive oxygen species as signals that modulate plant stress responses and programmed cell death. *BioEssays, 28*(11), 1091–1101.

Gibson, C. W. D., Brown, V. K., & Jepson, M. (1987). Relationships between the effects of insect herbivory and sheep grazing on seasonal changes in an early successional plant community. *Oecologia, 71*(2), 245–253.

Gill, R. S., Gupta, K., Taggar, G. K., & Taggar, M. S. (2010). Role of oxidative enzymes in plant defenses against herbivory. *Acta Phytopathologica et Entomologica Hungarica, 45*, 277–290.

Gulsen, O., Eickhoff, T., Heng-Moss, T., Shearman, R., Baxendale, F., Sarath, G., et al. (2010). Characterization of peroxidase changes in resistant and susceptible warm-season turfgrasses challenged by *Blissus occiduus*. *Arthropod-Plant Interactions, 4*, 45–55.

Halliwell, B. (2006). Reactive species and antioxidants: Redox biology is a fundamental theme of aerobic life. *Plant Physiology, 141*(2), 312–322.

Hanley, M. E., Fenner, M., & Edwards, P. J. (1995). An experimental field study of the effects of mollusc grazing on seedling recruitment and survival in grassland. *Journal of Ecology, 83*(4), 621–627.

Hartley, S. E., & Jones, T. H. (2003). Plant diversity and insect herbivores: Effects of environmental change in contrasting model systems. *Oikos, 101*(1), 6–17.

Havir, E. A., & McHale, N. A. (1989). Enhanced-peroxidatic activity in specific catalase isozymes of tobacco, barley, and maize. *Plant Physiology, 91*(3), 812–815.

Heil, M., & Kost, C. (2006). Priming of indirect defenses. *Ecology Letters, 9*(7), 813–817.

Heil, M., & Silva Bueno, J. C. (2007). Within-plant signaling by volatiles leads to induction and priming of an indirect plant defense in nature. *Proceedings of the National Academy of Sciences, 104*(13), 5467–5472.

Heng-Moss, T. M., Sarath, G., Baxendale, F., Novak, D., Bose, S., Ni, X., et al. (2004). Characterization of oxidative enzyme changes in buffalograsses challenged by *Blissus occiduus*. *Journal of Economic Entomology, 97*, 1086–1095.

Herrera, C. M., & Pellmyr, O. (2002). *Plant-Animal Interactions: An Evolutionary Approach*. Wiley.

Holopainen, J. K. (2004). Multiple functions of inducible plant volatiles. *Trends in Plant Science, 9*(11), 529–533.

Horiuchi, J. I., Arimura, G. I., Ozawa, R., Shimoda, T., Dicke, M., Takabayashi, J., et al. (2003). Lima bean leaves exposed to herbivore-induced conspecific plant volatiles attract herbivores in addition to carnivores. *Applied Entomology and Zoology (Japan), 38*, 365–368.

Houlahan, J. E., & Findlay, C. S. (2004). Effect of invasive plant species on temperate wetland plant diversity. *Conservation Biology, 18*(4), 1132–1138.

Huang, W., Zhikuan, J., & Qingfang, H. (2007). Effects of herbivore stress by *Aphis medicaginis* Koch on the malondialdehyde contents and activities of protective enzymes in different alfalfa varieties. *Acta Ecologica Sinica, 27*, 2177–2183.

Hulme, P. E. (1994). Seedling herbivory in grassland: Relative impact of vertebrate and invertebrate herbivores. *Journal of Ecology, 82*(4), 873–880.

Hulme, P. E. (1996). Herbivores and the performance of grassland plants: A comparison of arthropod, mollusc, and rodent herbivory. *Journal of Ecology, 84*(1), 43–51.

Huntly, N. (1991). Herbivores and the dynamics of communities and ecosystems. *Annual Review of Ecology and Systematics, 22*(1), 477–503.

Jacks, T. J., & Davidonis, G. H. (1979). Superoxide, hydrogen peroxide, and the respiratory burst of fungally infected plant cells. *Molecular and Cellular Biochemistry, 158*, 77–79.

Jiang, H., Cai, Y. L., Li, K., Wang, H., & Wang, L. (2005). Intensity and patterns of leaf area eaten of *Lithocarpus glaber* by insects at Tiantong Forest Park, Zhejiang. *Chinese Journal of Ecology, 24*(9), 989–993.

Johnson, M. T. J. (2011). Evolutionary ecology of plant defenses against herbivores. *Functional Ecology, 25*(2), 305–311.

Kaspari, M., & Welti, E. A. (2024). Nutrient dilution and the future of herbivore populations. *Trends in Ecology & Evolution, 39*(9), 809–820.

Kempema, L. A., Cui, X. P., Holzer, F. M., & Walling, L. L. (2007). Arabidopsis transcriptome changes in response to phloem-feeding silverleaf whitefly nymphs: Similarities and distinctions in responses to aphids. *Plant Physiology, 143*, 849–865. https://doi.org/10.1104/pp.106.090662.

Kunkel, B. N., & Brooks, D. M. (2002). Cross talk between signaling pathways in pathogen defense. *Current Opinion in Plant Biology, 5*, 325–331. https://doi.org/10.1016/S1369-5266(02)00275-3.

Kursar, T. A., & Coley, P. D. (2003). Convergence in defense syndromes of young leaves in tropical rainforests. *Biochemical Systematics and Ecology, 31*(8), 929–949.

Liu, M., Jiang, P., Chase, J. M., & Liu, X. (2024). Global insect herbivory and its response to climate change. *Current Biology, 34*(12), 2558–2569.

Lodge, D. M. (1991). Herbivory on freshwater macrophytes. *Aquatic Botany, 41*(1–3), 195–224.

Lu, Y., Xia, W., Yonggen, Y. L., & Jiaan, C. (2006). Role of ethylene signalling in the production of rice volatiles induced by the rice brown planthopper *Nilaparvata lugens*. *Chinese Science Bulletin, 51*, 2457–2465. https://doi.org/10.1007/s11434-006-2148-3.

Ludwig, A. A., Romeis, T., & Jones, J. D. (2004). CDPK-mediated signaling pathways: Specificity and cross-talk. *Journal of Experimental Botany, 55*, 181–188. https://doi.org/10.1093/jxb/erh008.

Maffei, M. E., Mithöfer, A., & Boland, W. (2007). Insects feeding on plants: Rapid signals and responses preceding the induction of phytochemical release. *Phytochemistry, 68*, 2946–2959. https://doi.org/10.1016/j.phytochem.2007.07.016.

Mao, Y. B., Cai, W. J., Wang, J. W., Hong, G. J., Tao, X. Y., Wang, L. J., & Chen, X. Y. (2007). Silencing a cotton bollworm P450 monooxygenase gene by plant-mediated RNAi impairs larval tolerance of gossypol. *Nature Biotechnology, 25*, 1307–1313. https://doi.org/10.1038/nbt1352.

Pauwels, L., Inzé, D., & Goossens, A. (2009). Jasmonate-inducible gene: What does it mean? *Trends in Plant Science, 14*, 87–91. https://doi.org/10.1016/j.tplants.2008.11.005.

Pazouki L and Niinemets Ü (2016) Multi-Substrate Terpene Synthases: Their Occurrence and Physiological Significance. Front. Plant Sci. 7:1019. doi: 10.3389/fpls.2016.01019

Pieterse, C. M. J., & Van Loon, L. C. (2004). NPR1: The spider in the web of induced resistance signaling pathways. *Current Opinion in Plant Biology, 7*, 456–464. https://doi.org/10.1016/j.pbi.2004.05.006.

Rayapuram, C., & Baldwin, I. T. (2007). Increased SA in NPR1-silenced plants antagonizes JA and JA-dependent direct and indirect defenses in herbivore-attacked *Nicotiana attenuata* in nature. *The Plant Journal, 52*, 700–715. https://doi.org/10.1111/j.1365-313X.2007.03267.x.

Reymond, P., Bodenhausen, N., Van Poecke, R. M. P., Krishnamurthy, V., Dicke, M., & Farmer, E. E. (2004). A conserved transcript pattern in response to a specialist and a generalist herbivore. *The Plant Cell, 16*, 3132–3147. https://doi.org/10.1105/tpc.104.026120.

Rhoads, D. M., Umbach, A. L., Subbaiah, C. C., & Siedow, J. N. (2006). Mitochondrial reactive oxygen species: Contribution to oxidative stress and interorganellar signaling. *Plant Physiology, 141*(2), 357–366.

Ribot, C., Zimmerli, C., Farmer, E. E., Reymond, P., & Poirier, Y. (2008). Induction of the *Arabidopsis PHO1;H10* gene by 12-oxo-phytodienoic acid but not jasmonic acid via a *CORONATINE INSENSITIVE1*-dependent pathway. *Plant Physiology, 147*, 696–706. https://doi.org/10.1104/pp.108.119321.

Rivas-San Vicente, M., & Plasencia, J. (2011). Salicylic acid beyond defence: Its role in plant growth and development. *Journal of Experimental Botany, 62*, 3321–3338. https://doi.org/10.1093/jxb/err031.

Robson, C. A., & Vanlerberghe, G. C. (2002). Transgenic plant cells lacking mitochondrial alternative oxidase have increased susceptibility to mitochondria-dependent and -independent pathways of programmed cell death. *Plant Physiology, 129*(4), 1908–1920.

Ruiz-Arocho, J., González-Salas, R., LeMay, G., Steinthal, N., Mastretta-Yanes, A., Wegier, A., et al. (2024). How is leaf herbivory related to agriculture? Insights from the Mexican center of crop origin. *Arthropod-Plant Interactions, 18*(1), 89–104.

Ruther, J. (2000). Retention index database for identification of general green leaf volatiles in plants by coupled capillary gas chromatography–mass spectrometry. *Journal of Chromatography A, 890*(2), 313–319.

Ruther, J., & Kleier, S. (2005). Plant-plant signaling: Ethylene synergizes volatile emission in *Zea mays* induced by exposure to (Z)-3-hexen-1-ol. *Journal of Chemical Ecology, 31*, 2217–2222. https://doi.org/10.1007/s10886-005-6413-8.

Sagi, M., & Fluhr, R. (2006). Production of reactive oxygen species by plant NADPH oxidases. *Plant Physiology, 141*(2), 336–340.

Sanchez-Casas, P., & Klessig, D. F. (1994). A salicylic acid-binding activity and a salicylic acid-inhibitable catalase activity are present in a variety of plant species. *Plant Physiology, 106*(4), 1675–1679.

Scheidel, U., & Bruelheide, H. (2005). Effects of slug herbivory on the seedling establishment of two montane Asteraceae species. *Flora, 200*(4), 309–320.

Sharma, P., & Dubey, R. S. (2005). Drought induces oxidative stress and enhances the activities of antioxidant enzymes in growing rice seedlings. *Plant Growth Regulation, 46*, 209–221.

Sheard, L. B., Tan, X., Mao, H., Withers, J., Ben-Nissan, G., Hinds, T. R., et al. (2010). Jasmonate perception by inositol-phosphate-potentiated COI1-JAZ co-receptor. *Nature, 468*, 400–405. https://doi.org/10.1038/nature09430.

Shivaji, R., Camas, A., Ankala, A., Engelberth, J., Tumlinson, J. H., Williams, W. P., et al. (2010). Plants on constant alert: Elevated levels of jasmonic acid and jasmonate-induced transcripts in caterpillar-resistant maize. *Journal of Chemical Ecology, 36*, 179–191. https://doi.org/10.1007/s10886-010-9752-z.

Søndergaard, M., Bruun, L., Lauridsen, T., Jeppensen, E., & Madsen, T. V. (1996). The impact of grazing waterfowl on submerged macrophytes: *In situ* experiments in a shallow eutrophic lake. *Aquatic Botany, 53*(1, 2), 73–84.

Stephensen, E., Sturve, J., & Förlin, L. (2002). Effects of redox cycling compounds on glutathione content and activity of glutathione-related enzymes in rainbow trout liver. *Comparative Biochemistry and Physiology Part C: Toxicology & Pharmacology, 133*(3), 435–442.

Stout, M. J., Riggio, M. R., & Yang, Y. (2009). Direct induced resistance in *Oryza sativa* to *Spodoptera frugiperda*. *Environmental Entomology, 38*, 1174–1181. https://doi.org/10.1603/022.038.0426.

Takemoto, D., Tanaka, A., & Scott, B. (2006). A p67Phox-like regulator is recruited to control hyphal branching in a fungal–grass mutualistic symbiosis. *The Plant Cell, 18*(10), 2807–2821.

Tanaka, A., Christensen, M. J., Takemoto, D., Park, P., & Scott, B. (2006). Reactive oxygen species play a role in regulating a fungus–perennial ryegrass mutualistic interaction. *The Plant Cell, 18*(4), 1052–1066.

Teixeira, F. K., Menezes-Benavente, L., Galvão, V. C., Margis, R., & Margis-Pinheiro, M. (2006). Rice ascorbate peroxidase gene family encodes functionally diverse isoforms localized in different subcellular compartments. *Planta, 224*, 300–314.

Thaler, J. S. (1999). Jasmonate-inducible plant defences cause increased parasitism of herbivores. *Nature, 399*(6737), 686–688.

Torres, M. A., Jones, J. D., & Dangl, J. L. (2006). Reactive oxygen species signaling in response to pathogens. *Plant Physiology, 141*(2), 373–378.

Tscharntke, T., Thiessen, S., Dolch, R., & Boland, W. (2001). Herbivory, induced resistance, and interplant signal transfer in *Alnus glutinosa*. *Biochemical Systematics and Ecology, 29*, 1025–1047. https://doi.org/10.1016/S0305-1978(01)00048-5.

Ulloa, R. M., Raíces, M., MacIntosh, G. C., Maldonado, S., & Téllez-Iñón, M. T. (2002). Jasmonic acid affects plant morphology and calcium-dependent protein kinase expression and activity in *Solanum tuberosum*. *Physiologia Plantarum, 115*, 417–427. https://doi.org/10.1034/j.1399-3054.2002.1150312.x.

Usha Rani, P., & Jyothsna, Y. (2010). Biochemical and enzymatic changes in rice as a mechanism of defense. *Acta Physiologiae Plantarum, 32*, 695–701. https://doi.org/10.1007/s11738-009-0449-2.

Valencia, E., Marin, A., & Hardy, G. (2001). Glutathione—Nutritional and pharmacological viewpoints: Part III. *Nutrition, 17*(7, 8), 696–697.

van Breusegem, F., Vranová, E., Dat, J. F., & Inzé, D. (2001). The role of active oxygen species in plant signal transduction. *Plant Science, 161*(3), 405–414.

van Donk, E. (1998). Switches between clear and turbid water states in a biomanipulated lake (1986–1996): The role of herbivory on macrophytes. In E. Jeppensen, M. Sondergaard, & K. Christoffersen (Eds.), *The Structuring Role of Macrophytes in Lakes* (pp. 290–297). Springer-Verlag.

van Loon, L. C., Geraats, B. P. J., & Linthorst, H. J. M. (2006). Ethylene as a modulator of disease resistance in plants. *Trends in Plant Science, 11*, 184–191. https://doi.org/10.1016/j.tplants.2006.02.005.

Verhage, A., van Wees, S. C. M., & Pieterse, C. M. J. (2010). Plant immunity: It's the hormones talking, but what do they say? *Plant Physiology, 154*, 536–540. https://doi.org/10.1104/pp.110.161570.

Voelckel, C., & Baldwin, I. T. (2004). Herbivore-induced plant vaccination. Part II. Array-studies reveal the transience of herbivore-specific transcriptional imprints and a distinct imprint from stress combinations. *Plant Journal, 38*, 650–663. https://doi.org/10.1111/j.1365-313X.2004.02077.x.

Voelckel, C., Weisser, W. W., & Baldwin, I. T. (2004). An analysis of plant-aphid interactions by different microarray hybridization strategies. *Molecular Ecology, 13*, 3187–3195. https://doi.org/10.1111/j.1365-294X.2004.02297.x.

Walter, A., Mazars, C., Maitrejean, M., Hopke, J., Ranjeva, R., Boland, W., et al. (2007). Structural requirements of jasmonates and synthetic analogues as inducers of Ca^{2+} signals in the nucleus and the cytosol of plant cells. *Angewandte Chemie International Edition, 46*, 4783–4785. https://doi.org/10.1002/anie.200604989.

War, A. R., Paulraj, M. G., War, M. Y., & Ignacimuthu, S. (2011a). Jasmonic acid-mediated induced resistance in groundnut (*Arachis hypogaea* L.) against *Helicoverpa armigera* (Hubner) (Lepidoptera: Noctuidae). *Journal of Plant Growth Regulation, 30*, 512–523. https://doi.org/10.1007/s00344-011-9213-0.

War, A. R., Paulraj, M. G., War, M. Y., & Ignacimuthu, S. (2011b). Role of salicylic acid in induction of plant defense system in chickpea (*Cicer arietinum* L.). *Plant Signaling & Behavior, 6*, 1787–1792. https://doi.org/10.4161/psb.6.11.17685.

Xiong, W., Yu, D., Wang, Q., Liu, C. H., & Wang, L. G. (2008). A snail prefers native over exotic freshwater plants: Implications for the enemy release hypothesis. *Freshwater Biology, 53*(11), 2256–2263.

Zhang, T. S. (2008). *Insect Diversity in Subtropical Evergreen Broad-Leaved Forests*. East China Normal University.

Zhao, L. Y., Chen, J. L., Cheng, D. F., Sun, J. R., Liu, Y., & Tian, Z. (2009). Biochemical and molecular characterizations of *Sitobion avenae*-induced wheat defense responses. *Crop Protection, 28*, 435–442. https://doi.org/10.1016/j.cropro.2009.01.005.

CHAPTER 10

The Employment of Endophytes Based on Their Volatile Organic Compounds in Pest Management: Mechanisms, Challenges, and Innovation

SHIMAA GHAREEB,[1,2,*] ADIL TONĞA,[3] SOZAN E. EL-ABEID,[4,5] RANA HUSSIEN,[1] YOMNA A. MOHAMED,[6] HOSSAM E. HARB,[6] and DINA S. S. IBRAHIM[7,*]

[1]*Plant Protection Research Institute, Agricultural Research Center, Giza, Egypt*

[2]*Faculty of Agriculture, Kyushu University, 744 Motooka, Fukuoka, Japan*

[3]*Laboratory of Pest Physiology, Plant Protection Research Institute, Agricultural Research Center, Giza, Egypt*

[4]*Nanotechnology and Advanced Nano-Materials Laboratory (NANML), Plant Pathology Research Institute, Agricultural Research Center, Giza, Egypt*

[5]*Department of Mycology and Disease Survey Research, Plant Pathology Research Institute, Agricultural Research Center, Giza, Egypt*

[6]*Department of Biotechnology, Faculty of Science, Cairo University, Cairo, Egypt*

[7]*Department of Nematode Diseases and Central Laboratory of Biotechnology, Plant Pathology Research Institute, Agricultural Research Center, Giza, Egypt*

*Corresponding author

Insect-Plant Interactions, Volume 1: Influence of Biotic and Abiotic Factors.
Jamin Ali, Adil Tonğa, Adnan Shakeel, and Nasheeman Ashraf (Eds.)

DOI: 10.1201/9781779644282-10

ABSTRACT

Understanding the ecological relationships between plants, pests, and endophytic microorganisms is crucial for developing sustainable pest management strategies. By harnessing these natural interactions, it has been possible to discover compounds that are both effective in their insecticidal potential and environmentally safe, minimizing harm to non-target organisms. Recently, research has increasingly focused on the insecticidal potential of secondary metabolites produced by endophytes. These endophytes, along with the volatile organic compounds (VOCs) they release, play a crucial role in agroecosystems by contributing to pest control and improving plant fitness. Through symbiosis with host plants, endophytes stimulate the production of secondary metabolites in plants while also producing their own secondary metabolites. This dual production strengthens plant health and offers natural pest resistance. This chapter explores the interactions between endophytes, plants, and pests through VOCs and discusses the challenges of utilizing VOCs in open field conditions. The integration of VOCs into pest management strategies offers a sustainable alternative to synthetic pesticides, fostering agricultural practices that promote environmental health.

10.1 INTRODUCTION

For a long time, agriculture and beneficial microorganisms have shared a relationship that has enhanced crop production and pest management systems globally (Orr and Lahiri, 2014) through the production of vital bioactive secondary metabolites (Sharma et al., 2022). A deeper understanding of the ecological dynamics of tripartite interactions, including distantly related organisms such as plants, microorganisms including fungi and bacteria, and insects, holds promise for sustainable advancements in crop management. Most of the studies on insect-plant–microorganism interactions employ terms like mutualism and symbiosis, reflecting the distinct advantages conferred to one or more of the interacting participants. However, identifying a unifying model for these complex systems remains challenging. The difficulty arises from the variability in insect behavior when orienting or avoiding microorganism-infected plants, which may be driven by multiple factors, including specific environmental conditions that foster unique tripartite relationships (Raman and Suryanarayanan, 2017).

Numerous fungus species, integral components of the plant microbiome, can establish endophytic relationships with their host plants, offering them a long-lasting shield against controversial organisms (Card et al., 2016). The endophytic organism-infected host plants can produce toxic secondary metabolites that are associated with the entomopathogenic potential of endophytes. Endophytes, which include fungi or bacteria mostly antagonistic to pest insects, have emerged as a potential tool for pest management practices (Berg and Hallmann, 2006). Endophytic fungi are unique varieties of fungi that are intimately associated with the host plant. They reside within various parts of plants, including roots, stems, and leaves (Suryanarayanan, 2013). The endophytic microorganisms can enhance the resilience of plants against biotic challenges using a range of strategies, including overall fitness effects and growth promotion achieved eventually in synergy with other components of the plant microbiome (Bamisile et al., 2023; Picciotti et al., 2023). Recent advancements have suggested that endophytic microorganisms could be exploited in pest management applications. For example, they could be employed as biocontrol agents (BCA) for controlling pests, and they may alter insect behavior and development through antixenotic and antibiotic effects. Moreover, these microorganisms can participate in the host plant defense against other biotic organisms by producing enzymes like cellulases, chitinases, amylases, and laccases (Samal et al., 2023). The same or structurally analogous compounds or active components can be produced by different endophytes and still may exhibit the same biological relevance to the biotic environment (Yan et al., 2014). Diverse types of host plants harbor endophyte microorganisms to produce metabolites in response to potential competitors. These compounds play a role in host competitive ability, fitness, and defense, exhibiting effects similar to those of natural plant hormones.

Specific strains of endophytic fungi, such as *Fusarium oxysporum, Cunninghamella echinulata,* and *Cladosporium sp.*, have shown insecticidal activities against aphids such as the black bean aphid, *Aphis fabae* Scop, and the green peach aphid, *Myzus persicae* Sulzer (Hemiptera: Aphididae). These fungal agents may reduce aphid populations and adversely affect their embryonic growth, which leads to decreased reproductive abilities (Bensaci et al., 2015, 2022). Furthermore, the larvae of *Spodoptera litura* (Lepidoptera: Noctuidae) feeding on leaves taken from endophyte-treated host plants were slow in movement and exhibited considerably enhanced mortality when compared with those feeding on control plants (Thakur et al., 2013). It is also known that endophytic microorganisms do not pose a danger to their host plants, which have been referred to as exhibiting no diseased symptoms in cauliflower plants that were infected with an endophytic strain of *C. uredinicola*, and there was no variation in the vigor of endophyte-infected plants from those of untreated ones.

The emission of low-molecular-weight VOCs from endophytic fungi can have multiple beneficial functions in agricultural ecosystems (Kaddes et al., 2019; Ling et al., 2024). They can trigger defense responses, suppress antagonistic pathogens, improve host plant growth, and act as insect attractants or repellents (Srikamwang et al., 2023; Wonglom et al., 2020; Zhang et al., 2020). Several VOCs, such as acids, alcohols, ketones, aldehydes, ethers, hydrocarbons, and esters, are emitted by endophytic fungi, such as *T. longibrachiatum* strain 2 (MK751759) (Rajani et al., 2021). Additionally, the host plant VOCs are essential in insect-herbivore interactions (Turlings and Erb, 2018).

This chapter offers a comprehensive overview of the roles and mechanisms of endophytes and their VOCs in pest management and plant health by documenting the suppressive potential on pest insects and their roles in indirect defense mechanisms. It also examines how recent advances in omics technologies have unveiled new opportunities for utilizing these microorganisms in sustainable agriculture and addresses the current challenges and future directions for optimizing the use of their bioactive compounds.

10.2 ENDOPHYTES AND THEIR VOLATILE ORGANIC COMPOUNDS (VOCS) IN AGRO-ECOSYSTEM

10.2.1 MECHANISMS OF ACTION IN PEST MANAGEMENT

10.2.1.1 DIRECT DEFENSE MECHANISM

Endophytes generate and secrete bioactive metabolites such as salicylic acid (SA), jasmonic acid (JA), phytoalexins, siderophores, and VOCs to efficiently suppress pests. It is well recognized that each of these bioactive metabolites has strong inhibitory effects on pests (Etminani and Harighi, 2018). In numerous ways, endophytic fungi reduce damage to their host plants caused by insects via different effects, such as reduced feeding, oviposition, growth rate, pest deterrence, avoidance, and survival (Vega, 2008). Some endophytes fill plants with substances that decrease herbivory by insects (Bagheri et al., 2021). *Metarhizium* and *Beauveria*, two genera of insect pathogenic fungi from the endophyte category, can infect insect pests through a combination of mechanical and enzymatic processes that disturb the cuticular structure of insect pests. These fungi release cuticle-degrading enzymes like proteases, chitinases, lipases, and esterases (Barelli et al., 2016; Pedrini et al., 2013; Schrank and Vainstein, 2010). Lipophilic conidia of endophytes attach to the insect cuticle, germinate, and produce appressoria. The penetration and infection processes are enhanced via the enzymatic degradation by proteases such as subtilisin-like protease (Pr1A), trypsin-like serine protease (Pr2), thermolysin-like metalloproteinase, as well as other peptidases (Semenova et al., 2020). Genomic analysis revealed that entomopathogens differ from other fungi by having many genes coding for subtilisin-like peptidases. For example, *Metarhizium robertsii* contains the highest number of these genes, whereas only one subtilisin gene is present in the case of *Saccharomyces kluyveri* (Li et al., 2017). Following the colonization of plants, this endophytic fungus alters the nutritional makeup of its host to enhance the production of secondary metabolites. In this process, insect herbivory and oviposition are hindered by changes in chemical constituents found within the host plants' tissues (Clay, 1990).

Secondary metabolites produced by microorganisms mediate interactions among organisms in different ecological conditions (Demain and Fang, 2000). For instance, antimicrobial compounds can be used as signaling mechanisms to protect against competitors and safeguard symbiotic organisms (Tyc et al., 2014). Antimicrobial agents work by interfering with basic cellular processes that are important for their survival; these include DNA replication, translation, transcription, membrane stability, and the synthesis of cell walls (Cotsonas and Wu, 2009). The relationship between endophytic fungi and host plants is a crucial mechanism in disease management; fungi can control phytopathogens through various mechanisms, such as activation of plant resistance resulting from the synthesis of fungal metabolites, hormonal signaling molecules, and gene expression (Adeleke et al., 2022).

The exploration of endophyte VOCs as replacements for pesticides is a topic that has attracted a great deal of focus (Razo-Belman and Ozuna, 2023). Agbessenou et al. (2020) investigated that secondary metabolites or microbial VOCs produced by *Beauveria bassiana* ICIPE 706, *Hypocrea lixii* F3ST1, and *Trichoderma asperellum* M2RT4 can be used as biopesticides for the control of *Tuta absoluta* (Lepidoptera: Gelechiidae). Additionally, a single endophyte may provide defense against insect pests and fungal pathogens. For instance, *Beauveria bassiana* inhibits insects by producing secondary metabolites. Some of the main secondary metabolites that *B. bassiana* produces are bassianin, oxalic acid, oosporein, beauvericin, and tenellin (Ávila-Hernández et al., 2020, 2022; Wang et al., 2021), and beauvericin is the most important one (Safavi, 2013; Wang et al., 2021). *Aphis gossypii* Glover (Hemiptera: Aphididae) had less reproduction and was killed when treated with these metabolites, which were removed *in vitro* from the mycelia of an endophytic isolate of *B. bassiana* (Gurulingappa et al., 2011). According to Russo et al. (2019), larvae of *Rachiplusia nu* (Order: Family) consumed a smaller proportion of maize leaves that were endophytically colonized by *B. bassiana. Trichoderma* spp. can manage insect pests directly by producing insecticidal secondary metabolites, repellent metabolites, and antifeedant compound synthesis. Moreover, it also works indirectly through the promotion of systemic plant defensive response activation (Poveda, 2021). When an isolate *of M. brunneum* endophytically colonized potato and tomato leaves, destruxin A was discovered in the leaves (Resquín-Romero et al., 2016). Likewise, destruxin A was found by Garrido-Jurado et al. (2017) in melon leaves that were endophytically colonized by an isolate of *M. brunneum* as well as in *B. tabaci* nymphs that were fed on the melon leaves.

A range of alkaloids produced by fungal endophytes in the genus *Epichloë* (Clavicipitaceae) prevents herbivores from feeding on plants, such as seeds, leaves, and culms. Similarly, the toxic alkaloid swainsonine, a potent anti-herbivore chemical and toxin, is produced by the endophytic fungus of the genus Undifilum in plants in the Fabaceae (White et al., 2019). There is a strong correlation between endophytes' capacity to ward off insects and the toxins they produce. Consequently, the production of different toxins by endophytic fungi provides the host defense against various herbivores. These toxins make the plant unfavorable to beetles, grasshoppers, and aphids, among other pests (Anamika et al., 2018; Bagheri et al., 2021). Indeed, endophytic fungi produce a variety of toxins that provide host plants with defense against various herbivores (Azevedo et al., 2000).

10.2.1.2 INDIRECT DEFENSE MECHANISM

Endophytes can produce secondary metabolites directly, affecting pests, or they can stimulate the host plant to produce these compounds (Li et al., 2022). González-Mas et al. (2019) confirmed that there were differences in the quality and quantity of volatiles released by endophytically colonized plants with *B. bassiana* and *M. brunneum* compared to control plants. These led to an impact on the production rate of aphid nymphs. Also, González-Mas et al. (2021) stated that plants colonized with *B. bassiana* release a different mixture of volatile compounds than uncolonized plants. Many of these compounds are known to contribute to the attraction of natural enemies and are related to herbivore-induced reactions. Another mechanism by which endophytes protect their host plant is by activating the plant's defense systems. This can occur by inducing systemic resistance (ISR) or systemic acquired resistance (SAR), both enhancing the plant's defenses against diseases and pests (Muthu Narayanan et al., 2022). The endophytic fungal *B. bassiana* strains YC and BB can significantly lower *Rhopalosiphum padi* (Hemiptera: Aphididae) population fitness, initiative, and feeding efficiency. The defense response in wheat plants was largely mediated by enhanced differentially expressed genes linked to defense-related metabolic pathways and by higher levels of protective phytohormones and secondary metabolites in wheat plants (Song et al., 2024). According to Shrivastava et al. (2015), tomato plants with endophytic colonization of *B. bassiana* showed a significant increase in monoterpene and sesquiterpene concentrations compared to non-inoculated plants, as well as the *S. exigua* larvae feeding on colonized ones having a lower weight than those feeding on non-inoculated plants. These findings imply that differences in terpenoids might be a mechanism of defensive response within fungus-inoculated plants' defense systems. As a result of inoculating the roots of sweet pepper plants (*Capsicum annuum* L.) with the *B. bassiana* ARSEF 3097 and *Akanthomyces muscarius* ARSEF 5128, the resistance of the plant was enhanced against the tobacco peach aphid *Myzus persicae* var. nicotianae by modification of plant volatile profiles. Compared to uninoculated plants, plants infected with *A. muscarius* produced significantly higher levels of β-pinene and indole than uninoculated plants. Additionally, there was a noticeable increase in terpinolene emission following inoculation with the fungal strains (Wilberts et al., 2022).

Thus, endophytes play dual roles in pest management: indirect defense mechanisms, which increase plant resistance and stimulate the release of secondary metabolites (Panwar and Szczepaniec, 2024), and direct defense mechanisms, which directly influence insect behavior or survival (Figure 10.1).

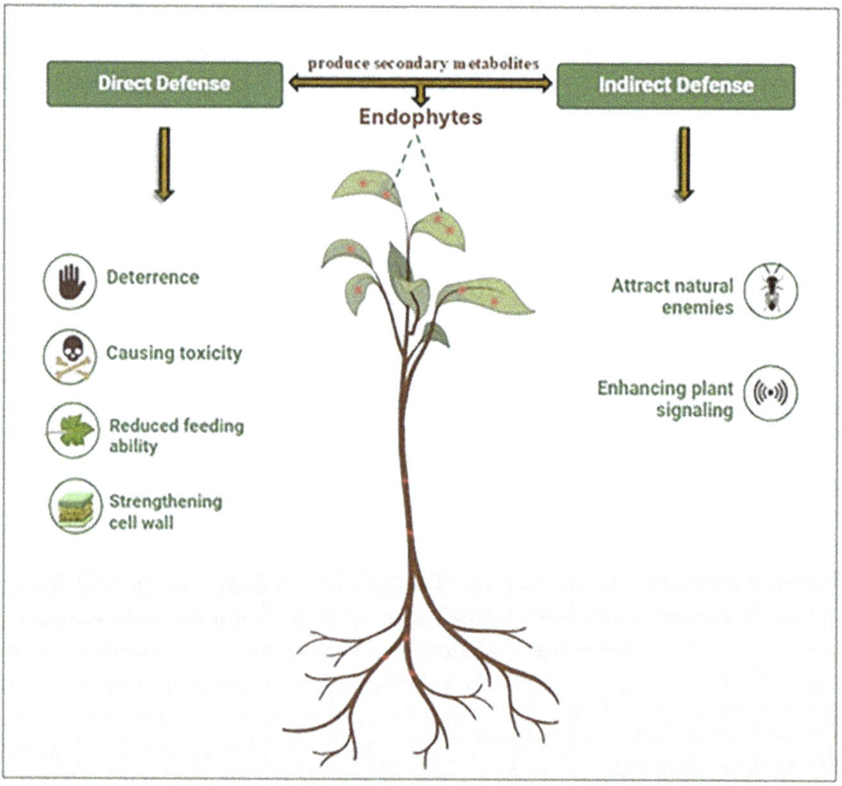

FIGURE 10.1 Defensive roles of endophytes in host plant resistance against insect pest.

10.2.2 THE ROLE IN PLANT-INSECT INTERACTIONS

Fungi endophytes can interact with the plants they inhabit and synthesize a combination of VOC molecules. This mixture facilitates complex signaling between plants and other organisms. VOCs can allow communication among various parts of the same plant or other plants and attract helpful organisms such as pollinators and seed dispersers (Delory et al., 2016; Farré-Armengol et al., 2016). Although plant protection and pollinator attraction may be the two most crucial roles for VOCs, they also play a variety of other roles in shaping the plants' interaction with their surroundings (Effah et al., 2019). They are considered initiator elicitors of defense responses in the plant (Ryu et al., 2003, 2004). Virtually all plant taxa naturally produce VOCs and provide an incredibly flexible option for integration into IPM systems (Vivaldo et al., 2017). Many studies have focused on the insecticidal properties of endophytes. For instance, some endophytes strengthen the endodermal cell wall to prevent pests from penetrating the stele (Gao et al., 2010). Fungal endophytes can act as bio-agents of insect disease by infecting *Lepidopteran larvae*, sucking insects, and others. They infect specific pest species, posing the least threat to other species, including beneficial insects via a host-based specification mechanism (Akutse et al., 2014). In sugarcane, *Bacillus* can cause abnormal development of giant borers, which lowers disease and insect infestations (Rocha et al., 2021). As we mentioned before, there are several toxic metabolites linked to endophytes, including ergovaline, pyrrolopyrazine alkaloid, pyrrolizidine, peramine ergot alkaloid, and alkaloids, among others (Wilkinson et al., 2000). Ergot alkaloids, such as ergotamine, ergocryptine, agroclavine, and elymoclavine, are produced by endophytic fungi in the tribe Balansiae. At concentrations of 77–100 mg/liter, these compounds reduce the weight of larvae and the amount of leaf area that *Spodoptera frugiperda* (Lepidoptera: Noctuidae) consumes (Kumar et al., 2008). *T. absoluta* larvae were affected by endophytic *B. bassiana* and *M. anisopliae* colonized tomato plants, which resulted in a shorter pupal weight and a shorter period between egg hatching and the first instar stage (Giannoulakis et al., 2023). Maize plants colonized by *Hypocrea lixii* F3ST1 and *Trichoderma asperellum* M2RT4 dramatically decreased the quantity of egg masses and the feeding rates of *S. frugiperda*. Furthermore, *Trichoderma harzianum* F2R41 exhibited a female-biased sex ratio and harmful effects on adult emergence and pupation (Kinyungu et al., 2023).

The effectiveness of VOCs against some of the most significant insect pests makes them an environmentally friendly alternative to pesticides (Conboy et al., 2019). A perfect example of this unique group of fungi is the *Metarhizium* species. In addition to their application as biological control agents to control various agricultural pests, they are known for their ability to perform endophytic relationships with host plants (Davis et al., 2012). They can suppress disease (Batta, 2013), enhance plant resistance toward different pests (Ahmad et al., 2019, 2020), and produce active VOCs that can manipulate insects' behavior and enhance plant growth and defense. The termites *Coptotermes lacteus* and *Macrotermes michaelseni* (Isoptera: Termitidae) exhibited avoidance behavior from VOCs emitted by virulent strains of *M. anisopliae* (Mburu et al., 2009; Staples and Milner, 2000). Similarly, VOCs produced by *Metarhizium brunneum* conidia provide it with antifeedant and repellent properties depending on the fungal strain and conidia concentration. Furthermore, some *M. brunneum* strains produce 3-octanone and 1-octen-3-ol with repellent, molluscicidal effects by direct contact or fumigation (Khoja et al., 2019). The most outstanding fungal VOCs tested as potential bio-pesticides for pest control are 3-methyl-1-butanol, 3-octanone, 1-octen-3-ol, and 1-octene. Various insects exhibit different behaviors (attraction or repulsion) toward these compounds depending on the compound concentration and fungal strain (Buda et al., 2016; Ferreira et al., 2020; Khoja et al., 2019). Davis et al. (2012) reported that VOCs emitted by *Aureobasidium pullulans* can attract eusocial wasps. For example, *Pinus taeda* L. barks or bolts colonized with blue-stain fungi (genus Ophiostoma) emitted attractant odors to two bark beetle parasitoids, Roptrocerus xylophagorum (Hymenoptera: Pteromalidae) and *Spathius pallidus* (Hymenoptera: Braconidae) (Sullivan et al., 2004). Also, *Cerrena unicolor* fungi attract the wood wasp *Tremex apicalis* (Hymenoptera: Siricidae) and its parasitoids, *Ibalia japonica* (Hymenoptera: Ibaliidae) (Kuramitsu et al., 2019). Herbivores can distinguish between host and non-host plants while navigating the sophisticated background and overlapping signals of the odor environment by utilizing their olfactory system (Bruce and Pickett, 2011). These organisms detect VOCs using olfactory receptor neurons (ORNs), which are primarily found on the antennae; the ORNs participate in the conversion of chemical signals into electrical signals, allowing for the advent of behavioral responses (Binyameen, 2013). The dual action of VOCs released from endophytic fungi represents promising bio-pesticides that alter insect behaviors, and active agents interact with plants, modulating plant growth and defense, and increasing plant resistance (Kanchiswamy et al., 2015a). Furthermore, these compounds not only affect plant growth but also increase fruit yields and significantly enhance the population dynamics of natural enemies (Song and Ryu, 2013). The progressive studies on fungal VOCs illustrate their critical role in interactions and their importance in both ecosystem and agricultural systems (Kanchiswamy et al., 2015b).

10.2.3 EFFECT ON PLANT GROWTH AND HEALTH IMPROVEMENT

Plant fitness is a major factor shaping its interaction with insects and is highly dependent on growth and health. Endophytes and their VOCs are critical in the growth and survival of plants, as well as regulating some defensive mechanisms, and they also play a crucial role in the trade-off between growth and defense. They form a composite balance to meet the host limit and establish a mutualistic relationship with the host (Alam et al., 2021). Previous studies have reported different evidence about the capability of endophytic fungi in promoting the development of plants, which is done both naturally and artificially by introducing fungi into hosts through diverse inoculation methods (Kushwaha et al., 2020).

Endophytic fungi in the different plant compartments encourage plant growth via direct and indirect mechanisms (Adeleke and Babalola, 2022). The direct mechanism involves the regulation of diverse plant hormones, such as auxins, cytokinins, and ET, increasing the availability of soil nutrients, including nitrogen fixation, iron, phosphorus solubilization, and siderophore production. The indirect mechanism encompasses the release of hydrogen cyanide, enzymes, volatile compounds, and antibiotics that impede pathogen activity, thus inducing systemic resistance in plants while preventing damage from them (Segaran and Sathiavelu, 2019). The presence of *Metarhizium* as endophytes promotes plant growth by improving nutrient uptake into the plant (nitrogen and phosphorus) (Behie et al., 2012; Elena et al., 2011). Nayak et al. (2020) confirmed that *Aspergillus clavatus* Y2H0002, which is a fungal endophyte isolated from *Nymphoides peltata*, can produce a large number of gibberellins, which are hormones that stimulate growth. *Trichoderma hamatum* (isolate DIS 219b), an endophytic fungus, encouraged root growth (Bae et al., 2009). Bacterial endophytes accelerate the release of mineral nutrients, produce different types of enzymes, phytohormones, and siderophores, and activate the plant's systemic resistance to pests, thus benefiting plant growth (Kushwaha et al., 2020). Several endophytic bacteria engaging in direct interaction with plants generate secondary metabolites serving as agents essential for the absorption of nutrients, especially siderophores associated with iron procurement (Barry and Challis, 2009). These metabolites interfere with plant hormone signaling (Glick, 2012; López et al., 2008). The crucial roles of auxins and ET in the development of plants are mediated by beneficial bacterial endophyte-plant interactions (Glick, 2012; Hardoim et al., 2008). Also, endophytic actinomycetes found in different plant parts, including leaves, roots, stems, fruits, and seeds, are recognized to enhance plant growth and stress resistance advantageously (Rutkowska et al., 2023). Through the studies of metabolomics, it has been found that the endophytic actinomycetes produce numerous secondary metabolites, including different BCA, stress protectants, and regulators of plant growth (Mishra et al., 2022). The effects of endophyte-derived VOCs on insect management are presented in Table 10.1.

TABLE 10.1 The Effects of Endophyte-Derived Volatile Organic Compounds (VOCs) on Insect Pests Suggesting Potential for Management Practices

Endophytes	Host Plant	Compounds	Effects
Neotyphodium uncinatum	A hybrid cultivar of *Festuca pratensis* × *Lolium perenne*	Green leaf volatiles (Hexanal, Monoterpenes, Tricyclene, α-Fenchene, Sabinen, β-Pinene, and Myrcene), Aromatic hydrocarbons (Phenylacetaldehyde), Monoterpene ketone (Camphor)	The existence of insect-toxic loline alkaloids in roots and shoots alters soil insects' *C. zealandica* behavior toward root volatiles and thus affects the distribution of soil insects (Rostás et al., 2015).
Fusarium solani	Tomato	sesquiterpenes, C6 VOCs and MeSA	Promote direct defense and indirect defense against spider mite *Tetranychus urticae* by attracted more predators (Pappas et al., 2018).
Trichoderma atroviride	maize	6-pentyl-2H-pyran-2-one and 1-octen-3-ol	decreased foliar tissue consumption and changed *S. frugiperda's* feeding pattern (Contreras-Cornejo et al., 2018).
Muscodor vitigenus	liana	Naphthalene	Act as insect repellent. It deters the *Cephus cinctus*, the adult wheat stem sawfly (Daisy et al., 2002).
Unidentified endophytic fungus	wintergreen	5-hydroxy-2-(1'-oxo-5'-methyl-4'-hexenyl) benzofuran and 5-hydroxy-2-(1'-hydroxy-5'-methyl-4'-hexenyl) benzofuran	Toxic to spruce budworm (Findlay et al., 1997).
Aspergillus niger and Ophiostoma minus	Mongolian pine	phenylethyl alcohol, hexadecane, acetophenone, isopropyl myristate, and acetylacetone	Repellent to female wood wasp *Sirex noctilio* (Wang et al., 2019).

10.3 OMICS APPROACHES TO UNVEIL THE HIDDEN POTENTIAL OF ENDOPHYTES IN PEST MANAGEMENT

Endophytes could become a significant source of bioactive natural compounds that can be used as bio-insecticides. Fungal genomes can be sequenced to discover unknown specialized metabolites and the effective methods to produce them. There are still many endophyte species and unknown bioactive agents to be discovered that may help in understanding the complex relationships between host plants, endophytes, and insects (Samal et al., 2023). Techniques such as genome sequencing, microarray analysis, metagenomics, comparative genomics, meta-transcriptomics, and next-generation sequencing (NGS) are essential for unraveling the interactions between plants and endophytes (Saini et al., 2024). Many studies have addressed the importance of metabolomics in finding bioactive compounds from certain endophytes under specific conditions and parameters (Fadiji and Babalola, 2020; Gakuubi et al., 2021; Rai et al., 2021). For instance, some bioactive metabolites like siderophore ferrirubin and 3-O-methylfunicone (an antibiotic) were identified via metabolomic analysis of the organic extracts acquired from the liquid culture of *Talaromyces pinophilus* (strain F36CF). Moreover, the former is used in iron transport and has antimicrobial properties, while the latter has an insecticidal effect against aphids (Vinale et al., 2017). It is possible to carry out metagenomic investigations of different organs of plants for a better understanding of the complex molecular pathways by which plants respond to endophytes. Long reads of metagenome sequence data could allow the exploration of the structures, roles, and phylogenetic construction of genetic relatedness in microbial genomes (Adeleke and Babalola, 2021). Genomic sequencing carried out on various microorganisms resulted in the identification of unknown microbial genomes, uncharacterized proteins, gene regulatory elements, disease-causing genes, and metabolic processes (Almeida and De Martinis, 2019). Thus, studying endophytes can aid in finding new biomolecules that can be employed in different applications of biotechnology in agriculture and may contribute to the development of sustainable agriculture (Bogas et al., 2022). NGS platforms are increasingly used in genomic research to provide useful insights into the taxonomy and phylogeny of endophytes, as well as the structural and functional characteristics of genes (Bosamia et al., 2020). These studies can be combined with other omics data to understand the impact of fungal endophytes on plant gene expression during interactions. For example, the differential expression of genes in *Festuca rubra* with and without colonization by the endophyte *Epichloë festucae* was assessed by Ambrose and Belanger (2012), revealing different expression levels of approximately 200 plant genes related to distinct physiological functions. The type of one secreted protein and the number of transcripts suggested that, in endophyte-infected *F. rubra*, proteins may play a role in disease resistance. Understanding how the host can acquire resistance due to the unique genes in *E. festucae* is crucial.

10.4 KEY CHALLENGES IN VOLATILE ORGANIC COMPOUND PRODUCTION AND APPLICATION

Volatile organic compounds (VOCs) provide innovative and environmentally friendly solutions for agricultural challenges with multiple advantages. Despite their potential, several factors limit their employment in practice. One significant obstacle is the lengthy and costly screening and development processes required to create efficient VOC formulations. Moreover, certain VOCs may adversely affect the quality of agricultural products (Razo-Belman and Ozuna, 2023). While Glare et al. (2012) highlight that the high biodegradability of VOCs helps mitigate long-term exposure risks, this same characteristic reduces their persistence and effectiveness in field applications. VOCs are highly reactive, which limits their dispersal over large areas, reducing their capacity to protect crops over extended distances. Lower concentrations of VOCs in the air can also weaken their effects on both pests and plants. Additionally, VOC performance is heavily influenced by climatic conditions such as humidity, temperature, and wind dynamics (Aksenov et al., 2014).

To address these limitations, encapsulation technology can be employed, providing a protective barrier around VOCs to shield them from environmental degradation, thereby enhancing their stability during application, transport, and storage. Encapsulation not only preserves the integrity of VOCs but also facilitates their controlled release, preventing rapid degradation due to environmental exposure (Paiva-Santos et al., 2022; Sharifi and Ryu, 2020). Among encapsulation techniques, nano-encapsulation has garnered particular attention for its superior stability, precise control over release rates, and higher encapsulation efficiency (Saifullah et al., 2019). For instance, polymeric and lipid nanoparticles have been successfully used to enhance VOC effectiveness (Dima and Dima, 2015).

However, before large-scale implementation, it is critical to evaluate the potential risks VOCs pose to human health and non-target organisms. Comprehensive studies in this area will help ensure their safe and responsible use, minimizing unforeseen hazards (Mangotra and Singh, 2024).

10.5 ADVANCED TECHNIQUES FOR VOLATILE ORGANIC COMPOUNDS ANALYSIS

Determining and quantifying VOCs presents a challenge (Epping and Koch, 2023). The most employed method for VOC measurement is gas chromatography (GC), which is followed by various other detection methods. This can be done through the direct introduction of samples into the GC system or through solvent or thermal desorption. Currently, a wide range of applications is accepted for this purpose (Lara-Ibeas et al., 2021; Yang et al., 2022). This method has strength and versatility, but it also poses some bottlenecks. The primary obstacle is maintaining the original sample's integrity during the collection phase. There might be difficulties in accurately collecting some compounds. The chemical structure of samples may not be successfully maintained during storage or cannot be accurately analyzed during collection without losses. Moreover, the analysis is also time-consuming, and this process requires much effort, qualified experts, and appropriate and modern equipment that renders the measurements complex and expensive. These challenges highlight the need for specific attention to the development of cost-effective techniques that can provide an accelerated process of generating reliable data (Epping and Koch, 2023). Apart from GC-based methods, several other techniques are available for VOC detection. A summary of methods that are employed in the identification and quantitation of VOCs, along with their advantages and disadvantages, is presented in Table 10.2.

TABLE 10.2 Tools That Have Been Employed to Qualitatively and Quantitatively Explore the Chemical Structure of Volatile Organic Compounds (VOCs)

Detection Methods	Advantages	Disadvantages	References
Gas chromatography–mass spectrometry (GC-MS)	• A wide range of analytes can be detected. • Analytes can be simultaneously quantified and confirmed. • High accuracy. • Analysis and resolution of overlapping peaks.	• Costly. • Expert user-required. • Complex measurements. • Methodological adjustments. • Difficulties in quantifications.	Dobrzyniewski et al. (2021); Epping and Koch (2023); Mastovska (2005).
Electronic nose	• Fast monitoring and sample detection for a range of gases; • Flexible tool for nondestructive; • Mobility; • Field applicable.	• Restricted detection sensitivity. • Long-term sensor drift and hardly achievable full calibration. • low chemical specificity. • Some sensors have short lifespan.	Cellini et al. (2017); Harper (2001); Tholl et al. (2021)
Proton transfer reaction mass spectrometry (PTR-MS)	• High sensitivity; • Quick analysis; • Non-renewable carrier gases and solvents are not needed.	• Insufficient to correctly recognize VOCs and inability to differentiate the isomer signal. • Expensive equipment	Mazzucotelli et al. (2022)
Selected ion flow tube-mass spectrometry (SIFT-MS)	• Allows for targeting a wide range of compounds. • High selectivity. • Real-time Detection. • Complex mixture analysis.	• Need skilled personnel to operate. • Expensive. • Due to the reliance on m/z ratios and insufficient compound libraries, it becomes challenging to identify VOCs in some cases.	Cellini et al. (2021); Roberts et al. (2022)

TABLE 10.2 *(Continued)*

Detection Methods	Advantages	Disadvantages	References
Photoionization detector (PID)	• High sensitivity. • Excellent dynamic range. • Ability to detect vapor without destruction	• False positive results happen frequently.	Stauffer et al. (2008); Zhu et al. (2015)
Surface acoustic wave (Saw) sensors	• High selectivity. • Reversibility. • Sensitivity. • A quick reaction time. • Inexpensive. • Simple manufacture. • Wireless operation	• May produce false positives or negatives, affecting the measurement's precision and dependability. • May have high sensitivity for specific parameters but poor sensitivity for others	Devkota et al. (2017); Mandal and Banerjee (2022); Viespe (2014)

10.6 CONCLUSION

In the agroecosystem, endophytes and their VOCs play an essential role in pest control, plant growth promotion, and the preservation of ecological balance. These microorganisms produce bioactive metabolites that modulate plant responses against herbivory threats and enhance host plant resistance against pests. These compounds possess great bioactivity, which attracts the interest of the industry to make VOC products commercially available for use in agricultural practices. However, there are still persistent challenges in finding the optimal techniques to isolate, extract, and preserve these compounds from their respective endophytic organisms. Commercializing endophyte-derived VOCs requires studies to reveal methods with convenient costs, as their limited emission hampers this process. Genetic tools can overcome this limitation by identifying and over-expressing key genes involved in the biosynthetic pathways of VOCs to improve transgenic plants, which could lead to large-scale production. Future research should focus on integrating endophyte secondary metabolites and volatiles into pest management strategies. This may result in innovative solutions that boost plant resistance against different stressors, reduce reliance on chemical pesticides, and promote sustainable agriculture.

KEYWORDS

- **biocontrol**
- **endophytes**
- **insecticidal potential**
- **pest management**
- **secondary metabolites**
- **volatile organic compounds**

REFERENCES

Adeleke, B. S., & Babalola, O. O. (2021). Biotechnological overview of agriculturally important endophytic fungi. *Horticulture, Environment, and Biotechnology, 62*(4), 507–520.

Adeleke, B. S., & Babalola, O. O. (2022). Meta-omics of endophytic microbes in agricultural biotechnology. *Biocatalysis and Agricultural Biotechnology, 42*, 102332.

Adeleke, B. S., Ayilara, M. S., Akinola, S. A., & Babalola, O. O. (2022). Biocontrol mechanisms of endophytic fungi. *Egyptian Journal of Biological Pest Control, 32*(1), 46.

Agbessenou, A., Akutse, K. S., Yusuf, A. A., Ekesi, S., Subramanian, S., & Khamis, F. M. (2020). Endophytic fungi protect tomato and nightshade plants against *Tuta absoluta* (Lepidoptera: Gelechiidae) through a hidden friendship and cryptic battle. *Scientific Reports, 10*(1), 22195. https://www.nature.com/articles/s41598-020-78898-8 (accessed on 3 April 2025).

Ahmad, I., Jiménez-Gasco, M. D. M., Luthe, D. S., Shakeel, S. N., & Barbercheck, M. E. (2020). Endophytic *Metarhizium robertsii* promotes maize growth, suppresses insect growth, and alters plant defense gene expression. *Biological Control, 144*, 104167. https://doi.org/10.1016/j.biocontrol.2019.104167.

Ahmad, I., Zaib, S., Alves, P. C. M. S., Luthe, D. S., Bano, A., & Shakeel, S. N. (2019). Molecular and physiological analysis of drought stress responses in *Zea mays* treated with plant growth-promoting rhizobacteria. *Biologia Plantarum, 63*, 536–547. https://doi.org/10.32615/bp.2019.092.

Aksenov, A. A., Martini, X., Zhao, W., Stelinski, L. L., & Davis, C. E. (2014). Synthetic blends of volatile, phytopathogen-induced odorants can be used to manipulate vector behavior. *Frontiers in Ecology and Evolution, 2*, 78.

Akutse, K. S., Fiaboe, K. K., Van den Berg, J., Ekesi, S., & Maniania, N. K. (2014). Effects of endophyte colonization of *Vicia faba* (Fabaceae) plants on the life history of leafminer parasitoids *Phaedrotoma scabriventris* (Hymenoptera: Braconidae) and *Diglyphus isaea* (Hymenoptera: Eulophidae). *PLoS ONE, 9*(10), e109965.

Alam, B., Lǐ, J., Gě, Q., Khan, M. A., Gōng, J., Mehmood, S., & Gǒng, W. (2021). Endophytic fungi: From symbiosis to secondary metabolite communications or vice versa? *Frontiers in Plant Science, 12*, 791033. https://doi.org/10.3389/fpls.2021.791033.

Almeida, O. G., & De Martinis, E. C. (2019). Bioinformatics tools to assess metagenomic data for applied microbiology. *Applied Microbiology and Biotechnology, 103*(1), 69–82.

Ambrose, K. V., & Belanger, F. C. (2012). SOLiD-SAGE of endophyte-infected red fescue reveals numerous effects on host transcriptome and an abundance of highly expressed fungal secreted proteins. *PLoS ONE, 7*(12), e53214.

Anamika, Joshi, S., Sahgal, M., Sahu, S., & Prakash, A. (2018). Fungal endophytes and their secondary metabolites: Role in sustainable agriculture. In *Fungi and Their Role in Sustainable Development: Current Perspectives* (pp. 121–146). Springer. https://link.springer.com/chapter/10.1007/978-981-13-0393-7_8.

Ávila-Hernández, J. G., Aguilar-Zárate, P., Carrillo-Inungaray, M. L., Michel, M. R., Wong-Paz, J. E., Muñiz-Márquez, D. B., & Martínez-Ávila, G. C. G. (2022). The secondary metabolites from *Beauveria bassiana* PQ2 inhibit the growth and spore germination of *Gibberella moniliformis* LIA. *Brazilian Journal of Microbiology, 53*(1), 143–152. https://doi.org/10.1007/s42770-021-00668-z.

Ávila-Hernández, J. G., Carrillo-Inungaray, M. L., De, R., La Cruz-Quiroz, J. E., Paz, D. B., Parra, R., & del Campo, R. (2020). *Beauveria bassiana* secondary metabolites: A review inside their production systems, biosynthesis, and bioactivities. *Mexican Journal of Biotechnology, 5*(4), 1–33. https://doi.org/10.29267/mxjb.2020.5.4.1.

Azevedo, J. L., Maccheroni Jr., W., Pereira, J. O., & De Araújo, W. L. (2000). Endophytic microorganisms: A review on insect control and recent advances on tropical plants. *Electronic Journal of Biotechnology, 3*(1), 15–16.

Bae, H., Sicher, R. C., Kim, M. S., Kim, S. H., Strem, M. D., Melnick, R. L., & Bailey, B. A. (2009). The beneficial endophyte *Trichoderma hamatum* isolate DIS 219b promotes growth and delays the onset of the drought response in *Theobroma cacao*. *Journal of Experimental Botany, 60*(11), 3279–3295.

Bagheri, A., Askari Seyahooei, M., & Fathipour, Y. (2021). The endophytes. In *Microbial Approaches for Insect Pest Management* (pp. 151–215).

Bamisile, B. S., Afolabi, O. G., Siddiqui, J. A., & Xu, Y. (2023). Endophytic insect pathogenic fungi-host plant-herbivore mutualism: Elucidating the mechanisms involved in the tripartite interactions. *World Journal of Microbiology and Biotechnology, 39*(12), 326.

Barelli, L., Moonjely, S., Behie, S. W., & Bidochka, M. J. (2016). Fungi with multifunctional lifestyles: Endophytic insect pathogenic fungi. *Plant Molecular Biology, 90*, 657–664.

Barry, S. M., & Challis, G. L. (2009). Recent advances in siderophore biosynthesis. *Current Opinion in Chemical Biology, 13*(2), 205–215.

Batta, Y. A. (2013). Efficacy of endophytic and applied *Metarhizium anisopliae* (Metch.) Sorokin (Ascomycota: Hypocreales) against larvae of *Plutella xylostella* L. (Yponomeutidae: Lepidoptera) infesting *Brassica napus* plants. *Crop Protection, 44*, 128–134.

Behie, S. W., Zelisko, P. M., & Bidochka, M. J. (2012). Endophytic insect-parasitic fungi translocate nitrogen directly from insects to plants. *Science, 336*, 1576–1577.

Bensaci, O. A., Daoud, H., Lombarkia, N., & Rouabah, K. (2015). Formulation of the endophytic fungus *Cladosporium oxysporum* Berk. and M. A. Curtis, isolated from *Euphorbia bupleuroides* subsp. *luteola*, as a new biocontrol tool against the black bean aphid (*Aphis fabae* Scop.). *Journal of Plant Protection Research, 55*(1).

Bensaci, O. A., Rouabah, K., Aliat, T., Lombarkia, N., Plushikov, V. G., Kucher, D. E., & Rebouh, N. Y. (2022). Biological pest management for sustainable agriculture: Understanding the influence of *Cladosporium*-bioformulated endophytic fungi application to control *Myzus persicae* (Sulzer, 1776) in potato (*Solanum tuberosum* L.). *Plants, 11*(15), 2055.

Berg, G., & Hallmann, J. (2006). Control of plant pathogenic fungi with bacterial endophytes. In *Microbial Root Endophytes* (pp. 53–69). Springer Berlin Heidelberg.

Binyameen, M. (2013). Olfactory mechanisms of host selection in phytophagous insects. *Acta Universitatis Agriculturae Sueciae, 2013*(11).

Bogas, A. C., Cruz, F. P. N., Lacava, P. T., & Sousa, C. P. (2022). Endophytic fungi: An overview on biotechnological and agronomic potential. *Brazilian Journal of Biology, 84*, e258557.

Bosamia, T. C., Barbadikar, K. M., & Modi, A. (2020). Genomic insights of plant-endophyte interaction: Prospective and impact on plant fitness. In *Microbial Endophytes* (pp. 227–249). Woodhead Publishing.

Bruce, T. J., & Pickett, J. A. (2011). Perception of plant volatile blends by herbivorous insects: Finding the right mix. *Phytochemistry, 72*(13), 1605–1611.

Buda, V., Apsegaite, V., Cereskiene, L. B., Butkiene, R., Nedveckyte, I., & Peciulyte, D. (2016). Response of moth *Plodia interpunctella* to volatiles of fungus-infected and uninfected wheat grain. *Journal of Stored Products Research, 69*, 152–158.

Cellini, A., Blasioli, S., Biondi, E., Bertaccini, A., Braschi, I., & Spinelli, F. (2017). Potential applications and limitations of electronic nose devices for plant disease diagnosis. *Sensors, 17*(11), 2596.

Cellini, A., Spinelli, F., Donati, I., Ryu, C. M., & Kloepper, J. W. (2021). Bacterial volatile compound-based tools for crop management and quality. *Trends in Plant Science, 26*(9), 968–983.

Clay, K. (1990). Fungal endophytes of grasses. *Annual Review of Ecology and Systematics, 21*, 275–297.

Conboy, N. J., McDaniel, T., Ormerod, A., George, D., Gatehouse, A. M., Wharton, E., & Tosh, C. R. (2019). Companion planting with French marigolds protects tomato plants from glasshouse whiteflies through the emission of airborne limonene. *PLOS ONE, 14*(3), e0213071.

Contreras-Cornejo, H. A., Macías-Rodríguez, L., del-Val, E., & Larsen, J. (2018). The root endophytic fungus *Trichoderma atroviride* induces foliar herbivory resistance in maize plants. *Applied Soil Ecology, 124*, 45–53.

Cotsonas King, A., & Wu, L. (2009). Macromolecular synthesis and membrane perturbation assays for mechanisms of action studies of antimicrobial agents. *Current Protocols in Pharmacology, 47*(1), 13A–7.

Daisy, B. H., Strobel, G. A., Castillo, U., Ezra, D., Sears, J., Weaver, D. K., & Runyon, J. B. (2002). Naphthalene, an insect repellent, is produced by *Muscodor vitigenus*, a novel endophytic fungus. *Microbiology, 148*(11), 3737–3741.

Davis, T. S., Boundy-Mills, K., & Landolt, P. J. (2012). Volatile emissions from an epiphytic fungus are semiochemicals for eusocial wasps. *Microbial Ecology, 64*, 1056–1063.

Delory, B. M., Delaplace, P., du Jardin, P., & Fauconnier, M. L. (2016). Barley (*Hordeum distichon* L.) roots synthesize volatile aldehydes with a strong age-dependent pattern and release (E)-non-2-enal and (E, Z)-nona-2,6-dienal after mechanical injury. *Plant Physiology and Biochemistry, 104*, 134–145.

Demain, A. L., & Fang, A. (2000). The natural functions of secondary metabolites. In *History of Modern Biotechnology, I* (pp. 1–39).

Devkota, J., Ohodnicki, P. R., & Greve, D. W. (2017). SAW sensors for chemical vapors and gases. *Sensors, 17*(4), 801.

Dima, C., & Dima, S. (2015). Essential oils in foods: Extraction, stabilization, and toxicity. *Current Opinion in Food Science, 5*, 29–35.

Dobrzyniewski, D., Szulczyński, B., Dymerski, T., & Gębicki, J. (2021). Development of gas sensor array for methane reforming process monitoring. *Sensors, 21*(15), 4983.

Effah, E., Holopainen, J. K., & McCormick, A. C. (2019). Potential roles of volatile organic compounds in plant competition. *Perspectives in Plant Ecology, Evolution and Systematics, 38*, 58–63.

Elena, G. J., Beatriz, P. J., Alejandro, P., & Lecuona, R. (2011). *Metarhizium anisopliae* (Metschnikoff) Sorokin promotes growth and has endophytic activity in tomato plants. *Advances in Biological Research, 5*, 22–27.

Epping, R., & Koch, M. (2023). On-site detection of volatile organic compounds (VOCs). *Molecules, 28*(4), 1598.

Etminani, F., & Harighi, B. (2018). Isolation and identification of endophytic bacteria with plant growth-promoting activity and biocontrol potential from wild pistachio trees. *The Plant Pathology Journal, 34*(3), 208.

Fadiji, A. E., & Babalola, O. O. (2020). Elucidating mechanisms of endophytes used in plant protection and other bioactivities with multifunctional prospects. *Frontiers in Bioengineering and Biotechnology, 8*, 467.

Farré-Armengol, G., Filella, I., Llusia, J., & Peñuelas, J. (2016). Bidirectional interaction between phyllospheric microbiotas and plant volatile emissions. *Trends in Plant Science, 21*(10), 854–860.

Ferreira, L. L., Filho, J. G. D., Silva, F. O., Ferraz, A. L. L., & Mascarin, G. M. (2020). Attract or repel *Amblyomma sculptum* ticks: Screening of semiochemicals. *Veterinary Parasitology, 278*, 109036.

Findlay, J. A., Buthelezi, S., Li, G., Seveck, M., & Miller, J. D. (1997). Insect toxins from an endophytic fungus from wintergreen. *Journal of Natural Products, 60*(11), 1214–1215.

Gakuubi, M. M., Munusamy, M., Liang, Z. X., & Ng, S. B. (2021). Fungal endophytes: A promising frontier for discovery of novel bioactive compounds. *Journal of Fungi, 7*(10), 786.

Gao, F. K., Dai, C. C., & Liu, X. Z. (2010). Mechanisms of fungal endophytes in plant protection against pathogens. *African Journal of Microbiology Research, 4*(13), 1346–1351.

Garrido-Jurado, I., Resquín-Romero, G., Amarilla, S. P., Ríos-Moreno, A., Carrasco, L., & Quesada-Moraga, E. (2017). Transient endophytic colonization of melon plants by entomopathogenic fungi after foliar application for the control of *Bemisia tabaci* Gennadius (Hemiptera: Aleyrodidae). *Journal of Pest Science, 90*, 319–330.

Giannoulakis, E., Mantzoukas, S., Lagogiannis, I., Dervisoglou, S., & Perdikis, D. (2023). Efficacy of endophytic wild strains of entomopathogenic fungi against the tomato leafminer *Tuta absoluta* (Lepidoptera: Gelechiidae) in tomato plants. *Egyptian Journal of Biological Pest Control, 33*(1), 19.

Glare, T., Caradus, J., Gelernter, W., Jackson, T., Keyhani, N., Köhl, J., & Stewart, A. (2012). Have biopesticides come of age? *Trends in Biotechnology, 30*(5), 250–258.

Glick, B. R. (2012). Plant growth-promoting bacteria: Mechanisms and applications. *Scientifica, 2012*, 963401.

González-Mas, N., Gutiérrez-Sánchez, F., Sánchez-Ortiz, A., Grandi, L., Turlings, T. C., Manuel Muñoz-Redondo, J., & Quesada-Moraga, E. (2021). Endophytic colonization by the entomopathogenic fungus *Beauveria bassiana* affects plant volatile emissions in the presence or absence of chewing and sap-sucking insects. *Frontiers in Plant Science, 12*, 660460.

González-Mas, N., Sánchez-Ortiz, A., Valverde-García, P., & Quesada-Moraga, E. (2019). Effects of endophytic entomopathogenic ascomycetes on the life-history traits of *Aphis gossypii* Glover and its interactions with melon plants. *Insects, 10*(6), 165.

Gurulingappa, P., McGee, P. A., & Sword, G. (2011). Endophytic *Lecanicillium lecanii* and *Beauveria bassiana* reduce the survival and fecundity of *Aphis gossypii* following contact with conidia and secondary metabolites. *Crop Protection, 30*(3), 349–353.

Hardoim, P. R., van Overbeek, L. S., & van Elsas, J. D. (2008). Properties of bacterial endophytes and their proposed role in plant growth. *Trends in Microbiology, 16*(10), 463–471.

Harper, W. J. (2001). The strengths and weaknesses of the electronic nose. *Headspace Analysis of Foods and Flavors: Theory and Practice, 488*, 59–71.

Kaddes, A., Fauconnier, M. L., Sassi, K., Nasraoui, B., & Jijakli, M. H. (2019). Endophytic fungal volatile compounds as a solution for sustainable agriculture. *Molecules, 24*(6), 1065.

Kanchiswamy, C. N., Malnoy, M., & Maffei, M. E. (2015a). Chemical diversity of microbial volatiles and their potential for plant growth and productivity. *Frontiers in Plant Science, 6*, 151.

Kanchiswamy, C. N., Malnoy, M., & Maffei, M. E. (2015b). Bioprospecting bacterial and fungal volatiles for sustainable agriculture. *Trends in Plant Science, 20*(4), 206–211.

Khoja, S., Khalifa, M. E., Baxter, I. K. M., Bull, C. J., Loveridge, E. J., & Butt, T. M. (2019). Fungal volatile organic compounds show promise as potent molluscicides. *Pest Management Science, 75*(12), 3392–3404.

Kinyungu, S. W., Agbessenou, A., Subramanian, S., Khamis, F. M., & Akutse, K. S. (2023). One stone for two birds: Endophytic fungi promote maize seedlings' growth and negatively impact the life history parameters of the fall armyworm, *Spodoptera frugiperda*. *Frontiers in Physiology, 14*, 1253305.

Korpi, A., Järnberg, J., & Pasanen, A. L. (2009). Microbial volatile organic compounds. *Critical Reviews in Toxicology, 39*(2), 139–193.

Kumar, S., Kaushik, N., Edrada-Ebel, R., Ebel, R., & Proksch, P. (2008). Endophytic fungi for pest and disease management. *Integrated Management of Diseases Caused by Fungi, Phytoplasma and Bacteria*, 365–387.

Kuramitsu, K., Ishihara, T., Sugita, A., Yooboon, T., Lustig, B., Matsumori, Y., Yamada, H., & Kinoshita, N. (2019). The attraction of *Tremex apicalis* (Hymenoptera, Siricidae, Tremecinae) and its parasitoid *Ibalia japonica* (Hymenoptera, Ibaliidae) to the fungus *Cerrena unicolor*. *Journal of Hymenoptera Research, 68*, 37–48.

Kushwaha, P., Kashyap, P. L., Bhardwaj, A. K., Kuppusamy, P., Srivastava, A. K., & Tiwari, R. K. (2020). Bacterial endophyte-mediated plant tolerance to salinity: Growth responses and mechanisms of action. *World Journal of Microbiology and Biotechnology, 36*(2), 26.

Lara-Ibeas, I., Cuevas, A. R., & Le Calvé, S. (2021). Recent developments and trends in miniaturized gas preconcentrators for portable gas chromatography systems: A review. *Sensors and Actuators B: Chemical, 346*, 130449.

Li, J., Gu, F., Wu, R., Yang, J., & Zhang, K. Q. (2017). Phylogenomic evolutionary surveys of subtilase superfamily genes in fungi. *Scientific Reports, 7*(1), 45456.

Li, Z., Wen, W., Qin, M., He, Y., Xu, D., & Li, L. (2022). Biosynthetic mechanisms of secondary metabolites promoted by the interaction between endophytes and plant hosts. *Frontiers in Microbiology, 13*, 928967.

Ling, L., Feng, L., Li, Y., Yue, R., Wang, Y., & Zhou, Y. (2024). Endophytic fungi volatile organic compounds as crucial biocontrol agents used for controlling fruit and vegetable postharvest diseases. *Journal of Fungi, 10*(5), 332.

López, M. A., Bannenberg, G., & Castresana, C. (2008). Controlling hormone signaling is a plant and pathogen challenge for growth and survival. *Current Opinion in Plant Biology, 11*(4), 420–427.

Mandal, D., & Banerjee, S. (2022). Surface acoustic wave (SAW) sensors: Physics, materials, and applications. *Sensors, 22*(3), 820.

Mangotra, A., & Singh, S. K. (2024). Volatile organic compounds: A threat to the environment and health hazards to living organisms—A review. *Journal of Biotechnology, 382*, 51–69.

Mastovska, K. (2005). Food and nutritional analysis: Pesticide residues. In *Encyclopedia of Analytical Science* (2nd ed., pp. 251-261).

Mazzucotelli, M., Farneti, B., Khomenko, I., Gonzalez-Estanol, K., Pedrotti, M., Fragasso, M., & Biasioli, F. (2022). Proton transfer reaction mass spectrometry: A green alternative for food volatilome profiling. *Green Analytical Chemistry, 3*, 100041.

Mburu, D. M., Ochola, L., Maniania, N. K., Njagi, P. G. N., Gitonga, L. M., Ndung'u, M. W., Wanjoya, A. K., & Hassanali, A. (2009). Relationship between virulence and repellency of entomopathogenic isolates of *Metarhizium anisopliae* and *Beauveria bassiana* to the termite *Macrotermes michaelseni*. *Journal of Insect Physiology, 55*, 774–780.

Mishra, S., Priyanka, & Sharma, S. (2022). Metabolomic insights into endophyte-derived bioactive compounds. *Frontiers in Microbiology, 13*, 835931.

Muthu Narayanan, M., Ahmad, N., Shivanand, P., & Metali, F. (2022). The role of endophytes in combating fungal- and bacterial-induced stress in plants. *Molecules, 27*(19), 6549.

Nayak, S., Samanta, S., & Mukherjee, A. K. (2020). Beneficial role of *Aspergillus* sp. in agricultural soil and environment. In *Frontiers in Soil and Environmental Microbiology* (pp. 17–36).

Orr, D., & Lahiri, S. (2014). Biological control of insect pests in crops. In *Integrated Pest Management* (pp. 531-548). Academic Press.

Paiva-Santos, A. C., Ferreira, L., Peixoto, D., Silva, F., Soares, M. J., Zeinali, M., & Veiga, F. (2022). Cyclodextrins as an encapsulation molecular strategy for volatile organic compounds—Pharmaceutical applications. *Colloids and Surfaces B: Biointerfaces, 218*, 112758.

Panwar, N., & Szczepaniec, A. (2024). Endophytic entomopathogenic fungi as biological control agents of insect pests. *Pest Management Science, 80*(12), 6033–6040.

Pappas, M. L., Liapoura, M., Papantoniou, D., Avramidou, M., Kavroulakis, N., Weinhold, A., Broufas, G. D., & Papadopoulou, K. K. (2018). The beneficial endophytic fungus *Fusarium solani* strain K alters tomato responses against spider mites to the benefit of the plant. *Frontiers in Plant Science, 9*, 1603.

Pedrini, N., Ortiz-Urquiza, A., Huarte-Bonnet, C., Zhang, S., & Keyhani, N. O. (2013). Targeting of insect epicuticular lipids by the entomopathogenic fungus *Beauveria bassiana*: Hydrocarbon oxidation within the context of a host-pathogen interaction. *Frontiers in Microbiology, 4*, 24.

Picciotti, U., Araujo Dalbon, V., Ciancio, A., Colagiero, M., Cozzi, G., De Bellis, L., & Porcelli, F. (2023). "Ectomosphere": Insects and microorganism interactions. *Microorganisms, 11*(2), 440.

Poveda, J. (2021). *Trichoderma* as a biocontrol agent against pests: New uses for a mycoparasite. *Biological Control, 159*, 104634.

Rai, N., Kumari Keshri, P., Verma, A., Kamble, S. C., Mishra, P., Barik, S., & Gautam, V. (2021). Plant-associated fungal endophytes as a source of natural bioactive compounds. *Mycology, 12*(3), 139–159.

Rajani, P., Rajasekaran, C., Vasanthakumari, M. M., Olsson, S. B., Ravikanth, G., & Shaanker, R. U. (2021). Inhibition of plant pathogenic fungi by endophytic *Trichoderma* spp. through mycoparasitism and volatile organic compounds. *Microbiological Research, 242*, 126595.

Raman, A., & Suryanarayanan, T. S. (2017). Fungus–plant interaction influences plant-feeding insects. *Fungal Ecology, 29*, 123–132.

Razo-Belman, R., & Ozuna, C. (2023). Volatile organic compounds: A review of their current applications as pest biocontrol and disease management. *Horticulturae, 9*(4), 441.

Resquín-Romero, G., Garrido-Jurado, I., Delso, C., Ríos-Moreno, A., & Quesada-Moraga, E. (2016). Transient endophytic colonizations of plants improve the outcome of foliar applications of mycoinsecticides against chewing insects. *Journal of Invertebrate Pathology, 136*, 23–31.

Roberts, I. J., Carpenter, L. J., Shaw, M. D., & Langford, V. S. (2022). Selected ion flow tube–mass spectrometry (SIFT-MS) study of the reactions of H_3O^+, NO^+ and O_2^+ with a range of oxygenated volatile organic carbons (OVOCs). *International Journal of Mass Spectrometry, 479*, 116892.

Rostás, M., Cripps, M. G., & Silcock, P. (2015). Aboveground endophyte affects root volatile emission and host plant selection of a belowground insect. *Oecologia, 177*(2), 487–497.

Russo, M. L., Scorsetti, A. C., Vianna, M. F., Cabello, M., Ferreri, N., & Pelizza, S. (2019). Endophytic effects of *Beauveria bassiana* on corn (*Zea mays*) and its herbivore, *Rachiplusia nu* (Lepidoptera: Noctuidae). *Insects, 10*(4), 110.

Rutkowska, N., Drożdżyński, P., Ryngajłło, M., & Marchut-Mikołajczyk, O. (2023). Plants as the extended phenotype of endophytes—the actual source of bioactive compounds. *International Journal of Molecular Sciences, 24*(12), 10096.

Ryu, C. M., Farag, M. A., Hu, C. H., Reddy, M. S., Kloepper, J. W., & Paré, P. W. (2004). Bacterial volatiles induce systemic resistance in *Arabidopsis*. *Plant Physiology, 134*(3), 1017–1026.

Ryu, C. M., Farag, M. A., Hu, C. H., Reddy, M. S., Wei, H. X., Paré, P. W., & Kloepper, J. W. (2003). Bacterial volatiles promote growth in *Arabidopsis*. *Proceedings of the National Academy of Sciences, 100*(8), 4927–4932.

Safavi, S. A. (2013). *In vitro* and *in vivo* induction, and characterization of beauvericin isolated from *Beauveria bassiana* and its bioassay on *Galleria mellonella* larvae. *Journal of Agricultural Science and Technology, 15*(1), 1–10.

Saifullah, M., Shishir, M. R. I., Ferdowsi, R., Rahman, M. R. T., & Van Vuong, Q. (2019). Micro and nano encapsulation, retention and controlled release of flavor and aroma compounds: A critical review. *Trends in Food Science & Technology, 86*, 230–251.

Saini, P., Verma, A., Tiwari, H., Mishra, V., & Gautam, V. (2024). Omics-based approaches in studying fungal endophytes and their associated secondary metabolites. In *Endophytic Fungi: The Hidden Sustainable Jewels for the Pharmaceutical and Agricultural Industries* (pp. 209–227). Cham: Springer International Publishing.

Samal, I., Bhoi, T. K., Majhi, P. K., Murmu, S., Pradhan, A. K., Kumar, D., & Balawi, T. A. (2023). Combatting insects mediated biotic stress through plant-associated endophytic entomopathogenic fungi in horticultural crops. *Frontiers in Plant Science, 13*, 1098673.

Schrank, A., & Vainstein, M. H. (2010). *Metarhizium anisopliae* enzymes and toxins. *Toxicon, 56*(7), 1267–1274.

Segaran, G., & Sathiavelu, M. (2019). Fungal endophytes: A potent biocontrol agent and a bioactive metabolites reservoir. *Biocatalysis and Agricultural Biotechnology, 21*, 101284.

Semenova, T. A., Dunaevsky, Y. E., Beljakova, G. A., & Belozersky, M. A. (2020). Extracellular peptidases of insect-associated fungi and their possible use in biological control programs and as pathogenicity markers. *Fungal Biology, 124*(1), 65–72.

Sharifi, R., & Ryu, C. M. (2020). Formulation and agricultural application of bacterial volatile compounds. In *Bacterial Volatile Compounds as Mediators of Airborne Interactions* (pp. 317–336).

Sharma, A., Kaushik, N., Sharma, A., Marzouk, T., & Djébali, N. (2022). Exploring the potential of endophytes and their metabolites for bio-control activity. *3 Biotech, 12*(10), 277.

Shrivastava, G., Ownley, B. H., Augé, R. M., Toler, H., Dee, M., Vu, A., & Chen, F. (2015). Colonization by arbuscular mycorrhizal and endophytic fungi enhanced terpene production in tomato plants and their defense against a herbivorous insect. *Symbiosis, 65*, 65–74.

Song, G. C., & Ryu, C. M. (2013). Two volatile organic compounds trigger plant self-defense against a bacterial pathogen and a sucking insect in cucumber under open field conditions. *International Journal of Molecular Sciences, 14*(5), 9803–9819.

Song, Y., Cui, H., Guo, W., Sindhu, L., Lv, S., Li, L., & Men, X. (2024). Endophytic fungi improved wheat resistance to *Rhopalosiphum padi* by decreasing its feeding efficiency and population fitness. *Ecotoxicology and Environmental Safety, 270*, 115865.

Srikamwang, C., Onsa, N. E., Sunanta, P., Sangta, J., Chanway, C. P., Thanakkasaranee, S., & Sommano, S. R. (2023). Role of microbial volatile organic compounds in promoting plant growth and disease resistance in horticultural production. *Plant Signaling & Behavior, 18*(1), 2227440.

Staples, J. A., & Milner, R. J. (2000). A laboratory evaluation of the repellency of *Metarhizium anisopliae* conidia to *Coptotermes lacteus* (Isoptera: Rhinotermitidae). *Sociobiology, 36*, 133–146.

Stauffer, E., Dolan, J. A., & Newman, R. (2008). Detection of ignitable liquid residues at fire scenes. In *Fire Debris Analysis* (pp. 131–161).

Sullivan, B. T., & Berisford, C. W. (2004). Semiochemicals from fungal associates of bark beetles may mediate host location behavior of parasitoids. *Journal of Chemical Ecology, 30*, 703–717.

Suryanarayanan, T. S. (2013). Endophyte research: Going beyond isolation and metabolite documentation. *Fungal Ecology, 6*(6), 561–568.

Thakur, A., Kaur, S., Kaur, A., & Singh, V. (2013). Enhanced resistance to *Spodoptera litura* in endophyte-infected cauliflower plants. *Environmental Entomology, 42*(2), 240–246.

Tholl, D., Hossain, O., Weinhold, A., Röse, U. S., & Wei, Q. (2021). Trends and applications in plant volatile sampling and analysis. *The Plant Journal, 106*(2), 314–325.

Turlings, T. C., & Erb, M. (2018). Tritrophic interactions mediated by herbivore-induced plant volatiles: Mechanisms, ecological relevance, and application potential. *Annual Review of Entomology, 63*(1), 433–452.

Tyc, O., van den Berg, M., Gerards, S., van Veen, J. A., Raaijmakers, J. M., De Boer, W., & Garbeva, P. (2014). Impact of interspecific interactions on antimicrobial activity among soil bacteria. *Frontiers in Microbiology, 5*, 115901.

Vega, F. E. (2008). Insect pathology and fungal endophytes. *Journal of Invertebrate Pathology, 98*(3), 277–279.

Viespe, C. (2014). Surface acoustic wave sensors based on nanoporous films for hydrogen detection. *Key Engineering Materials, 605*, 331–334.

Vinale, F., Nicoletti, R., Lacatena, F., Marra, R., Sacco, A., Lombardi, N., & Woo, S. L. (2017). Secondary metabolites from the endophytic fungus *Talaromyces pinophilus*. *Natural Product Research, 31*(15), 1778–1785.

Vivaldo, G., Masi, E., Taiti, C., Caldarelli, G., & Mancuso, S. (2017). The network of plant volatile organic compounds. *Scientific Reports, 7*(1), 1–18.

Wang, H., Peng, H., Li, W., Cheng, P., & Gong, M. (2021). The toxins of *Beauveria bassiana* and the strategies to improve their virulence to insects. *Frontiers in Microbiology, 12*, 705343.

Wang, L. X., Ren, L. L., Liu, X. B., Shi, J., Wang, J. Z., & Luo, Y. Q. (2019). Effects of endophytic fungi in Mongolian pine on the selection behavior of woodwasp (*Sirex noctilio*) and the growth of its fungal symbiont. *Pest Management Science, 75*(2), 492–505.

White, J. F., Kingsley, K. L., Zhang, Q., Verma, R., Obi, N., Dvinskikh, S., & Kowalski, K. P. (2019). Endophytic microbes and their potential applications in crop management. *Pest Management Science, 75*(10), 2558–2565.

Wilberts, L., Vuts, J., Caulfield, J. C., Thomas, G., Birkett, M. A., Herrera-Malaver, B., & Lievens, B. (2022). Impact of endophytic colonization by entomopathogenic fungi on the behavior and life history of the tobacco peach aphid *Myzus persicae var. nicotianae*. *PLOS ONE, 17*(9), e0273791.

Wilkinson, H. H., Siegel, M. R., Blankenship, J. D., Mallory, A. C., Bush, L. P., & Schardl, C. L. (2000). Contribution of fungal loline alkaloids to protection from aphids in a grass-endophyte mutualism. *Molecular Plant-Microbe Interactions, 13*(10), 1027–1033.

Yan, J., Qi, N., Wang, S., Gadhave, K., & Yang, S. (2014). Characterization of secondary metabolites of an endophytic fungus from *Curcuma wenyujin*. *Current Microbiology, 69*, 740–744.

Yang, Z., Ren, Z., Cheng, Y., Sun, W., Xi, Z., Jia, W., & Li, D. (2022). Review and prospect on portable mass spectrometer for recent applications. *Vacuum, 199*, 110889.

Zhu, H., Nidetz, R., Zhou, M., Lee, J., Buggaveeti, S., Kurabayashi, K., & Fan, X. (2015). Flow-through microfluidic photoionization detectors for rapid and highly sensitive vapor detection. *Lab on a Chip, 15*(14), 3021–3029.

CHAPTER 11

Biocontrol Agents for Insect Pests and Plant Protection

MUBASHER AHMAD MALIK,[1,*] GOZDE BUSRA EROGLU,[2] JORGE POVEDA,[3] AZHAR ABBAS KHAN,[1] and JAVAD KARIMI[4]

[1]*Department of Entomology, College of Agriculture, University of Layyah, Layyah, Pakistan*

[2]*Department of Molecular Biology and Genetics, Erzurum Technical University, Erzurum, Türkiye*

[3]*Recognized Research Group AGROBIOTECH, UIC-370 (JCyL), Department of Plant Production and Forest Resources, Higher Technical School of Agricultural Engineering of Palencia, University Institute for Research in Sustainable Forest Management (iuFOR), University of Valladolid, Palencia, Spain*

[4]*Department of Plant Protection, School of Agriculture, Ferdowsi University of Mashhad, Mashhad, Iran*

*Corresponding author

ABSTRACT

Agricultural insect pests represent a major source of losses to agricultural production, necessitating the development of new sustainable strategies for their control. This chapter focuses on the intersection of biocontrol agents (BCAs) and plant protection within the intricate relationships between insects and plants. BCAs are naturally occurring living organisms, such as viruses, bacteria, fungi, protozoa, nematodes, mites, and insects, whose biological activity includes the direct or indirect attack on

Insect-Plant Interactions, Volume 1: Influence of Biotic and Abiotic Factors.
Jamin Ali, Adil Tonğa, Adnan Shakeel, and Nasheeman Ashraf (Eds.)

DOI: 10.1201/9781779644282-11

populations and individuals of plant pathogens and pests. Among BCAs, microbial control is important in managing pests that threaten plant health. Potential microbial agents isolated must be commercialized through a series of steps, including mass production, formulation, field testing, and safety evaluation. The various methods used in mass production, such as liquid fermentation, solid-state fermentation, and diphasic liquid-solid fermentation, are discussed. The main aim of this chapter is to discuss the importance of BCAs in plant protection, to optimize their potential to reduce reliance on chemical insecticides, and to ensure efficient pest management for sustainable agriculture. By exploring the complex relationship between insects, plants, and BCAs, this chapter develops an environmentally friendly strategy for plant protection; ultimately, the complex intricate relationship shapes ecological systems.

11.1 INTRODUCTION

Diverse biological communities are critical for maintaining ecosystem services (ES) (Millennium Ecosystem Assessment, 2005). As the largest animal biomass, insects play an important role in agriculture and the ecosystem by performing roles such as pollination, pest control, herbivory, and recycling of soil nutrients (Jankielsohn et al., 2023). About three-quarters of the world's wild plant species and two-thirds of all plant species depend on insects for pollination (Schoonhoven, 2005). Different control methods, such as physical, chemical, and biocontrol, have been used to manage insect pests. Among these, chemical methods have been preferred by farmers due to convenience and the visible negative outcomes of pesticide applications, which lead to an immediate reduction in pests and benefits. The continuous use of pesticides has created resurgences, insecticide resistance, and concerns about environmental and human health. Therefore, it is necessary to use biological control methods to minimize future reliance on chemicals, biodiversity, and population dynamics. Arthropod pests cause about 18% of annual yield losses in different agricultural crop production (Culliney, 2014), while only 0.5% of insect species are pests.

The biological controls of insect pests have been used since ancient civilizations. For example, Chinese farmers employed predatory ants (*Oecophylla smaragdina* Fabricius) (Hymenoptera: Formicidae) to manage leaf-feeding pests of citrus orchards (Yang, 2002). The era of modern biocontrol began in the late 18th century when Charles V. Riley introduced the shipment of predatory mites, *Tyroglyphus phylloxera* Riley (Acari: Acaridae), as a biological control in France to manage grapevine phylloxera, *Daktulosphaira vitifoliae* (Fitch) (Phylloxeridae: Hemiptera). In 1888–1889, the Vedalia beetle, *Rodolia cardinalis* Mulsant (Coleoptera: Coccinellidae), was imported from Australia to the U.S., where it managed the cottony cushion scale (*Icerya purchasi*) Maskell (Hemiptera: Monophlebidae) and significantly contributed to pest management in California's citrus cultivation. This success is considered a landmark event for the further introduction of beneficial insects in the U.S. (Coulson et al., 2000). IPM strategies include biological, cultural, physical, and chemical methods to manage pest populations in both economically viable and environmentally sustainable ways. The development and commercialization of biocontrol products have been promoted by the Food and Agriculture Organization (FAO) with the help of countries such as the United States and members of the European Union (van Lenteren, 2007).

Biological control agents are living organisms that are used to manage insect pest populations below damaging levels. The three main categories of insect pests include predators, parasites, and pathogens. Predators are important BCA that hunt and consume a variety of prey throughout their life cycles (Joshi et al., 2024). These organisms maintain the natural balance of pest populations, which leads to the health of both agricultural and natural ecosystems. Predatory insects, mites, and entomopathogens are effectively used as biocontrol agents (BCA) and key components of IPM. Out of 53 recorded species, all are classified into one of 9 orders, with a majority (30 species) belonging to the Coccinellidae family. The ladybird beetle belongs to the family Coccinellidae (lady beetles or ladybirds) within the superfamily Coccinellidae, which is composed of more than 6000 known species worldwide, representing 360 genera (Ślipiński and Tomaszewska, 2010). They are probably the most widely recognized beneficial insect of field and horticultural crops (Che et al., 2022).

The family of green lacewing (*Chrysoperla carnea*), Steph (Neuropteran: Chrysopidae), is composed of approximately 1200 identified species in the most frequently studied order, Neuroptera, because of their value in biological control (Brook, 1997). The larvae of many chrysopid species feed on nymphs, larvae, and adults of soft-bodied insects and mites in crops or horticultural plantings because of their value in biological control (Rash and Carde, 2009). Chrysalides are the most frequently studied of the Neuroptera (Gerth et al., 2017). The Phytoseiidae family of predatory mites (Acari: Mesostigmata) constitutes 60% of the biological market worldwide (Helyer et al., 2014; Knapp et al., 2018). They play a leading role in commercial augmentative biological control, especially in greenhouse and open-field ornamental cultivation systems (McMurtry et al., 2015). Ground beetles (Coleoptera, Carabidae) are one of the largest groups, with 40,000 species in the order Coleoptera. They play a significant role in biocenoses as entomophagists or bioindicators in terrestrial invertebrates (Bondarenko et al., 2020). Both adults and larvae feed on economically important pests of forestry and agriculture.

Parasitoids comprise 68,000 named species. Parasitoids feed on various stages, including egg, larva, pupa, and adult, and are distinguished by their parasitic development within or on a single host individual (Bruke and Sharanowski, 2024). They lay their eggs in or on the body of the host insect, and as the larvae develop, they feed on the host's body content, ultimately leading to the host's death. Mostly, they belong to the Hymenopteran order (Tables 11.1 and 11.2), which includes 15 families; among them, the major families are Trichogrammatidae, Mymaridae, Braconidae, Ichneumonidae, and Sciaenidae. There are 230 recorded species of *Trichogramma* from the USA, India, Brazil, China, and Russia, parasitizing eggs of over 200 insect species belonging to 70 different families (Jalali et al., 2016). The genus *Encarsia* (family Aphelinidae) was first described in 1875. It has more than 450 species that target a diversity of agricultural pest insects, ranging from whiteflies to armored scale insects (Polaszek et al., 2023). Among the 130 species of the genus *Aphidius, Aphidius colemani* (Hymenoptera: Braconidae) is commercially available as a biological control for many aphid species of greenhouse crops. *Aphidius colemani* is thought to originate from India or Pakistan (Messing and Rabasse, 1995). It was first used in biological programs in 1971 (Starý, 1975). The genus *Cotesia* is the largest genus within the subfamily Microgastrinae, which comprises approximately 320 species (Fernandez-Triana et al., 2020). Among the natural enemies of the large cabbage white butterfly, *Cotesia glomerata* is a generalist parasitoid of several species of *Pieris* butterflies but prefers to feed on *P. brassicae* in cabbage plantations (Gols et al., 2019; Ullah et al., 2016).

TABLE 11.1 Important Parasitoid Species Characteristics and Example

Important Parasitoid Species	Parasitoid Order	Target Pest	Number of Species	Target Life Stages of Pest	Example Application	References
Trichogramma. chilonis Ishii, Trichogramma. japonicum Ashmead, Trichogramma. pretiosum Riley *Trichogramma achaeae* Nagaraja and Nagarkatti, *Trichogramma brassicae* (Bezdenko), *Trichogramma minutum* (Riley), *Trichogramma nubilale* Ertle and Davis, *Trichogramma platneri* Nagarkatti, *Trichogramma fuentesi* Torre	*Hymenoptera*	Lepidopteron egg	230	Egg	Control of different borer	Dodiya et al. (2023)
Encarsia Formosa Gahan Encarsia luteola Howard, Encarsia sophia Girault and Dodd	*Hymenoptera*	Whiteflies	>450	Female nymphal parasitoid and Male hyperparasitoid of conspecific species	Management of whitefly in green house	Polaszek et al. (2023)
Aphidius colemani	*Hymenoptera*	Aphid	130	Adult stage	Management of aphid population in controlled condition	Messing and Rabasse (1995); Woolay et al. (2021)
Cotesia glomerate Linnaeus, *Cotesia plutellae.* Kurdjumov	*Hymenoptera*	Cabbage caterpillar	320	Larval, pupal stage	Control of cabbage butterfly	Fernandez-Triana et al. (2020); Lazervic et al. (2023)

TABLE 11.2 Important Predator Species Characteristics and Example

Predator	Important Species of Predator	Order	Total Species	Target Pest	Key Characteristics	Example Application	References
Lady beetle	*Hippodamia convergens Guérin-Méneville; Coccinella septempunctata* Linnaeus. *Adalia bipunctata* Linnaeus. *Coleomegilla maculata lengi; Harmonia axyridis* Pallas *Epilachna varivestis Mulsant*	*Coleoptera*	6,000	Aphid, Mites	Feed on soft bodies insect	Control of aphid on vegetables and horticulture crops	Che et al. (2021)
Lace wing	*Chrysoperla carnea* Stephens, *Chrysopa oculata* Say, *Chrysoperla rufilabris* Burmeister, *Micromus posticus* Walker, *Hemerobius humulinus* Linnaeus, *Sympherobius amiculus* Fitch	*Neuroptera*	1415	Aphid, Caterpillar	Feed on nymph, larva, and adult of insect	Management of soft boded field and greenhouse insect pest	Oswald et al. (2012)
Predatory mite	*Phytoseiulus persimilis* Athias-Henriot, *Neoseiulus californicus* McGregor, *Amblyseius andersoni* Chant, *Neoseiulus cucumeris* Oudemans, *Typhlodromus pyri Scheuten, Amblydromalus limonicus* (Garman and McGregor)	*Mesostigmata*	2700	Spider, Mite	Highly active in humid conditions	Control of mites, thrips, and whiteflies	Demite et al. (2023)
Ground beetle	*Calosoma sycophanta* Linnaeus, *Lebia grandis* Hentz, *Pterostichus nigra* Schaller, *Bembidion spp. Scaphinotus spp.*	*Carabidae*	40000	Bioindicator pest	Feed on underground caterpillar soil-dwelling, gypsy moth, pupae of the Colorado potato beetle.	Soil pest management in field crops	Sahito et al. (2021); Yurievich and Lutsenko (2022)

Pathogens, such as fungi, bacteria, viruses, and nematodes, play a crucial role in inducing diseases and subsequent mortalities in pest populations (Table 11.3). This process results in substantial decreases in pest numbers, contributing to healthier ecosystems. Among microbial agents, entomopathogenic fungi are the most widely utilized pathogens in biocontrol strategies. They infect their host when spores in the conidia come in contact with the insect cuticle, germinate, and subsequently penetrate the host's body. Notable examples include fungi such as *Beauveria bassiana* and *Metarhizium anisopliae*, both of which are effective against a diverse range of insect pests, including aphids, whiteflies, and beetles (Butt, 2000). This process leads to the host's death through tissue destruction and nutrient depletion.

Bacteria can cause lethal infections in pests, with *Bacillus thuringiensis* (*Bt*) being one of the most widely used bacterial BCAs. *Bt* produces toxins that, when ingested by pests, disrupt their gut cells, ultimately leading to the pests' death. *Bt* produces protein crystals that are capable of causing cell lysis in the midgut of insects, ultimately resulting in the death of the susceptible insect. This mechanism of action is highly effective against caterpillars and other pests (Ruiu, 2015).

Viruses, particularly baculoviruses, are capable of infecting and killing insect pests. These viruses are highly specific to their host insects, making them ideal BCAs. One notable example is the *Autographa californica* multiple nucleopolyhedrovirus (AcMNPV), which is used to control various Lepidopteran pests (Miller, 1998).

Entomopathogenic nematodes (EPNs), including species from the *Steinernema* and *Heterorhabditis* genera, are highly effective BCAs. These nematodes target and kill a range of soil-dwelling insect pests. Upon infecting their host, the nematodes release symbiotic bacteria that induce septicemia, ultimately leading to the host's death (Forst, 1997). This symbiotic relationship significantly enhances the nematodes' effectiveness as BCAs, allowing them to efficiently target and eliminate pest populations (Table 11.3).

BCAs utilize both direct and indirect mechanisms to control pest populations. Direct mechanisms include predation, parasitism, and infection, while indirect mechanisms involve altering the environment or influencing the behavior of pests to reduce their impact (Joshi et al., 2024). Various factors, such as environmental conditions, the characteristics of the agents, and the interactions between pests and their natural enemies, alter the efficacy of BCAs (Abd-Elgawad et al., 2020).

TABLE 11.3 Important Pathogen Species, Characteristics and Example

Type of Entomopathogen	Target Insect Pest Order	Species	Mode of Action	Example Application	References
Bacteria	*Lepidopteran*, *Coleopteran*, and *Hemipteran*	*Bacillus thuringiensis Pseudomonas syringae*, *Serratia ficaria*, *Clostridium bifermentans* and Yersinia species	Produce toxins that disrupt gut cells and cause septicemia	*Bt* spray for *Lepidopteran*, *Coleopteran*, and *Hemipteran* pest	Ruiu (2015).
Fungi	*Lepidopterans*, *Coleopterans*, and *Hemipterans*	*Metarhizium anisopliae*, *Beauveria bassiana*, *Hirsutella*, *Isaria*, *Lecanicillium*, *Paecilomyces*, and *Verticillium* species	Produce spores that Penetrate and proliferate within the cuticle	Biopesticide for general pest control	Butt (2000)
Viruses	*Lepidopteran*	Baculoviruses (Nucleopolyhedrovirus and Granulovirus)	Release Occlusion bodies that bound to the peritrophic membrane	Biopesticide spray for *Lepidopteran* pest	Miller (1998)
Nematodes	*Lepidopteran*	*Steinernema* and *Heterorhabditis*	Release the bacteria into host hemolymph that produces a toxin that kills the host	Biopesticide for general pest control	Forst (1997)

11.2 IDENTIFICATION AND COMMERCIALIZATION OF POTENTIAL MICROBIAL BIOCONTROL AGENTS

The initial step includes the isolation of potential bacteria, fungi, viruses, or nematodes that disrupt the life cycle of plant pests. The probability of obtaining useful microorganisms can be significantly increased through the processes of sampling. For this purpose, samples of beneficial microorganisms are taken from dead cadavers of arthropods, disease-free soil, and healthy plants in epidemic areas. These beneficial microorganisms are isolated in suitable culture media, in synthetic media, or living biological systems like cellular cultures and trap organisms (Montesinos, 2003).

After identification through sequencing, the efficacy of the submitted isolate against the target insect pest is evaluated under controlled laboratory conditions. This is accomplished using various methods such as *in vitro, ex vivo*, or *planta* tests. Pilot trials of selected strains are conducted under natural conditions in which biological control is expected to be employed for practice. Unfortunately, the majority of BCAs discovered during the initial assessments, such as pilot trials, are hardly maintained due to a narrow host range that limits the employment of alternative hosts, contradictory results between studies, or low efficacy under field conditions (Harman, 2000).

Apart from the aforementioned steps, implementing biosafety measures is important to avoid the nontoxic effect of microbial pesticides on humans, plants, animals, and the environment. Rigorous toxicological assessments in mammal models are essential to ensure the safety and health of consumers and handlers exposed to microbial pesticide products. Biopesticide patents are a separate category of biotechnological inventions because they consist of living organisms or natural components derived from these organisms. However, obtaining any patent for BCAs does not fully provide authorization for commercial use. Specific analysis methods are necessary not only to assess the quality of the microbial pesticide active ingredient (the BCA) but also to study its traceability, residue analysis, and environmental impact.

11.3 VARIOUS METHODS FOR MASS PRODUCTION OF MICROBIAL BIOCONTROL AGENTS

Microbial pesticides are produced at an industrial scale through the processes of fermentation, which are further preserved, stored, and formulated (Powell and Justsum, 1993). Based on the nature of the agent (bacteria, fungi or yeast, nematodes, or viruses), there are generally two methods for the industrial-scale production of microbial agents: solid-phase or liquid-phase fermentation, which can benefit from the advanced technology in the pharmaceutical and food industries. Mostly, liquid fermentation is used for the production of bacteria and yeast, in which tanks are continuously stirred, while many fungi are fermented using the solid-state technique (Fravel, 1999).

Independent of the method used for fermentation, the aim is to achieve the highest possible yield with the lowest cost of culture medium. This is achieved on a small scale by using molasses, peptones, yeast extract, and tryptone. Industrial-grade whey, corn steep liquor, soybean bran, wheat bran, barley fiber, fishmeal, molasses, composts, or wastewater sludge may all serve as good low-cost alternative carbon and nitrogen sources (Bejarano and Puopolo, 2020). Besides culture media, optimal culture parameters, including dissolved oxygen, temperature, pH, and stirring speed, need to be identified. Dissolved oxygen in liquid fermentation systems is often between 90% and 100%, with a temperature of 25°C–28°C for biocontrol bacteria production.

11.4 DEVELOPMENT OF VARIOUS BIOCONTROL AGENT FORMULATIONS AND THEIR IMPLEMENTATION

There are only two commercial formulations of BCAs: powder/granulated or liquid forms (Keswani et al., 2016). For each of these formulations, a variety of carriers and adjuvants are available.

11.4.1 POWDER AND GRANULAR FORMULATIONS

The powder formulations of microorganisms are created by mixing microorganisms with additional ingredients (carriers and adjuvants) in a blender until a uniform mixture is obtained. The particle size of the powder is about 100 micrometers. In addition to this, granules can be ground into a fine powder by lyophilization and spray drying. Granular formulations of microorganisms are usually made by incorporating microorganisms into a powder carrier, followed by pressing through roller compactors (Bejarano and Puopolo, 2020). Granule particle sizes usually range between 0.1 and 2.5 mm. Wettable powders and water-dispersible granules are formulated in the same way as mentioned above, but only anionic dispersants and wetting agents must be added to the formulation (Brar et al., 2006).

11.4.2 LIQUID FORMULATIONS

Liquid formulations of BCAs (Figure 11.1) consist of BCAs suspended in water, mineral oil, or organic oil. Suspension concentrations of microorganisms are typically produced by combining microbial cells or immobilized microbial cells with an aqueous solution. Usually, emulsifiers are used to mix two (or more) immiscible liquids, so that one of the liquids, the dispersed phase, results in the production of small droplets in the other phase. Two types can be produced: oil in water (O/W) or water in oil (W/O), which is the so-called invert emulsion (Bejarano and Puopolo, 2020).

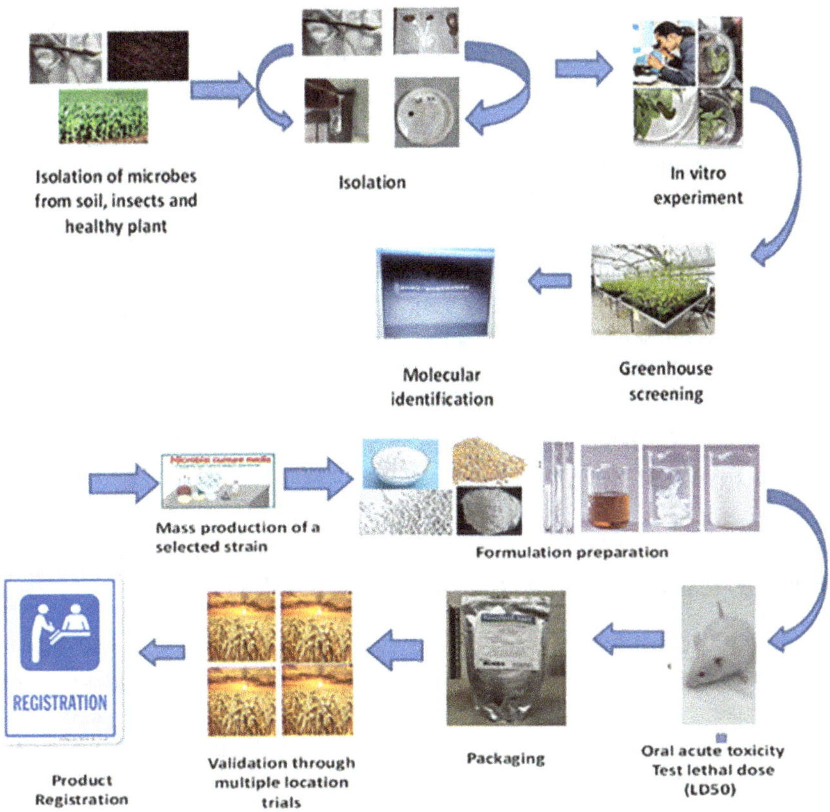

FIGURE 11.1 Identification and mass production of biocontrol agent.

11.5 INTERACTION BETWEEN BIOCONTROL AGENTS AND INSECT PESTS

The relationship between BCAs and insect pests is intricate, involving the mechanism of action of microorganisms, ecological considerations, and the integration of semiochemicals. BCAs, like microorganisms, are employed to control pest populations. These microorganisms have different modes of action to control insect pests. Entomopathogenic fungi attach to the host cuticle, germinate, and penetrate within the host to cause death by the production of toxins. Other microorganisms, such as bacteria, viruses, and nematodes, also act as BCAs by entering the host and causing septicemia. Semiochemicals play an important role in controlling pest populations by attracting natural predators and parasitoids to pest-infected areas, thereby enhancing the effectiveness of BCAs. For example, repellents push insects away from the main stage of crops, while attractants pull them toward the wing of trap crops. The incorporation of semiochemicals with natural enemies and microbial control agents is an important part of integrated pest management (IPM). The interspecific combination between BCAs and insect pests (Sharma, 2019) is significant.

BCAs are highly recognized due to their host-specific nature, lower toxicity to the environment, and potential for self-propagation. A field bioassay of different ecological entomopathogenic microorganisms is conducted to unlock their full potential and deployment as successful BCAs.

11.6 CASE STUDIES AND SUCCESS STORIES OF BIOCONTROL AGENT IMPLEMENTATION

11.6.1 CASE STUDY 1: BIOLOGICAL CONTROL OF TOMATO LEAF MINER WITH ENTOMOPATHOGENIC NEMATODES

The tomato leaf miner, *Tuta absoluta* (Lep., Gelechidae), is a key threat to tomatoes in the world (Deseneux et al., 2022). Currently, growers in Europe use a combined application of a predator agent like *Phytoseiulus persimilis* and an entomopathogen like species of *Steinernema spp.* or *B. thuringiensis* to manage the density of the pest larvae. The predator is active on the moth eggs or initial larvae, while *S. feltiae*/*Bt* actively kills other larval stages of this gelechiid (Deseneux et al., 2022; Kamali et al., 2017).

11.6.2 CASE STUDY 2: CONTROL OF COTTONY CUSHION SCALE WITH VEDALIA BEETLE

The vedalia beetle, *Rodolia cardinalis* Mulsant (Coleoptera: Coccinellidae), is a successful BCA for controlling cottony cushion scale. The beetle was introduced from Australia to control cottony cushion scale in California in 1888–1889, making it a well-documented example of biological control. The cottony cushion scale has four to five generations per year. It lays 600–2000 red eggs, which hatch and feed on the sap of leaves and twigs, ultimately reducing plant vigor and causing dieback of citrus groves. The female vedalia beetle lays eggs near the scale egg sac. The newly hatched larvae feed on the scale eggs. The predation is more effective when it feeds on different life stages of the scale (Heimpel, 2017).

11.6.3 CASE STUDY 3: MANAGEMENT OF DIAMONDBACK MOTH WITH PARASITOID WASPS

The diamondback moth (*Plutella xylostella*) Linnaeus (Lepidoptera: Plutellidae) is considered a major threat to cruciferous crops all over the world. The introduction of parasitoid wasps, such as *Cotesia plutellae* (Hymenoptera, Braconidae) and *Diadegma insulare* Cresson (Hymenoptera: Ichneumonidae), has significantly reduced moth populations in various regions. These parasitoids lay their eggs inside the caterpillars, leading to the death of the host. The use of parasitoids has decreased the need for chemical pesticides, promoting sustainable pest management practices (Sarfraz et al., 2005). The effectiveness of parasitoids against diamondback moths increases when used in conjunction with planting insectary plants that attract these beneficial insects.

11.6.4 CASE STUDY 4: ANTICARSIA GEMMATALIS (LEPIDOPTERA: NOCTUIDAE) (AGMNPV) FOR THE CONTROL OF THE ANTICARSIA GEMMATALIS VELVETBEAN CATERPILLAR IN SOYBEAN CROPS

There is a case for the use of AgMNPV to suppress the population of *A. gemmatalis* in the soybean crops of Brazil. Farmers can mass rear the virus as a cottage industry at a very low cost and apply it as a bioinsecticide.

In 1980/1981 and 1981/1982, a pilot use of AgMNPV was conducted under the supervision of the famous Brazilian entomologist, Moscardi, and it remains a typical and brilliant example (Moscardi, 1999).

11.6.5 CASE STUDY 5: BT IN LEPIDOPTERAN PEST MANAGEMENT

The bacterium *Bt* has been widely used to control Lepidopteran pests in various crops, including corn, cotton, and vegetables. *Bt* produces crystal proteins that are toxic to caterpillars when ingested. The use of *Bt*-based biopesticides has reduced the reliance on chemical insecticides and minimized environmental contamination. *Bt* crops, genetically engineered to express *Bt* toxins (vegetative insecticidal proteins and crystal inclusions), have also been developed. These crops, such as *Bt* cotton, have been engineered to produce Cry proteins, providing protection against key agricultural Lepidopteran pests like the cotton bollworm (*Helicoverpa armigera*) Hübner (Lepidoptera: Noctuidae) and pink bollworm (*Pectinophora gossypiella*) Saunders (Gelechiidae; Lepidoptera) and adopted globally, providing effective pest control and enhancing crop productivity (James, 2016). *Bt*-based bioinsecticide formulations such as Dipel and Xentari contain *Bt* toxins of various strains that have shown effectiveness against a range of Lepidopteran species, including *Spodoptera* and *Diatraea* species.

11.7 INTEGRATION OF BIOCONTROL AGENT WITH IPM PROGRAM

11.7.1 CASE STUDY 1: INTEGRATED MANAGEMENT OF DIAMONDBACK MOTH IN CRUCIFEROUS CROPS

The diamondback moth (*P. xylostella*) is a severe pest of cruciferous crops, causing extensive damage to cabbage, broccoli, and other brassicas. In various areas of the world, parasitoid wasps (*C. plutellae*) and microbial insecticides such as *Bt* are incorporated into IPM programs. These BCAs intercept different life stages of the diamondback moth, reducing its population and minimizing crop damage. The combination of biocontrol with cultural practices, such as crop rotation and resistant varieties, has led to sustainable pest management and reduced pesticide use. In these integrated programs, adult parasitoid wasps deposit their ovipositor inside the larvae of the diamondback moth and lay eggs. The emerging larvae consume the internal contents of the host, thereby reducing moth populations. Insecticides containing a toxin from *Bt* are applied in a targeted manner to control caterpillar larvae. By using this integrated approach, the mechanism of resistance to *Bt* toxins has been reduced, and farmers have noted a substantial lowering of pest damage and increased yields (Shelton et al., 2012).

11.7.2 CASE STUDY 2: IPM FOR SOYBEAN APHID IN NORTH AMERICA

The soybean aphid (*Aphis glycines*) Matsumura (Hemiptera: Aphididae) is a destructive pest in North American soybean production. BCAs, like the Asian ladybird (*Harmonia axyridis*) and parasitic wasps (*Aphelinus certus*, Yasnosh) (Hymenoptera: Aphelinidae), are included in IPM programs. These living organisms keep aphid populations below economic thresholds, which has led to the reduction of chemical insecticides. Habitat manipulation practices, such as planting strips of flowering plants, have been implemented to enhance the habitat for these beneficial insects. As a result, the IPM programs have led to increased soybean yields and reduced environmental impact (Ragsdale et al., 2011). In these programs, lady beetles and parasitoid wasps are released or conserved to control aphid populations. Lady beetles feed on large quantities of aphids, while parasitic wasps parasitize the aphid colonies. The growing of flowering plants along the main crop provides nectar and pollen, increases the populations of natural enemies, and enhances their effectiveness. The incorporation of BCAs into IPM programs not only achieves sustainable pest control but also improves crop performance. The decrease in chemical insecticide use has a positive impact on environmental and economic benefits (Hajek et al., 2018).

11.7.3 CASE STUDY 3: IPM IN GREENHOUSE TOMATO PRODUCTION

The production of tomatoes in greenhouses is susceptible to pests, such as whiteflies, thrips, and spider mites. In these environments, BCAs, such as predatory mites (*P. persimilis*), parasitic wasps (*E. formosa*), and predatory mites (*Amblyseius swirskii*), Athias-Henriot (Acari: Phytoseiidae), have been successfully integrated into IPM programs. These BCAs provide effective control of key pests, reducing the need for chemical pesticides.

The integration of biocontrol with cultural practices such as sanitation and environmental control has led to improved pest management and higher tomato yields (Van Lenter et al., 1988).

In IPM programs, predatory mites are released to control spider mite populations, while parasitic wasps target whitefly nymphs. Predatory beetles are used to manage thrips and other small pests. These BCAs are introduced into the greenhouse environment in a systematic manner, often through regular releases or augmentation. By combining biocontrol with practices such as maintaining optimal humidity levels and implementing strict sanitation measures, greenhouse tomato producers have achieved effective and sustainable pest control. This has resulted in healthier plants, higher yields, and reduced reliance on chemical pesticides (Van Lenter et al., 1988).

11.8 CHALLENGES AND FUTURE PROSPECTS

One of the primary hurdles in the use of BCAs is their inconsistent efficacy under natural conditions. Environmental factors, both biotic and abiotic, significantly influence the performance of these agents, leading to variable results across different environments. Additionally, regulatory requirements pose another challenge in the commercialization of microbial control agents. These regulations necessitate comprehensive documentation regarding the identity, efficacy, and potential environmental impact of BCAs, often resulting in delays in their approval. Furthermore, many farmers express uncertainty about the effectiveness of low-risk plant protection products (PPPs), which leads them to prefer traditional chemical insecticides. In the end, there is a knowledge gap in understanding the mechanism by which BCAs interact with plant pathogens.

Future research should be focused on the identification of novel strains by using genomics and metagenomics that can effectively control plant diseases and pests. Integrating control agents with resistant crop varieties and cultural practices could lead to more sustainable and effective pest control solutions. Cost-effectiveness, as compared to conventional insecticides, may aid in their adoption. Recent regulatory changes in the EU, designed to simplify the approval process for microbial PPPs, may pave the way for the registration of more biocontrol products. Understanding how these agents can mitigate the impacts of environmental changes on plant health will be crucial for ensuring future agricultural sustainability.

KEYWORDS

- **biocontrol**
- ***Bacillus thuringiensis***
- **entomopathogenic fungi**
- **immiscible liquids**
- **lady beetles**
- **pest management**
- **plant protection products**
- **soybean**

REFERENCES

Abd-Elgawad, M. M. (2020). Managing nematodes in Egyptian citrus orchards. *Bulletin of the National Research Centre, 44*(1), 41.

Bejarano, A., & Puopolo, G. (2020). Bioformulation of microbial biocontrol agents for sustainable agriculture: How research can stimulate the development of commercial biological control. *Plant Disease, 104*(2), 275–293.

Bondarenko, O. V., Utescher, T., Blokhina, N. I., Evstigneeva, T. A., & Kezina, T. V. (2020). Temporal climate and vegetation gradient of the Paleocene in the Amur Region (Far East of Russia). *Botanica Pacifica, 9*(2), 13–35.

Brar, S. K., Verma, M., Tyagi, R. D., & Valéro, J. R. (2006). Recent advances in downstream processing and formulations of *Bacillus thuringiensis*-based biopesticides. *Process Biochemistry, 41*(2), 323–342.

Brooks, S. J. (1997). An overview of the current status of *Chrysopidae* (*Neuroptera*) systematics. *Deutsche Entomologische Zeitschrift, 44*(2), 267–275.

Butt, T. M., & Copping, L. G. (2000). Fungal biological control agents. *Pesticide Outlook, 11*(5), 186–191.

Che, Y., Liu, X., Pu, Y., Zhou, M., Zhao, Z., Jiang, R., & Li, Q. (2021). Randomized, double-blinded, placebo-controlled phase 2 trial of an inactivated severe acute respiratory syndrome coronavirus 2 vaccine in healthy adults. *Clinical Infectious Diseases, 73*(11), e3949–e3955.

Coulson, J. R. (2000). Chapter IV: 1973–1993. In J. R. Coulson, P. V. Vail, M. E. Dix, D. A. Nordlund, & W. C. Kauffman (Eds.), *110 Years of Biological Control Research in the United States Department of Agriculture 1883–1993* (pp. 44–99). United States Department of Agriculture, Agriculture Research Service.

Culliney, T. W. (2014). Crop losses to arthropods. In *Integrated Pest Management: Pesticide Problems* (Vol. *3*, pp. 201–225).

Demite, P. R., Santos, J. C., & Holtz, A. M. (2023). Rediscovery and complementary description of *Amblyseius riodocei* El-Banhawy, 1984 (*Acari: Phytoseiidae*). *Entomological Communications, 5*, ec05007.

Desneux, N., Han, P., & Mansour, R. (2022). Integrated pest management of *Tuta absoluta*: Practical implementations across different world regions. *Journal of Pest Science, 95*, 17–39.

Dicke, M. (2018). Insects as feed and the Sustainable Development Goals. *Journal of Insects as Food and Feed, 4*(3), 147–156.

Dodiya, H. D., Patel, D. B., & Patel, S. M. (2023). Technological gap in adoption of pomegranate cultivation practices by the pomegranate growers. *Gujarat Journal of Extension Education, 36*(1), 109–112.

Fernandez-Triana, J., Shaw, M. R., Boudreault, C., Beaudin, M., & Broad, G. R. (2020). Annotated and illustrated world checklist of *Microgastrinae* parasitoid wasps (*Hymenoptera, Braconidae*). *ZooKeys, 920*, 1–107.

Forst, S., Dowds, B., Boemare, N., & Stackebrandt, E. (1997). *Xenorhabdus* and *Photorhabdus* spp.: Bugs that kill bugs. *Annual Review of Microbiology, 51*(1), 47–72.

Fravel, D. R., Rhodes, D. J., & Larkin, R. P. (1999). Production and commercialization of biocontrol products. In *Integrated Pest and Disease Management in Greenhouse Crops* (pp. 365–376). Springer Netherlands.

Gerth, M., Wolf, R., Bleidorn, C., Richter, J., Sontowski, R., Unrein, J., & Gruppe, A. (2017). Green lacewings (*Neuroptera: Chrysopidae*) are commonly associated with a diversity of *Rickettsial* endosymbionts. *Zoological Letters, 3*, 1–7.

Gols, R., Ros, V. I., Ode, P. J., Vyas, D., & Harvey, J. A. (2019). Varying degree of physiological integration among host instars and their endoparasitoid affects stress-induced mortality. *Entomologia Experimentalis et Applicata, 167*(5), 424–432.

Hajek, A. E., Hannam, J. J., Nielsen, C., Bell, A. J., & Liebherr, J. K. (2007). Distribution and abundance of *Carabidae* (*Coleoptera*) associated with soybean aphid (*Hemiptera: Aphididae*) populations in central New York. *Annals of the Entomological Society of America, 100*(6), 876–886.

Harman, G. E. (2000). Myths and dogmas of biocontrol: Changes in perceptions derived from research on *Trichoderma harzianum* T-22. *Plant Disease, 84*(4), 377–393.

Heimpel, G. E., Abram, P. K., Causton, C. E., Celis, S. L., Coll, M., Hardy, I. C., & Segoli, M. (2024). A benefit–risk analysis for biological control introductions based on the protection of native biodiversity. *Ecological Applications, 34*(6), e3012.

Helyer, N., Cattlin, N. D., & Brown, K. C. (2014). *Biological Control in Plant Protection: A Color Handbook.* CRC Press.

Jalali, S. K., Mohanraj, P., & Lakshmi, B. L. (2016). Trichogrammatids. In *Ecofriendly Pest Management for Food Security* (pp. 139–181). Academic Press.

Jankielsohn, A. (2023). Sustaining insect biodiversity in agricultural systems to ensure future food security. *Frontiers in Conservation Science, 4*, 1195512.

Joshi, M. D., Srivastava, A. K., Ashaq, M., Jaggi, S., Gupta, P., Hasan, W., & Gupta, S. (2024). Biocontrol agents and plant protection. *Uttar Pradesh Journal of Zoology, 45*(16), 109–131.

Kamali, S., Karimi, J., & Koppenhöfer, A. M. (2018). New insight into the management of the tomato leaf miner, *Tuta absoluta* (*Lepidoptera: Gelechiidae*) with entomopathogenic nematodes. *Journal of Economic Entomology, 111*(1), 112–119.

Keswani, C., Bisen, K., Singh, V., Sarma, B. K., & Singh, H. B. (2016). Formulation technology of biocontrol agents: Present status and future prospects. In *Bioformulations: For Sustainable Agriculture* (pp. 35–52).

Knapp, S., & van der Heijden, M. G. (2018). A global meta-analysis of yield stability in organic and conservation agriculture. *Nature Communications, 9*(1), 3632.

Lazarevic, N., Smurthwaite, K. S., Batterham, P. J., Lane, J., Trevenar, S. M., D'Este, C., & Kirk, M. D. (2023). Psychological distress in three Australian communities living with environmental per- and polyfluoroalkyl substances contamination. *Science of the Total Environment, 874*, 162503.

Lenteren, J. V. (1988). Biological and integrated pest control in greenhouses. *Annual Review of Entomology, 33*, 239–269.

Mahr, D. L., Whitaker, P., & Ridgway, N. M. (2008). *Biological Control of Insects and Mites: An Introduction to Beneficial Natural Enemies and Their Use in Pest Management.* University of Wisconsin Cooperative Extension.

McMurtry, B. (2015). *An Economic Analysis of High-Intensity, Short-Duration Grazing Systems in South Dakota and Nebraska.* Paper presented at the 2015 Agricultural and Applied Economics Association Annual Meeting, San Francisco, CA.

Messing, R., & Rabasse, J. M. (1995). Oviposition behavior of the polyphagous aphid parasitoid *Aphidius colemani* Viereck (*Hymenoptera: Aphidiidae*). *Agriculture, Ecosystems & Environment, 52*, 13–17.

Miller, L. K., & Ball, L. A. (1998). *The Insect Viruses*. Springer Science & Business Media.

Montesinos, E. (2003). Development, registration, and commercialization of microbial pesticides for plant protection. *International Journal of Microbiology, 6*, 245–252.

Moscardi, F. (1999). Assessment of the application of baculoviruses for the control of *Lepidoptera*. *Annual Review of Entomology, 44*, 257–289.

Polaszek, A., & Vilhemsen, L. (2023). Biodiversity of hymenopteran parasitoids. *Current Opinion in Insect Science, 56*, 101026.

Polaszek, A., Hernández-Suárez, E., Kresslein, R. L., Hanson, P., Linton, Y. M., MacKenzie-Dodds, J., & Schmidt, S. (2023). A revision of the *Encarsia mexicana* species-group (= *Dirphys* Howard) (*Hymenoptera: Aphelinidae*), gregarious endoparasitoids of whiteflies (*Hemiptera: Aleyrodidae*) in the Neotropical region. *Insects, 14*(6), 570.

Powell, K. A., & Jutsum, A. R. (1993). Technical and commercial aspects of biocontrol products. *Pesticide Science, 37*(4), 315–321.

Ragsdale, D. W., Landis, D. A., Brodeur, J., & Heimpel, G. E. (2004). Ecology and management of the soybean aphid in North America. *Annual Review of Entomology, 49*, 207–233.

Ruiu, L. (2015). Insect pathogenic bacteria in integrated pest management. *Insects, 6*(2), 352–367.

Sahito, Z. A., Zehra, A., Tang, L., Ali, Z., Bano, N., Ullah, M. A., & Yang, X. (2021). Arsenic and mercury uptake and accumulation in oilseed sunflower accessions selected to mitigate co-contaminated soil coupled with oil and bioenergy production. *Journal of Cleaner Production, 291*, 125226.

Sarfraz, M., Keddie, A. B., & Dosdall, L. M. (2005). Biological control of the diamondback moth, *Plutella xylostella*: A review. *Biocontrol Science and Technology, 15*(8), 763–789.

Schoonhoven, L. M., Van Loon, J. J., & Dicke, M. (2005). *Insect-Plant Biology*. Oxford University Press.

Sharma, A., Sandhi, R. K., & Reddy, G. V. (2019). A review of interactions between insect biological control agents and semiochemicals. *Insects, 10*(12), 439.

Shelton, A. M., Olmstead, D. L., Burkness, E. C., Hutchison, W. D., Dively, G., Welty, C., & Sparks, A. N. (2013). Multi-state trials of Bt sweet corn varieties for control of the corn earworm (*Lepidoptera: Noctuidae*). *Journal of Economic Entomology, 106*(5), 2151–2159.

Slipinski, A., & Tomaszewska, W. (2010). Revision of the family *Cavognathidae* (*Coleoptera: Cucujoidea*). *Australian Journal of Entomology, 49*(3), 256–267.

Starý, P. (1975). *Aphidius colemani* Viereck: Its taxonomy, distribution, and host range (*Hymenoptera, Aphidiidae*). *Acta Entomologica Bohemoslovaca, 72*, 156–163.

Ullah, M. I., Arshad, M., Ali, S., Iftikhar, Y., & Mahmood, S. U. (2016). Effect of thiamethoxam and some botanical extracts on *Cotesia glomerata* L. (*Braconidae: Hymenoptera*): An endoparasitoid of *Pieris brassicae* (L.) (*Pieridae: Lepidoptera*). *Egyptian Journal of Biological Pest Control, 26*(3), 545.

van Lenteren, J. V. (2007). Biological control for insect pests in greenhouses: An unexpected success. In *Biological Control: A Global Perspective* (pp. 105–117). CAB International, Wallingford.

Wilson, M., Nitzsche, P., & Shearer, P. W. (1999). Entomopathogenic nematodes to control black vine weevil (Coleoptera: Curculionidae) on strawberry. *Journal of Economic Entomology, 92*(3), 651–657.

CHAPTER 12

Battle Between Underground Insects and Plants

MUBASHER AHMAD MALIK,[1,*] JORGE POVEDA,[2] JAVAD KARIMI,[3] and SERGIO RASMANN[4]

[1]*Department of Entomology, College of Agriculture, University of Layyah, Layyah, Pakistan*

[2]*Recognized Research Group Agrobiotech, UIC-370 (JCyL), Department of Plant Production and Forest Resources, Higher Technical School of Agricultural Engineering of Palencia, University Institute for Research in Sustainable Forest Management (iuFOR), University of Valladolid, Palencia, Spain*

[3]*Department of Plant Protection, School of Agriculture, Ferdowsi University of Mashhad, Mashhad, Iran*

[4]*Functional Ecology Laboratory, Institute of Biology, University of Neuchâtel, Neuchâtel, Switzerland*

*Corresponding author

ABSTRACT

This chapter investigates the relationship between soil-dwelling insect herbivores and plants. Root-feeding insects, which include highly destructive pests, present substantial threats to plant health. In response, plant roots have developed a variety of adaptive strategies to counter root herbivory, such as induced defenses and tolerance mechanisms. These root-based systemic responses can extend to the shoots, influencing aboveground herbivores and higher trophic levels. Tolerance, a physiological response

Insect-Plant Interactions, Volume 1: Influence of Biotic and Abiotic Factors.
Jamin Ali, Adil Tonğa, Adnan Shakeel, and Nasheeman Ashraf (Eds.)

DOI: 10.1201/9781779644282-12

that mitigates the impact of herbivory, is shaped by the surrounding biotic and abiotic conditions. Insects, conversely, have evolved detoxification mechanisms to neutralize and even sequester plant chemicals. The complex relationship between plants and root herbivores is influenced by numerous factors, including evolutionary adaptations, climate change, and human agricultural practices. Ultimately, there is a critical need to integrate multiple control strategies—such as biological control, genetically modified crops, and resilient plant varieties—into a comprehensive and scientifically robust integrated pest management (IPM) framework to effectively manage soil-dwelling insects.

12.1 INTRODUCTION

Root herbivores have been shown to play crucial ecological roles, such as influencing plant abundance, species diversity, and processes of succession in natural vegetation (van der Putten, 2003). Indeed, while root tissues are considered inferior food quality for heterotrophic organisms, as they have a higher carbon to nitrogen ratio than shoots (Hol et al., 2003; Hunter, 2001; Tjoelker et al., 2005), a large diversity of organisms have been observed to feed partially or exclusively on roots (Whittaker, 2003). Therefore, the natural selection pressure exerted by belowground herbivores can be expected to have contributed to the evolution of root defenses in the same way that aboveground herbivores have selected for the large diversity of chemical defenses and defense levels in shoots (Macel et al., 2004; van der Meijden, 1996). Yet, still to date, despite the substantial impact on plant survival and fitness, the interactions between plants and belowground herbivores, as well as the consequences for the evolution of chemical plant defenses, remain understudied when compared to the broad range of studies on the interactions between plants and the herbivores that feed on leaves, stems, and flowers (Hunter, 2001). This chapter aims to explore the breadth of literature that has focused on root herbivore-plant interactions, both from a general mechanistic understanding of these interactions, as well as from a broader ecological, evolutionary, and applied perspective. While the aim is not to be fully comprehensive, our goal is to highlight key fields that await further exploration.

Insect herbivores represent most of the animal diversity damaging plants, both in natural and agricultural ecosystems (Meehan et al., 2011).

Modern intensified agricultural landscapes, specifically due to their ecological simplification and genetic oversimplification (Vagge et al., 2024), are particularly prone to being attacked by insect pests (Bernal and Medina, 2018). Indeed, insect damage accounts for more than 15% of annual crop losses worldwide (Schoonhoven, 2007). In this regard, it is important to consider that below-ground herbivores account for a large fraction of crop losses worldwide, feeding on multiple underground parts of the plants, such as roots, rhizomes, and storage organs (Hunter, 2001; Koppenhofer, 2000). For instance, the root-feeding aphid species *Daktulosphaira vitifoliae,* commonly known as grape phylloxera, almost destroyed the entire European wine industry in the late 19th century (Johnson et al., 2016). Root pests such as wireworms (Coleoptera: Elateridae) feed on a wide range of crops, including cereals, potatoes, carrots, sugar beets, and fruit orchards (Barsics, 2013). Annual losses due to the western corn rootworm (*Diabrotica virgifera*) in Europe and the USA could be much greater than $1 billion annually (Gray, 2009). In the southern hemisphere, the damage caused by the greyback canegrubs (*Dermolepida albohirtum*) can cost more than $10 million to sugarcane producers (Chandler, 2002). Despite the economic importance of roots and the impact of several root-feeding insects, research on root herbivores and how plants defend against them has always lagged behind research on aboveground plant-insect interactions (van Dam, 2009), mostly because of the technical challenges associated with working underground (Hunter, 2001). Yet, it has been estimated that about 17% of all 257 families of insects contain species that feed on the roots of plants (Rasmann and Agrawal, 2008). Therefore, on a broader level, insect root herbivory should be accounted for when trying to explain population and community-level dynamics in natural systems (van der Putten, 2009).

12.2 DEFENSE MECHANISMS AGAINST ROOT FEEDING INSECTS

Upon insect attack, inducible defense could be a cost-effective strategy to protect plants. In the absence of insect attack, plants can instead allocate their resources for growth and competition with their neighbors (van Dam and Baldwin, 2001). In addition, theory suggests that if the chance of attack is high, then constitutive defenses should be considered more advantageous than inducible defenses (Agrawal and Karban, 1999). Roots play an important role in absorbing water and nutrients from the soil, which are essential for plant growth and survival. It is reasonable to assume that infestation by root herbivores can generate substantial detrimental effects on plant fitness (Kaplan et al., 2008b). Because of their economic and ecological importance as pests, there has been substantial interest in identifying how plants can defend themselves against root pests. While aboveground, long-standing literature has shown that plants have evolved sophisticated defense mechanisms against insect pests, including physical structures like trichomes or spines, and chemical defenses, like toxins or chemical deterrents (War et al., 2018), belowground this remains to be fully addressed. For instance, it is still not clear how physical defenses act on root herbivores. Recent evidence indicates, for instance, that roots can also exude latex (plants release a milky fluid when they undergo a mechanical injury caused by various biological aggressors (Konno, 2011)), which has been shown to reduce the impact of root-feeding beetles in diverse systems such as the *Asclepias* (Rasmann et al., 2009), in dogbanes (*Apocynum* spp.) (Rasmann, 2014), or in dandelions (*Taraxacum* spp.) (Huber et al., 2016). It is also likely that root tissues, rich in lignin for example, can constitute natural barriers against insect attack; however, to our knowledge, we don't yet have full evidence of how root physical structures can inhibit herbivory. One interesting approach in this regard was to show that root architecture per se can influence the performance of the western corn rootworm larvae (*Diabrotica virgifera*) (Schumann et al., 2018). Now, whether such root architecture effects can also be mediated by changes in root structural defenses needs to be further researched.

Moreover, as the aboveground organs, roots can also defend themselves using chemical defenses, which can be produced at all times (i.e., constitutive defenses) or only induced upon herbivory (i.e., inducible defenses) (Touw and Vandam, 2023). Roots do not display constitutive chemical defenses stronger than those of shoots. The concentration and profile of constitutive chemical defenses differ among plant families, species, and genotypes (Agerbirk et al., 2003; Collantes et al., 1999; Hol et al., 2004), the age of the plant (Frolich et al., 2006; Hara et al., 2000), and the developmental stage of the tissues (Ohnmeiss and Baldwin, 2000). In *Cynoglossum officinale* (Boraginaceae), younger leaves contain 190 times higher concentrations of pyrrolizidine alkaloids compared to older senescent leaves (van Dam et al., 1995). Likewise, young leaves of *Arabidopsis thaliana* (Brassicaceae) contain roughly three times higher glucosinolate

(GSL) levels than old leaves (Hunziker et al., 2021). The molecular mechanisms of inducible chemical defense in roots are unclear so far. However, upon root herbivory attack, significant changes in root (GSL) profiles of *Brassica* spp. occurred. The concentration of GSL increases regardless of the location of herbivores feeding in the root system (Tsunoda et al., 2018). In *Brassica rapa*, indole GSL was four times higher in the induction site compared to the undamaged part. The production of indole GSLs is related to GTRs and biosynthesis genes (Touw et al., 2020). Different well-known secondary metabolites having defensive properties, such as alkaloids, GSLs, phenolics, terpenoids, furanocoumarins, and cardenolides, have been identified in plants. These compounds confer resistance to root herbivores. For example, when an artificial diet enriched with phenolic ester was fed to the African sweet potato weevil, it decreased the survival of the weevil (Stevenson et al., 2009). Similar to aboveground tissue, roots of plants deploy herbivore-induced plant volatiles (HIPVs) to recruit natural enemies for their own defense (Tonelli et al., 2016; Turlings and Erb, 2018). One well-studied example is the emission of root herbivore-induced plant volatiles (RHIPVs) of maize, which increases the performance of *Diabrotica virgifera* and recruits entomopathogenic nematodes (Rasmann et al., 2005). Sometimes, secondary metabolites reduce the performance of herbivores' natural enemies because herbivores selectively accumulate and sequester these metabolites for their own defense (Erb et al., 2019). Engineering plant defense resistance to natural enemies of herbivores is a promising approach. For instance, when endosymbiotic bacteria of *Photorhabdus luminescens* were isolated and exposed to benzoxazinoids for 25 days, it resulted in the evolution of benzoxazinoid resistance. Bringing these bacteria back into the nematodes resulted in improved benzoxazinoid-dependent predation success (Machado et al., 2020). Hence, roots have also evolved a range of defenses against root-feeding pests, both direct and indirect, and both constitutive and inducible defenses (Erb et al., 2008; Rasmann and Agrawal, 2008).

12.3 PHYTOHORMONES REGULATING PLANT DEFENSES

The expression of defenses in plants is intricately controlled by a complex network of signaling pathways, where plant hormones play a crucial role. These signaling cascades and the interactions between various plant hormones, known as cross-talk, determine how plants respond to different stresses, whether they are biotic (such as attacks by pathogens or herbivores) or abiotic (like drought or nutrient deficiency). The integration of signals from hormones like salicylic acid (SA), jasmonic acid (JA), and ethylene (ET) allows plants to finely tune their defensive responses, optimizing their survival and fitness in a constantly changing environment (Pieterse et al., 2009). Such dynamic regulation enables plants to balance growth and defense, ensuring that resources are allocated efficiently based on the specific type and intensity of stress encountered.

Induction of the JA pathway in root herbivory tissue is different from that of the above-ground-induced JA pathway; however, they do not induce the SA pathway (Acosta et al., 2013; Erb et al., 2012). The application of JA and MeJA reduced rice water weevil (*Lissorhoptrus oryzophilus*) oviposition and larval numbers feeding on rice (*Oryza sativa*) (Hamm et al., 2010), pupation rates of cabbage fly (*Delia radicum*) larvae feeding on broccoli (*Brassica oleracea* ssp. Italic) (Pierre et al., 2012), numbers of phylloxera (*Daktulosphaira vitifoliae*) eggs on grape (*Vitis vinifera*) vine roots, and reduced feeding of root-feeding chafer (*Dermolepida albohirtum*) on *Microlaena stipoides*. JA and MeJA were found to induce a variety of root defenses, including sesquiterpenes in maize (*Zea mays*) (Erb et al., 2011). It was shown that JA-deficient antisense 13-lipoxygenase (asLOX) and mutant allene oxide cyclase Habiba plants of rice lost more root biomass under the attack of cucumber beetles and rice water weevils. Shoot herbivores induce the production of phytohormones, which reduce the fitness of root herbivores. For example, the performance of cabbage root fly larvae (*Delia radicum*) was reduced on cabbage plants (*Brassica oleracea*) when previously exposed to leaf-chewing caterpillars (Karssemeijer et al., 2020, 2024). Similar to shoots, the JA pathway in plants can be induced in roots to a lower extent following root herbivore attack (Lu et al., 2016). After root herbivore attack, JA signaling can be enhanced in root organs by priming seeds or foliar application with JA, SA, or aminobutyric acid (BABA) (Kesseler and Mueller, 2024; Rasmann et al., 2012). For example, it has been shown that the belowground signaling molecule (−)-Loliolide induces JA signaling in diverse plant species in response to *Holotrichia oblita* Faldermann (Li et al., 2022). Artificial priming induces two types of responses in plants against root herbivore attacks: systemic acquired resistance (SAR) and induced systemic resistance (ISR). ISR is an important defense mechanism in plants that produces resistance against biotrophic pathogens. In contrast, plants activate ISR in response to biotic and abiotic factors, which further activate the JA and SA signaling pathways. ISR can be artificially induced by certain chemicals (e.g., acibenzolar-S-methyl, ASM).

In addition to producing chemical and physical defenses, plants can also use tolerance strategies to mitigate the adverse effects of belowground herbivory (Núñez-Farfán et al., 2007). Tolerance is the extent to which a unit increase in damage reduces fitness (Simms, 2000). Initially, it was argued that both resistance and tolerance mechanisms are alternate strategies (Núñez-Farfán et al., 2007), but later experiments have shown that these two modes of resistance can simultaneously occur within one plant. In the lab, after removing the artificially induced mechanism of resistance, different root classes have been observed in beans (Johnson et al., 2016). Despite the excision of a root class, the remaining root classes were able to compensate, mitigating the impact on shoot biomass. There were significant differences in root classes to overcome root loss. For example, when primary roots were removed, the plant could not compensate for the root removal (Rubio and Lynch, 2007). The common dandelion (*Taraxacum officinale*) is generally attacked by the larvae of *Melolontha melolontha* carabid beetles. Yet, *T. officinale* plants have been shown to be able to compensate for tissue loss through induced regrowth (Herve and Erb, 2019). The sustainability of the tolerance strategy against root herbivory may depend on the specific environmental conditions and ecological context. For instance, root losses due to herbivory are compensated for in the presence of high nitrogen and low phosphorus (Steinger and Muller-Schärer, 1992). Another strategy to adopt is to escape. Roots exhibit remarkable plasticity, allowing them to grow towards favorable niches, such as areas rich in nutrients. Roots also possess the physiological capability to redirect their growth in response to chemical signals released by soil organisms, including pathogens. Roots can grow up to two centimeters per day in order to escape from small enemies that attempt to invade at the root tip (Farrar and Jones, 2003).

12.4 PLANT–MICROBE–INSECT INTERACTIONS

Plants and their roots are almost ubiquitously associated with soil-borne microbes, including bacteria, fungi, and nematodes (Hartmann, 2009).

Among soil-dwelling microorganisms, arbuscular mycorrhizal fungi (AMF) can indirectly affect the performance of underground insects by modulating the defense chemistry of plants (Pozo and Azcón-Aguilar, 2007; Vannette and Hunter, 2009). Belowground herbivores can affect plant growth and nutrient uptake by modifying AMF-plant symbioses (Frew, 2022). Firstly, colonization of plant roots by AMF seems to enhance direct plant defenses, such as the production of induced secondary metabolites that deter herbivores (Borowicz, 2001). Frew suggested that specific genera of AMF may help agricultural crops tolerate root-feeding insects (NG, 2023). In another study, inoculation of rice plants with AMF enhanced the susceptibility of rice plants to both insect herbivores and pathogens, without significantly altering nutrient concentrations (Bernaola et al., 2018). Secondly, the volatiles produced by plants in combination with AMF can promote indirect plant defenses (i.e., the attraction of natural enemies of the herbivore) (Yu et al., 2022). Belowground herbivory by striped cucumber beetle larvae induces the systemic emission of HIPVs, specifically the volatile compound (E)-β-ocimene, which primes neighboring plants to better defend against squash bugs, demonstrating a novel form of cross-compartment volatile signaling between plants (Thompson et al., 2023).

12.5 PLANT MEDIATED ABOVE GROUND (AG) AND BELOW GROUND (BG) INTERACTIONS

In nature, plants provide shelter to many herbivores simultaneously, leading to competition for resource acquisition (Denno et al., 1995). However, direct competition does not apply to below- and above-ground insect herbivores. Two mechanisms, such as variation in primary and secondary metabolism, drive the plant-mediated AG–BG interactions, which are associated with the "stress response hypothesis" (i.e., predominantly driven by plant primary metabolism) and the "defense induction hypothesis" (i.e., predominantly driven by plant secondary metabolism) (Bezemer et al., 2002). The effect of root herbivores on aboveground herbivores can be positive or negative. Regarding alterations in primary metabolites, it has been widely believed that the attack of root insects on BG tissue leads to root-herbivore-induced drought stress (Biere and Goverse, 2016; Johnson et al., 2012; Papadopoulou and van Dam, 2017). Consequently, drought stress increases water and nutrient levels in the leaves, thereby enhancing the performance of foliar-sucking insects such as aphids. In another study, the grub of root feeders (*Otiorhynchus sulcatus* beetle larvae) improved the performance of phloem feeders, an effect influenced by changes in carbon, nitrogen, and phosphorus (Johnson et al., 2013). Leaf volatiles induced by the root-feeding larvae attracted above-ground conspecific adults in laboratory and field experiments (Sun et al., 2019). In contrast, for instance, the larvae of the western corn rootworm (*Diabrotica virgifera*) attacking corn plants (*Zea mays*) are responsible for the activation of genes related to abscisic acid in leaves, thereby increasing both plant defense and plant priming and limiting the development of AG chewers such as *Spodoptera littoralis* caterpillars. It has been shown that AG chewer armyworm (*S. littoralis*) and cabbage butterfly (Pieris brassicae) can affect root biomass or root mineral content (e.g., C, S), consequently benefiting root chewer (*Agriotes* spp.) performance. Several studies have shown that the abundance or performance of aboveground natural enemies depends on belowground insects in various systems. Concerning abundance, there is an increase or negative effect on AG natural enemies following root herbivore attack (Johnson et al., 2013; White and Andow, 2006). For instance, root feeding by *Phyllopertha horticola* or *O. sulcatus* led to an increase in *Terellia ruficauda* or *N. olfaciens*, which subsequently increased the population of natural enemies (*Torymus chloromerus, Aphidius spp*., respectively). In the second case, the negative correlation between root herbivory and the abundance of AG parasitoids (*Macrocentrus grandii*) regulating the European corn borer (*Ostrinia nubilalis*) relied on plant-habitat modification by BG herbivores (*Diabrotica virgifera*). As a result, parasitoids tend to avoid the volatile compounds emitted by plants damaged by root herbivores (Rasmann and Turlings, 2007). Regarding changes in primary and secondary plant metabolites, Soler et al. (2005) conducted an experiment in which root herbivory negatively affected the development time of the leaf feeder (*P. brassicae*) and significantly increased the parasitoid (*Cotesia glomerata*). Conversely, root herbivores can have positive effects on beneficial above-ground insects. For instance, feeding by click beetles on *Sinapis arvensis* plants has been shown to attract more pollinators (Poveda et al., 2003). On the other hand, only one example addressed the effect of AG insect herbivores on BG natural enemies and tri-trophic interactions. For example, systemic changes and accumulation of toxic GSLs in the roots of *B. nigra* in response to shoot-feeding herbivores (*P. brassicae*) lead to a reduction in the performance of both root feeders (*D. radicum*) and their natural enemies, such as *Trybliographa rapae* parasitoid wasps (Soler et al., 2007). For instance, biotic and abiotic factors can differ dramatically along elevational gradients. At low elevation, biotic stress is expected to be maximal, while at high elevation, abiotic stress tends to be at its peak. This results in a decline in nutrient availability and aboveground herbivore pressure, but an increase in root colonization by herbivorous nematodes. At mid-elevation, the strength of AG–BG interactions is high, whereas it is low at high and low elevations. Therefore, the AG-BG constitutive state versus inducibility should vary along elevation in a predictable manner (Kergunteuil et al., 2018) (Figure 12.1).

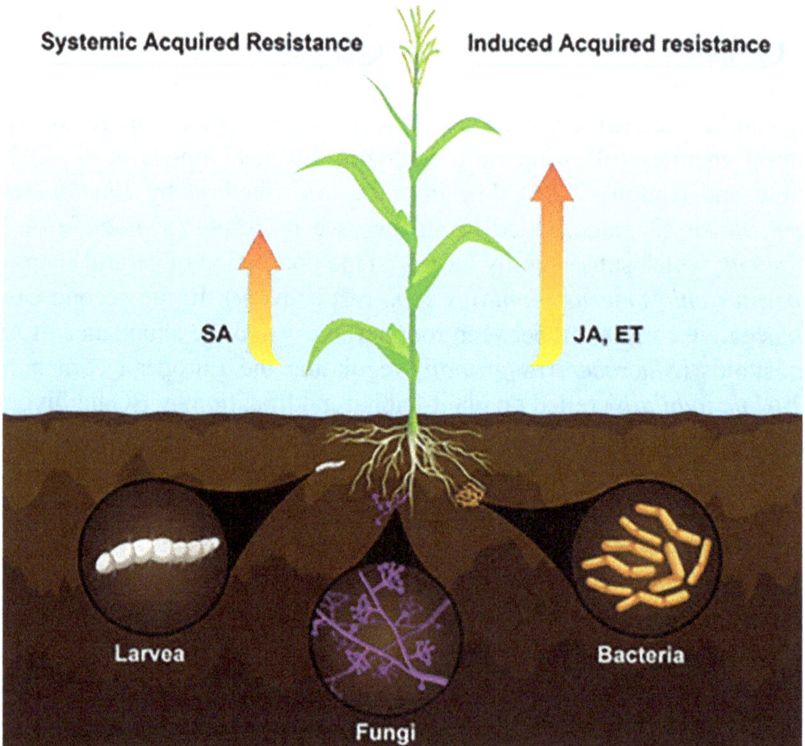

FIGURE 12.1 Systemic acquired resistance (SAR) is induced by biotrophic pathogen like fungi (3); induced systemic resistance (ISR) is induced by biotrophic and abiotic factor 1. insect; 2. fungi, 3. bacteria; SA, salicylic acid; JA, jasmonic acid; ET, ethylene.

12.6 INSECT COUNTER ADAPTATION AND STRATEGIES

Insects cope with plant defenses through counter-adaptations that can be genetically determined or as a result of behavioral plasticity (Fox et al., 2004). Insects have adapted external and internal strategies to overcome plant defenses (Despres et al., 2007). Upon ingestion of host tissues by an insect, it can excrete, sequester, or detoxify chemical defenses internally. Such post-ingestive mechanisms and enzymes in the insect midgut, such as cytochromes P450 and glutathione S-transferases, are well studied in relation to chemical defenses (Despres et al., 2007; Enyati et al., 2005; Feyereisen, 2006). On the other hand, insect counter-adaptations to physical defense traits are not well studied.

12.6.1 INTERNAL STRATEGIES OF ROOT FEEDING INSECTS TO COPE WITH PLANT DEFENSES

Insects use internal counter-adaptations to circumvent the toxic effects of specialized plant metabolites. These internal counter-adaptations against plant metabolites include tasting, avoidance of toxic excretion, detoxification, and sequestration. Sequestration is the ability of insects to accumulate plant defensive compounds to avoid the attack of predators (Nishida, 2002; Petschenka and Agrawal, 2016). In plants, precursors (so-called protoxins) and the enzymes that activate them are stored in different compartments. Toxic substances are formed when plant tissues are damaged, and their cells rupture, causing protoxins and activating enzymes to come together (Hopkins et al., 2009; Wouters et al., 2016). Sequestration is best studied in corn rootworm (*D. virgifera*), where western corn rootworm sequesters plant protoxins and produces its own activating enzymes to release the toxins, thereby protecting itself against natural enemies (Robert et al., 2017). In another study, larvae of banded cucumber beetles (*Diabrotica balteata*) sequestered and metabolized cucurbitacins, which reduced the efficacy of biological control agents (Bruo et al., 2023). On the other hand, sucking insects, like aphids, inject proteins into plants that neutralize and limit their defense responses, making the plant more prone to colonization (Guo et al., 2023). The production of secondary metabolites in plants is often characterized as a means of producing toxins and deterrents to protect against insect herbivores; however, there is a wealth of literature describing insects feeding on plants that have abundant specialized metabolites without suffering negative effects (Kshatriya and Gershenzon, 2024). Excretion is the simple elimination of ingested toxic plant metabolites from the insect gut with the feces (Zagrobelny et al., 2004). Insects that feed on diverse chemical defenses have the ability to excrete potentially toxic metabolites. Detoxification refers to biochemical processes that eliminate the toxic compounds that have been ingested. Insects have developed mechanisms of detoxification to defend against the chemical defenses of plants that include oxidation, hydrolysis, or reduction, as well as conjugation to molecules that can be excreted with feces (Despres et al., 2007). Different enzymes involved in the detoxification of plant metabolites are cytochrome P450 monooxygenases (CYP450s), glutathione S-transferases, and carboxylesterases. So far, several studies explain the mechanisms of excretion in aboveground herbivores, where they developed physiological strategies to overcome the glucosinolate-isothiocyanate (GSL-ITC) plant defenses (Crespo et al., 2012), but how belowground herbivores deal with these toxic secondary compounds remains unclear. Sontowski et al. (2022) investigated two underground Brassica specialists, *D. radicum* (Linnaeus) and *D. floralis* (Fallén) (Diptera: Anthomyiidae), to analyze the way they detoxify the GSL-ITC defenses of their host plants. Both species detoxify ITCs to ITC-cysteine conjugates and amines using glutathione S-transferase (GST) and a hydrolytic pathway, respectively, with different efficacies and substrate processes, where performance assays confirmed that ITCs negatively affected the performance of both species. Agrawal and Husting (2023) demonstrate that the four-eyed milkweed beetle *Tetraopas tetrophthalmus* tolerates cardenolide toxins secreted by milkweed *Asclepias syriaca* L through specific amino acid substitutions in its Na+/K+–ATPase. The same detoxification mechanisms could be related to the fact that the cabbage root fly *D. radicum* prefers Chinese cabbage over broccoli (Lamy et al., 2018).

Insects' different counter-adaptations to plant protein inhibitors (PIs) include the production of protease-insensitive isomers (Zhu-Salzman and Zeng, 2015), overproduction of PIs (De Leo et al., 1998), and hydroxylation and detoxification of PIs by alternative proteases (Ahn and Zhu-Salzman, 2009). *Agrotis ipsilon* produces proteases insensitive to soybean Kunitz trypsin inhibitor (Volpicella et al., 2003).

12.6.2 EXTERNAL STRATEGIES OF BELOW GROUND INSECT TO COPE WITH PLANT DEFENSES

Before ingestion, behavioral mechanisms are adopted to overcome plant defense traits. In behavioral mechanisms, insects partially or completely ignore the negative impacts of plant defenses. Such avoidance strategies include timing (phenology) or feeding on tissues that are less well defended. Some root herbivores feed on less defended parts of the plant by avoiding or ingesting high concentrations of plant toxins, for example. The clover root weevil (CRW) *Sitona obsoletus* exhibited low feeding on white clover (*Trifolium repens*) due to lignified and toxic tap roots (McNeil et al., 2016). Some root-feeding insects have evolved specialized mouthparts and digestive systems to cope with tough, lignified plant tissues and sequestered plant toxins. For example, the cotton root aphid has stylet-like mouthparts that allow it to feed on the vascular tissues of cotton plants without ingesting high concentrations of gossypol (Brożek et al., 2015).

12.7 IMPLICATION OF HIDDEN BATTLEFIELD

Plant health and yield could be enhanced by the interaction of root-feeding insects and plant roots. Root-feeding larvae can inject proteins into the growing roots of plants that may stop the plants from inducing defense mechanisms, thereby allowing for easier feeding and colonization (REF). These intricate relationships highlight how plants modify their defense mechanisms to counter underground insects (Zhou et al., 2023). The role of mycorrhizal fungi should not be overlooked in the underground battle because fungi either protect plants by enhancing plant defense or provide food to insects. Overall, the hidden battlefield beneath the soil significantly impacts plant health, agricultural productivity, and the evolutionary trajectories of both plants and their insect herbivores, emphasizing the need for further research into these subterranean interactions. Plants produce chemical defenses in response to herbivore attacks. These chemicals have medicinal potential, waiting to be discovered by researchers. By studying these chemicals, researchers may discover new drugs for treating human diseases. Additionally, studying the complex mechanisms and survival tactics of these subterranean insects could lead to innovative solutions in fields like agriculture, where farmers must tackle these destructive pests. Understanding how insects overcome plant defenses and use their roots for feeding will be helpful in developing environmentally friendly pest control methods. Perhaps most significantly, the existence of this unseen war underscores the incredible complexity and dynamism of the natural world, even in the most unassuming of environments. By shedding light on the hidden battles taking place beneath our feet, we gain a deeper appreciation for the delicate balance that sustains all life on our planet. The safeguarding of subterranean ecosystems plays an important role in the delicate balance of our soil ecosystem.

12.8 FUTURE CHALLENGES AND RESEARCH DIRECTION

The future challenges in the ongoing battle between underground insects and plants are numerous and multifaceted. The intricate relationships between underground insects and plants are important for maintaining good ecological conditions, food security, and agricultural sustainability. Soil-dwelling insects can damage crops, thereby decreasing crop yield, which can lead to increased vulnerability to diseases. Understanding the complex relationships between insects and plants is crucial for developing effective pest management strategies while minimizing damage to aboveground beneficial insects and the surrounding environment. The relationship of plants is shaped by several factors, including evolutionary adaptations, climate change, and human agricultural practices. Plants have developed various responses to the attack of root herbivores, such as changes in root architecture and the production of secondary metabolites. Further research is needed to clarify these mechanisms and how they can be harnessed for plant protection. Climate change is another important factor that affects the relationships between insects and their plant hosts. High temperatures increase the insect reproductive rate, which puts pest pressure on crops. Therefore, further research is needed to explore in depth how these changes are impacting agriculture and insect pest management.

Defenses lead to a complex evolutionary arms race. Many insect species are specialized to exploit a limited range of host plants, which allows them to maximize their feeding efficiency. However, some insects have evolved the ability to attack multiple plant species, posing significant threats to agricultural diversity and crop yields. Human agricultural practices have always prioritized crop yield over plant defense. As a result, many field crops have reduced their chemical defenses. For example, through breeding, the removal of GSLs, which is an important plant defense compound, has become a most undesirable trait for customers. The removal of plant defenses not only increases the attack of insects but also increases reliance on pesticide use.

Instead of relying on a single solution, there is a need to combine multidisciplinary methods, including ecology, molecular biology, and soil science, in order to understand the intricate relationship between plants, insects, and their shared environment. By using cutting-edge technologies like whole-genome sequencing and metabolomics, we can gain insight into the chemical and genetic bases of plant defense against root insects. These molecular tools can be helpful in identifying the genes associated with resistant traits, combining different control strategies such as biological control, genetically modified crops, and environmentally resistant varieties into a scientifically sound IPM approach to combat soil insects. Uncovering the role of soil microbes with plants and insects will help in exploring new strategies that enhance plant health.

KEYWORDS

- **benzoxazinoid resistance**
- **counteradaptation**
- **detoxification**
- **grape phylloxera**
- **pest management**
- **plant defenses**
- **root herbivores**
- **sequestration**
- **soil-dwelling insects**
- **systemic acquired resistance**

REFERENCES

Acosta, J. C., Banito, A., Wuestefeld, T., Georgilis, A., Janich, P., Morton, J. P., & Gil, J. (2013). A complex secretory program orchestrated by the inflammasome controls paracrine senescence. *Nature Cell Biology, 15*(8), 978–990.

Agerbirk, N., Olsen, C. E., Bibby, B. M., Frandsen, H. O., Brown, L. D., Nielsen, J. K., & Renwick, J. A. A. (2003). A saponin correlated with variable resistance of *Barbarea vulgaris* to the diamondback moth *Plutella xylostella*. *Journal of Chemical Ecology, 29*, 1417–1433.

Agrawal, A. A., & Hastings, A. P. (2023). Tissue-specific plant toxins and adaptation in a specialist root herbivore. *Proceedings of the National Academy of Sciences, 120*(22), 2302251120.

Agrawal, A. A., & Karban, R. (1999). Why induced defences may be favored over constitutive strategies in plants. *The Ecology and Evolution of Inducible Defences, 10*, 000331745.

Ahn, J. E., & Zhu-Salzman, K. (2009). *CmCatD*, a cathepsin D-like protease has a potential role in insect defence against a phytocystatin. *Journal of Insect Physiology, 55*(8), 678–685.

Barsics, F., Haubruge, E., & Verheggen, F. J. (2013). Wireworms' management: An overview of the existing methods, with particular regards to *Agriotes* spp. (Coleoptera: Elateridae). *Insects, 4*(1), 117–152.

Bernal, J. S., & Medina, R. F. (2018). Agriculture sows' pests: How crop domestication, host shifts, and agricultural intensification can create insect pests from herbivores. *Current Opinion in Insect Science, 26*, 76–81.

Bernaola, L., Cosme, M., Schneider, R. W., & Stout, M. (2018). Belowground inoculation with arbuscular mycorrhizal fungi increases local and systemic susceptibility of rice plants to different pest organisms. *Frontiers in Plant Science, 9*, 747.

Bezemer, T. M., Wagenaar, R., Van Dam, N. M., & Wäckers, F. L. (2003). Interactions between above- and belowground insect herbivores as mediated by the plant defence system. *Oikos, 101*(3), 555–562.

Bezemer, T. M., Wagenaar, R., van Dam, N. M., Van Der Putten, W. H., & Wäckers, F. L. (2004). Above- and belowground terpenoid aldehyde induction in cotton, *Gossypium herbaceum*, following root and leaf injury. *Journal of Chemical Ecology, 30*, 53–67.

Biere, A., & Goverse, A. (2016). Plant-mediated systemic interactions between pathogens, parasitic nematodes, and herbivores above and belowground. *Annual Review of Phytopathology, 54*(1), 499–527.

Borowicz, V. A. (2001). Do arbuscular mycorrhizal fungi alter plant–pathogen relations? *Ecology, 82*(11), 3057–3068.

Brożek, J., Mróz, E., Wylężek, D., Depa, Ł., & Węgierek, P. (2015). The structure of extremely long mouthparts in the aphid genus *Stomaphis* Walker (Hemiptera: Sternorrhyncha: Aphididae). *Zoomorphology, 134*, 431–444.

Bruno, P., Arce, C. C., Machado, R. A., Besomi, G., Spescha, A., Glauser, G., & Turlings, T. C. (2023). Sequestration of cucurbitacins from cucumber plants by *Diabrotica balteata* larvae provides little protection against biological control agents. *Journal of Pest Science, 96*(3), 1061–1075.

Chandler, K. (2002). *Strategies to Control Greyback Canegrub in Early Harvested Ratoon Crops*. Bureau of Sugar Experiment Stations: Queensland, Australia.

Collantes, H. G., Gianoli, E., & Niemeyer, H. M. (1999). Defoliation affects chemical defences in all plant parts of rye seedlings. *Journal of Chemical Ecology, 25*, 491–499.

De Leo, F., Bonadé-Bottino, M. A., Ceci, L. R., Gallerani, R., & Jouanin, L. (1998). Opposite effects on *Spodoptera littoralis* larvae of high expression level of a trypsin proteinase inhibitor in transgenic plants. *Plant Physiology, 118*(3), 997–1004.

Denno, R. F., McClure, M. S., & Ott, J. R. (1995). Interspecific interactions in phytophagous insects: Competition reexamined and resurrected. *Annual Review of Entomology, 40*(1), 297–331.

Després, L., David, J. P., & Gallet, C. (2007). The evolutionary ecology of insect resistance to plant chemicals. *Trends in Ecology & Evolution, 22*(6), 298–307.

Erb, M., & Reymond, P. (2019). Molecular interactions between plants and insect herbivores. *Annual Review of Plant Biology, 70*(1), 527–557.

Erb, M., Meldau, S., & Howe, G. A. (2012). Role of phytohormones in insect-specific plant reactions. *Trends in Plant Science, 17*(5), 250–259.

Erb, M., Robert, C. A., Hibbard, B. E., & Turlings, T. C. (2011). Sequence of arrival determines plant-mediated interactions between herbivores. *Journal of Chemical Ecology, 99*(1), 7–15.

Erb, M., Robert, C. A., Marti, G., Lu, J., Doyen, G. R., Villard, N., & Gershenzon, J. (2015). A physiological and behavioral mechanism for leaf herbivore-induced systemic root resistance. *Plant Physiology, 169*(4), 2884–2894.

Erb, M., Ton, J., Degenhardt, J., & Turlings, T. C. (2008). Interactions between arthropod-induced aboveground and belowground defences in plants. *Plant Physiology, 146*(3), 867–874.

Eyles, A., Jones, W., Riedl, K., Cipollini, D., Schwartz, S., Chan, K., & Bonello, P. (2007). Comparative phloem chemistry of Manchurian (*Fraxinus mandshurica*) and two North American ash species (*Fraxinus americana* and *Fraxinus pennsylvanica*). *Journal of Chemical Ecology, 33*, 1430–1448.

Farrar, J. F., & Jones, D. L. (2003). The control of carbon acquisition by and growth of roots. In *Root Ecology* (pp. 91–124). Berlin, Heidelberg: Springer.

Fornoni, J. (2011). Ecological and evolutionary implications of plant tolerance to herbivory. *Functional Ecology, 25*(2), 399–407.

Fox, C. W., Stillwell, R. C., Amarillos, A. R., Czesak, M. E., & Messina, F. (2004). Genetic architecture of population differences in oviposition behavior of the seed beetle *Callosobruchus maculatus*. *Journal of Evolutionary Biology, 17*(5), 1141–1151.

Frew, A., Antunes, P. M., Cameron, D. D., Hartley, S. E., Johnson, S. N., Rillig, M. C., & Bennett, A. E. (2022). Plant herbivore protection by arbuscular mycorrhizas: A role for fungal diversity? *New Phytologist, 233*(3), 1022–1031.

Frölich, C., Hartmann, T., & Ober, D. (2006). Tissue distribution and biosynthesis of 1,2-saturated pyrrolizidine alkaloids in *Phalaenopsis* hybrids (*Orchidaceae*). *Phytochemistry, 67*(14), 1493–1502.

Gray, M. E., Sappington, T. W., Miller, N. J., Moeser, J., & Bohn, M. O. (2009). Adaptation and invasiveness of western corn rootworm: Intensifying research on a worsening pest. *Annual Review of Entomology, 54*(1), 303–321.

Guo, H., Zhang, Y., Tong, J., Ge, P., Wang, Q., Zhao, Z., & Sun, Y. (2020). An aphid-secreted salivary protease activates plant defence in phloem. *Current Biology, 30*(24), 4826–4836.

Hamm, J. C., Stout, M. J., & Riggio, R. M. (2010). Herbivore- and elicitor-induced resistance in rice to the rice water weevil (*Lissorhoptrus oryzophilus* Kuschel) in the laboratory and field. *Journal of Chemical Ecology, 36*, 192–199.

Hanley, M. E., Lamont, B. B., Fairbanks, M. M., & Rafferty, C. M. (2007). Plant structural traits and their role in anti-herbivore defence. *Perspectives in Plant Ecology, Evolution and Systematics, 8*(4), 157–178.

Hartmann, A., Schmid, M., Tuinen, D. V., & Berg, G. (2009). Plant-driven selection of microbes. *Plant and Soil, 321*(1), 235–257.

Hervé, M. R., & Erb, M. (2019). Distinct defence strategies allow different grassland species to cope with root herbivore attack. *Oecologia, 191*(1), 127–139.

Hol, W. H. G., Vrieling, K., & van Veen, J. A. (2003). Nutrients decrease pyrrolizidine alkaloid concentrations in *Senecio jacobaea*. *New Phytologist, 158*, 175–181.

Hopkins, R. J., van Dam, N. M., & van Loon, J. J. A. (2009). Role of glucosinolates in insect-plant relationships and multitrophic interactions. *Annual Review of Entomology, 54*, 57–83.

Huber, M., Bont, Z., Fricke, J., Brillatz, T., Aziz, Z., Gershenzon, J., & Erb, M. (2016). A below-ground herbivore shapes root defensive chemistry in natural plant populations. *Proceedings of the Royal Society B: Biological Sciences, 283*(1827), 20160285.

Hunter, M. D. (2001). Out of sight, out of mind: The impacts of root-feeding insects in natural and managed systems. *Agricultural and Forest Entomology, 3*, 3–9.

Hunter, M. D., & McNeil, J. N. (1997). Host plant quality influences diapause and voltinism in a polyphagous insect herbivore. *Ecology, 78*(4), 977–986.

Hunziker, P., Lambertz, S. K., Weber, K., Crocoll, C., Halkier, B. A., & Schulz, A. (2021). Herbivore feeding preference corroborates optimal defence theory for specialized metabolites within plants. *Proceedings of the National Academy of Sciences, 118*(47), e2111977118.

Johnson, S. N., & Rasmann, S. (2015). Root-feeding insects and their interactions with organisms in the rhizosphere. *Annual Review of Entomology, 60*(1), 517–535.

Johnson, S. N., Clark, K. E., Hartley, S. E., Jones, T. H., McKenzie, S. W., & Koricheva, J. (2012). Aboveground–belowground herbivore interactions: A meta-analysis. *Ecology, 93*(10), 2208–2215.

Johnson, S. N., Erb, M., & Hartley, S. E. (2016). Roots under attack: Contrasting plant responses to below- and aboveground insect herbivory. *New Phytologist, 210*(2), 413–418.

Johnson, S. N., Mitchell, C., McNicol, J. W., Thompson, J., & Karley, A. J. (2013). Downstairs drivers: Root herbivores shape communities of aboveground herbivores and natural enemies via changes in plant nutrients. *Journal of Animal Ecology, 82*(5), 1021–1030.

Kaplan, I., Halitschke, R., Kessler, A., Rehill, B. J., Sardanelli, S., & Denno, R. F. (2008a). Physiological integration of roots and shoots in plant defence strategies links above- and belowground herbivory. *Ecology Letters, 11*, 841–851.

Kaplan, I., Halitschke, R., Kessler, A., Sardanelli, S., & Denno, R. F. (2008b). Constitutive and induced defences to herbivory in above- and belowground plant tissues. *Ecology, 89*(2), 392–406.

Karssemeijer, P. N., Croijmans, L., Gajendiran, K., Gols, R., van Apeldoorn, D. F., van Loon, J. J., & Poelman, E. H. (2024). Diverse cropping systems lead to higher larval mortality of the cabbage root fly (*Delia radicum*). *Journal of Pest Science, 97*(1), 337–353.

Karssemeijer, P. N., Reichelt, M., Gershenzon, J., van Loon, J., & Dicke, M. (2020). Foliar herbivory by caterpillars and aphids differentially affects phytohormonal signalling in roots and plant defence to a root herbivore. *Plant, Cell & Environment, 43*(3), 775–786.

Kergunteuil, A., Bakhtiari, M., & Rasmann, S. (2018). Eco-evolutionary factors driving plant-mediated above-belowground invertebrate interactions along elevation gradients. In *Aboveground–Belowground Community Ecology* (pp. 223–245).

Kessler, A., & Mueller, M. B. (2024). Induced resistance to herbivory and the intelligent plant. *Plant Signaling & Behavior, 19*(1), 2345985.

Konno, K. (2011). Plant latex and other exudates as plant defence systems: Roles of various defence chemicals and proteins contained therein. *Phytochemistry, 72*(13), 1510–1530.

Koppenhöfer, A. M., & Kaya, H. K. (1997). Additive and synergistic interaction between entomopathogenic nematodes and *Bacillus thuringiensis* for scarab grub control. *Biological Control, 8*, 131–137.

Kshatriya, K., & Gershenzon, J. (2024). Disarming the defences: Insect detoxification of plant defence-related specialized metabolites. *Current Opinion in Plant Biology, 81*, 102577.

Lamy, F., Dugravot, S., Cortesero, A. M., Chaminade, V., Faloya, V., & Poinsot, D. (2018). One more step toward a push-pull strategy combining both a trap crop and plant volatile organic compounds against the cabbage root fly *Delia radicum*. *Environmental Science and Pollution Research, 25*, 29868–29879.

Liu, H., Carvalhais, L. C., Kazan, K., & Schenk, P. M. (2016). Development of marker genes for jasmonic acid signaling in shoots and roots of wheat. *Plant Signaling & Behavior, 11*(5), e1176654.

Macel, M., Vrieling, K., & Klinkhamer, P. G. L. (2004). Variation in pyrrolizidine alkaloid patterns of *Senecio jacobaea*. *Phytochemistry, 65*, 865–873.

Mertens, D., Fernandez de Bobadilla, M., Rusman, Q., Bloem, J., Douma, J. C., & Poelman, E. H. (2021). Plant defence to sequential attack is adapted to prevalent herbivores. *Nature Plants, 7*(10), 1347–1353.

Ng, A., Wilson, B. A., & Frew, A. (2023). Belowground crop responses to root herbivory are associated with the community structure of native arbuscular mycorrhizal fungi. *Applied Soil Ecology, 185*, 104797.

Nishida, R. (2014). Chemical ecology of insect–plant interactions: Ecological significance of plant secondary metabolites. *Bioscience, Biotechnology, and Biochemistry, 78*(1), 1–13.

Ode, P. J. (2006). Plant chemistry and natural enemy fitness: Effects on herbivore and natural enemy interactions. *Annual Review of Entomology, 51*(1), 163–185.

Ohnmeiss, T. E., & Baldwin, I. T. (2000). Optimal defence theory predicts the ontogeny of an induced nicotine defence. *Ecology, 81*(7), 1765–1783.

Papadopoulou, G. V., & van Dam, N. M. (2017). Mechanisms and ecological implications of plant-mediated interactions between belowground and aboveground insect herbivores. *Ecological Research, 32*, 13–26.

Petschenka, G., & Agrawal, A. A. (2016). How herbivores coopt plant defences: Natural selection, specialization, and sequestration. *Current Opinion in Insect Science, 14*, 17–24.

Pierre, S. P., Dugravot, S., Hervé, M. R., Hassan, H. M., Van Dam, N. M., & Cortesero, A. M. (2013). Belowground induction by *Delia radicum* or phytohormones affect aboveground herbivore communities on field-grown broccoli. *Frontiers in Plant Science, 4*, 305.

Pieterse, C. M. J., Leon-Reyes, A., Van der Ent, S., & Van Wees, S. C. M. (2009). Networking by small-molecule hormones in plant immunity. *Nature Chemical Biology, 5*, 308–316.

Poelman, A., Mojet, J., Lyon, D., & Sefa-Dedeh, S. (2008). The influence of information about organic production and fair trade on preferences for and perception of pineapple. *Food Quality and Preference, 19*(1), 114–121.

Poveda, K., Steffan-Dewenter, I., Scheu, S., & Tscharntke, T. (2005a). Effects of decomposers and herbivores on plant performance and aboveground plant-insect interactions. *Oikos, 108*(3), 503–510.

Poveda, K., Steffan-Dewenter, I., Scheu, S., & Tscharntke, T. (2005b). Floral trait expression and plant fitness in response to below- and aboveground plant–animal interactions. *Perspectives in Plant Ecology, Evolution and Systematics, 7*(2), 77–83.

Pozo, M. J., & Azcón-Aguilar, C. (2007). Unraveling mycorrhiza-induced resistance. *Current Opinion in Plant Biology, 10*, 393–398.

Rasmann, S. (2014). Fine-tuning of defences and counter-defences in a specialised plant–herbivore system. *Ecological Entomology, 39*(3), 382–390.

Rasmann, S., & Agrawal, A. A. (2008). In defence of roots: A research agenda for studying plant resistance to belowground herbivory. *Plant Physiology, 146*(3), 875–880.

Rasmann, S., & Turlings, T. C. (2007). Simultaneous feeding by aboveground and belowground herbivores attenuates plant-mediated attraction of their respective natural enemies. *Ecology Letters, 10*(10), 926–936.

Rasmann, S., Agrawal, A. A., Cook, S. C., & Erwin, A. C. (2009). Cardenolides, induced responses, and interactions between above- and belowground herbivores of milkweed (*Asclepias* spp.). *Ecology, 90*(9), 2393–2404.

Rasmann, S., De Vos, M., Casteel, C. L., Tian, D., Halitschke, R., Sun, J. Y., & Jander, G. (2012). Herbivory in the previous generation primes plants for enhanced insect resistance. *Plant Physiology, 158*(2), 854–863.

Rasmann, S., Köllner, T. G., Degenhardt, J., Hiltpold, I., Toepfer, S., Kuhlmann, U., & Turlings, T. C. (2005). Recruitment of entomopathogenic nematodes by insect-damaged maize roots. *Nature, 434*(7034), 732–737.

Robert, C. A., Zhang, X., Machado, R. A., Schirmer, S., Lori, M., Mateo, P., & Gershenzon, J. (2017). Sequestration and activation of plant toxins protect the western corn rootworm from enemies at multiple trophic levels. *eLife, 6*, e29307.

Rubio, G., & Lynch, J. P. (2007). Compensation among root classes in *Phaseolus vulgaris* L. *Plant and Soil, 290*, 307–321.

Schausberger, P., Peneder, S., Jürschik, S., & Hoffmann, D. (2012). Mycorrhiza changes plant volatiles to attract spider mite enemies. *Functional Ecology, 26*(2), 441–449.

Schoonhoven, L. M., van Loon, J. J. A., & Dicke, M. (2005). *Insect-Plant Biology*. Oxford University Press.

Simms, E. L. (2000). Defining tolerance as a norm of reaction. *Evolutionary Ecology, 14*, 563–570.

Soler, R., Bezemer, T. M., Van Der Putten, W. H., Vet, L. E., & Harvey, J. A. (2005). Root herbivore effects on above-ground herbivore, parasitoid, and hyperparasitoid performance via changes in plant quality. *Journal of Animal Ecology, 74*(6), 1121–1130.

Sontowski, R., Guyomar, C., Poeschl, Y., Weinhold, A., van Dam, N. M., & Vassão, D. G. (2022). Mechanisms of isothiocyanate detoxification in larvae of two belowground herbivores, *Delia radicum* and *D. floralis* (Diptera: Anthomyiidae). *Frontiers in Physiology, 13*, 874527.

Stam, J. M., Kroes, A., Li, Y., Gols, R., van Loon, J. J., Poelman, E. H., & Dicke, M. (2014). Plant interactions with multiple insect herbivores: From community to genes. *Annual Review of Plant Biology, 65*(1), 689–713.

Steinger, T., & Müller-Schärer, H. (1992). Physiological and growth responses of *Centaurea maculosa* (Asteraceae) to root herbivory under varying levels of interspecific plant competition and soil nitrogen availability. *Oecologia, 91*, 141–149.

Stevenson, P. C., Muyinza, H., Hall, D. R., Porter, E. A., Farman, D. I., Talwana, H., & Mwanga, R. O. (2009). Chemical basis for resistance in sweetpotato *Ipomoea batatas* to the sweetpotato weevil *Cylas puncticollis*. *Pure and Applied Chemistry, 81*(1), 141–151.

Sun, X., Siemann, E., Liu, Z., Wang, Q., Wang, D., Huang, W., & Ding, J. (2019). Root-feeding larvae increase their performance by inducing leaf volatiles that attract above-ground conspecific adults. *Journal of Ecology, 107*(6), 2713–2723.

Thompson, M. N., Arriaga, J., Bradford, B. J., Kurian, R., Strozier, G., & Helms, A. M. (2024). Belowground insect herbivory induces systemic volatile emissions that strengthen neighboring plant resistance aboveground. *Plant, Cell & Environment, 47*(2), 714–725.

Tjoelker, M. G., Craine, J. M., Wedin, D., Reich, P. B., & Tilman, D. (2005). Linking leaf and root trait syndromes among 39 grassland and savannah species. *New Phytologist, 167*, 493–508.

Tonelli, M., Peñaflor, M. F. G. V., Leite, L. G., Silva, W. D., Martins, F., & Bento, J. M. S. (2016). Attraction of entomopathogenic nematodes to sugarcane root volatiles under herbivory by a sap-sucking insect. *Chemoecology, 26*, 59–66.

Touw, A. J., & van Dam, N. M. (2023). Optimal chemical defence allocation in roots: Where, why and how? *Phytochemistry Reviews, 24*(1), 27–36.

Tsunoda, T., Makoto, K., Suzuki, J. I., & Kaneko, N. (2018). Warming increased feeding of a root-chewing insect at the soil surface and enhanced its damage on a grass. *Soil Biology and Biochemistry, 126*, 213–218.

Turlings, T. C., & Erb, M. (2018). Tritrophic interactions mediated by herbivore-induced plant volatiles: Mechanisms, ecological relevance, and application potential. *Annual Review of Entomology, 63*(1), 433–452.

Vagge, I., Sgalippa, N., & Chiaffarelli, G. (2024). The role of agroforestry in solving the agricultural landscape vulnerabilities in the Po Plain district. *Community Ecology, 25*(3), 1–27.

van Dam, N. M. (2009). Belowground herbivory and plant defences. *Annual Review of Ecology, Evolution, and Systematics, 40*, 373–391.

Van Dam, N. M., & Baldwin, I. T. (2001). Competition mediates costs of jasmonate-induced defences, nitrogen acquisition, and transgenerational plasticity in *Nicotiana attenuata*. *Functional Ecology, 15*(3), 406–415.

van Dam, N. M., Vuister, L. W., Bergshoeff, C., de Vos, H., & Van der Meijden, E. D. (1995). The "Raison D'être" of pyrrolizidine alkaloids in *Cynoglossum officinale*: Deterrent effects against generalist herbivores. *Journal of Chemical Ecology, 21*, 507–523.

van der Meijden, E. (1996). Plant defense, an evolutionary dilemma: Contrasting effects of (specialist and generalist) herbivores and natural enemies. In *Proceedings of the 9th International Symposium on Insect-Plant Relationships* (pp. 307–310). Springer Netherlands.

Van der Putten, W. H. (2003). Plant defence belowground and spatiotemporal processes in natural vegetation. *Ecology, 84*(9), 2269–2280.

Van Der Putten, W. H., Bardgett, R. D., de Ruiter, P. C., Hol, W. H. G., Meyer, K. M., Bezemer, T. M., & Wardle, D. A. (2009). Empirical and theoretical challenges in aboveground–belowground ecology. *Oecologia, 161*, 1–14.

Vannette, R. L., & Hunter, M. D. (2009). Mycorrhizal fungi as mediators of defence against insect pests in agricultural systems. *Agricultural and Forest Entomology, 11*, 351–358.

Volpicella, M., Ceci, L. R., Cordewener, J., America, T., Gallerani, R., Bode, W., & Beekwilder, J. (2003). Properties of purified gut trypsin from *Helicoverpa zea*, adapted to proteinase inhibitors. *European Journal of Biochemistry, 270*(1), 10–19.

War, A. R., Taggar, G. K., Hussain, B., Taggar, M. S., Nair, R. M., & Sharma, H. C. (2018). Plant defence against herbivory and insect adaptations. *AoB Plants, 10*(4), ply037.

White, J. A., & Andow, D. A. (2006). Habitat modification contributes to associational resistance between herbivores. *Oecologia, 148*, 482–490.

Whittaker, J. B. (2003). Root-animal interactions. In *De Kroon & Visser* (Eds.), *Root Ecology* (pp. 363–385). Springer.

Wouters, F. C., Blanchette, B., Gershenzon, J., & Vassão, D. G. (2016). Plant defense and herbivore counter-defense: Benzoxazinoids and insect herbivores. *Phytochemistry Reviews, 15*, 1127–1151.

Zhu-Salzman, K., & Zeng, R. (2015). Insect response to plant defensive protease inhibitors. *Annual Review of Entomology, 60*(1), 233–252.

CHAPTER 13

From Bacteria to Beetles: Microbes' Impact on Plant-Feeding Insects

NOMAN WAHEED,[1,2*]† SABAHAT SAEED,[3,4]† ALEENA ALAM,[5] SOHAIL ABBAS,[5] ZEESHAN AHMAD,[4] USAMA NAWAZ,[4] and NOORAL AMIN[6]

[1]*The United Graduate School of Agricultural Sciences, Kagoshima University, Kagoshima, Japan*

[2]*Department of Zoology, Hazara University Mansehra KPK, Pakistan*

[3]*College of Animal Sciences and Technology, Jilin Agricultural University, Changchun, Jilin, China*

[4]*Department of Zoology, Hazara University, Mansehra, Khyber Pakhtunkhwa, Pakistan*

[5]*School of Agricultural and Pest Control, College of Plant Protection, Jilin Agricultural University, Changchun, Jilin, China*

[6]*Plant Biotechnology Center, College of Agronomy, Jilin Agricultural University, Changchun, Jilin, China.*

†*The authors equally share first authorship and are entitled to claim it as first authors on their resumes*

*Corresponding author

ABSTRACT

Microbes play a significant role in shaping the ecology and behavior of plant-feeding insects, such as aphids and beetles, which are critical to both ecosystems and agriculture. This chapter explores how microorganisms, particularly symbiotic bacteria within insect bodies, influence these insects' nutrition, fitness, and ecological interactions. These endosymbionts supply

Insect-Plant Interactions, Volume 1: Influence of Biotic and Abiotic Factors.
Jamin Ali, Adil Tonğa, Adnan Shakeel, and Nasheeman Ashraf (Eds.)

DOI: 10.1201/9781779644282-13

essential nutrients, enabling insects to thrive on nutrient-poor plant diets, while gut microbes help break down complex plant compounds, aiding digestion. Microbes also modulate insect-plant interactions by affecting herbivore resistance and susceptibility to plant defenses. Additionally, microbial communities within insects can vary based on host plants, influencing insect behavior and host preferences. The chapter further examines the broader ecological impact of these microbial interactions on insect populations and community dynamics, as well as their potential in pest management and crop yield enhancement. By highlighting the often-overlooked role of microbes, this chapter underscores their critical influence on insect physiology and ecology. Understanding these microbial relationships offers new perspectives on sustainable pest management and could lead to more ecologically balanced approaches in agriculture.

13.1 MICROBES ROLE IN LIFE OF LIVING ORGANISMS

Microbes are a vital and ubiquitous part of life on our planet. They include a wide variety of bacteria, viruses, fungi, archaea, and protozoa (Kaur et al., 2023). Microbes are prevalent and essential to the health and sustainability of ecosystems, found in the highest mountain peaks, the depths of the ocean, and even the cells of complex animals. Their accomplishments are significant and diversified. Microbes play a critical role in the global cycling of nutrients (Shu and Huang, 2022). They are the force behind degradation, fixation of nitrogen, and photosynthesis. A considerable quantity of oxygen is produced by photosynthetic microbes, such as phytoplankton and certain bacteria, which also serve as the base of the food chain and support life in all sizes, from tiny zooplankton to enormous whales. Nitrogen-fixing bacteria are essential for turning atmospheric nitrogen into plant-usable forms, promoting plant development and productivity. Decomposers, on the other hand, break down organic matter, restoring important nutrients to the soil and promoting plant growth and health (Dong et al., 2022).

Apart from their ecological duties, microorganisms have a substantial effect on the well-being of humans. The human microbiome, including a variety of microorganisms that live on and within our bodies, has a significant impact on our health (Panthee et al., 2022). Gut microorganisms promote digestion, manufacture critical vitamins, and contribute to the formation and function of our immune system. Imbalances in the gut microbiota have been related to a variety of health problems, emphasizing the importance of these microorganisms in our overall health. Microbes also play a role in the manufacturing of food and beverage products (Patil et al., 2023). Fermentation, an intricate procedure overseen by a variety of microorganisms, serves as the method for generating fundamental food items such as bread, yogurt, beer, and cheese (Baumgarthuber, 2021). In addition, microorganisms have been instrumental in driving significant progress in the fields of healthcare and biotechnology (Wohlgemuth et al., 2021). They play a crucial role in manufacturing antibiotics, insulin, and vaccines, thus contributing to the preservation of innumerable lives. The realms of genetic engineering and synthetic biology, grounded in our comprehension of microbial genetics and physiology, present exciting opportunities for fresh medical therapies, sustainable energy generation, and environmental restoration (Timmis et al., 2019).

Microorganisms, although having a dual nature, present both benefits and risks. Pathogenic microorganisms possess the potential to inflict severe illnesses in humans, animals, and plants. Grasping the intricacies of these detrimental microorganisms is essential for formulating approaches to address infectious diseases and uphold the security of our food and surroundings (Salehi et al., 2018). Essentially, microorganisms play a significant role in the drama of existence, molding our environments, impacting our well-being, and propelling innovations in science and technology that could revolutionize our prospects (Council, 2007). Microbiology, the investigation of microorganisms, persists in uncovering their mysteries, revealing the profound influence these minuscule organisms exert on all aspects of life (Prajapati et al., 2023).

13.2 PLANT FEEDING INSECTS

Insects that feed on plants, ranging from small aphids to large beetles, have developed intricate relationships with the plant world (Whitehill et al., 2023). Nevertheless, the role of microorganisms in these relationships has received notable focus in recent times (Jacoby et al., 2017). Microbes linked to plant-feeding insects have the potential to directly influence insect herbivory, nutrient acquisition, defensive strategies, and can also have ramifications for plant well-being and development (Whitehill et al., 2023). Within the expansive and intricate web of ecological interactions, the connection between plants and insects serves as a fundamental cornerstone. This connection, however, transcends a straightforward narrative of herbivores consuming plants. Instead, it embodies a complex interplay, a subtle choreography influenced by an imperceptible realm—microbes (Lange et al., 2023). Exploring the intricate mutual interactions between plants, insects, and microorganisms, we uncover the significant effects of these minuscule organisms on the ecosystem of herbivorous insects (Turbé et al., 2010).

The vast and intricate microbial world is home to a wide range of symbiotic relationships with both plants and insects. These symbiotic connections, which go beyond what is immediately observable, play a pivotal role in influencing the behavior, physiology, and evolutionary paths of insects that rely on plant matter for sustenance (Pouliot, 2018). The microbial realm wields significant influence over the dynamics of insect populations that primarily feed on plants, ranging from gut microorganisms that facilitate digestion to endosymbiotic bacteria that supply vital nutrients (Xu et al., 2023). To grasp the complex interaction described, we begin by exploring the small-scale world of bacteria and other microorganisms found in the digestive systems of herbivorous insects. These often-neglected microorganisms have a crucial function in decomposing intricate plant compounds and aiding in the absorption of nutrients. Consequently, they have a significant impact on the feeding choices and survival strategies of these insects (Magdoff and Van Es, 2021).

Moreover, the noteworthy symbiotic relationships between insects and particular microorganisms provide insights into how these partnerships empower insects to flourish on otherwise demanding plant-based diets (Wernegreen, 2012b). From the nitrogen-fixing bacteria that improve the availability of nutrients for plants to the endophytic fungi that provide resistance against herbivores, these collaborations illustrate the complex methods through which microorganisms influence the ecological relationships within this dynamic trio (Chewning, 2018). Exploring the captivating realm of plant-facilitated interactions, wherein plants serve as intermediaries, directing mutually beneficial connections between microorganisms and insects (Kang, 2019). Plants play a dual role in shaping their microbiome and affecting insect behavior through chemical signaling, thereby enhancing their defense mechanisms against herbivores. This interaction blurs the boundaries of the ecological relationship between these three elements (Trivedi et al., 2020).

13.3 AIM OF CHAPTER

While in this chapter, we will reveal the captivating tale of how microorganisms serve as a vital link between plants and insects, intricately exerting their impact across the entire web of life. This symbiotic association, ranging from bacteria to beetles, embodies the remarkable intricacy of the natural world and the profound interconnectedness among all living entities within it. The aim of this chapter is to explore the intricate relationships between microbes, plants, and plant-feeding insects, examining the diverse impacts that microbial communities, ranging from bacteria to fungi, have on the ecology, behavior, and physiology of these insects. Through a comprehensive review of current research, this chapter aims to elucidate the complex interplay between microorganisms and plant-feeding insects, shedding light on the ecological significance and potential applications of understanding these interactions.

13.4 PLANT PARTS AND STRUCTURES PLAY A CRUCIAL ROLE IN THE INTERACTIONS BETWEEN MICROBES AND PLANT-FEEDING INSECTS

The components and configurations within plants are anticipated to assume a vital function in the dynamics involving microorganisms, herbivorous insects, and their respective host vegetation (Noman et al., 2020). The subsequent section enumerates some of these plant components and structures that could potentially participate in these processes.

13.4.1 FLOWERS

The interaction between microbes and plants is significantly influenced by the presence of flowers. Flowers play a pivotal role in facilitating this interaction through various mechanisms. While flowers are crucial for reproduction, they can also face challenges from plant-feeding insects. Flower-feeding insects play a crucial role in pollination (Pegoraro et al., 2020). Insects such as bees and butterflies have the potential to introduce microorganisms when gathering nectar and pollen. Some insects, such as bees and butterflies, are attracted to flowers for nectar and pollen (Liu et al., 2021).

13.4.2 FRUITS

The interaction between fruits, microbes, and plants is complex and plays a crucial role in various ecological processes (Hardoim et al., 2015). Insects that primarily consume fruit have the potential to affect the quality of the fruit and can also shape the composition of the microbial communities that inhabit the plant (Behar et al., 2008). Fruit-feeding insects encompass a diverse group of organisms that feed on various parts of fruits, including the pulp, seeds, or other tissues. These insects play significant roles in ecosystems, contributing to processes such as pollination, seed dispersal, and nutrient cycling (Nair, 2007). Some insects, such as the Mediterranean Fruit Fly (*Ceratitis capitata*), Codling Moth (*Cydia pomonella*), and Brown Marmorated Stink Bug (*Halyomorpha halys*), are examples of fruit-feeding insects (Ekramirad et al., 2016).

13.4.3 LEAVES

Leaves play a vital role in facilitating the interaction between microorganisms and plants. Leaf-feeding insects, known as folivores, comprise a diverse group of herbivores that play significant roles in ecosystems and agricultural settings (Ghosh, 2023). Caterpillars, the larval stage of butterflies and moths, are prominent leaf feeders (Hazra et al., 2021). For instance, the cabbage white butterfly (*Pieris rapae*) caterpillars voraciously consume the leaves of cruciferous plants, impacting crops like cabbage and broccoli (Talbot, 2017). Leafhoppers, small insects with the ability to jump great distances, also feed on leaves by puncturing plant tissues and extracting sap. Aphids are another common group of leaf-feeding insects (Tholt et al., 2018).

13.4.4 STEM

The stem of a plant plays a crucial role in the complex interaction between plants and microbes. Insect species that feed on plant stems constitute a diverse category with notable implications for the overall well-being and vitality of plants. These insects display a range of feeding behaviors, including burrowing into stems and extracting sap or nutrients (Stephens and Westoby, 2015). One notable example is the corn stalk borer (*Diatraea saccharalis*), a moth larva that infests the stems of corn plants. These larvae tunnel into the stalks, weakening the structural integrity of the plant and potentially causing lodging (Hodgson and Dean, 2020). Another example is the stem-boring weevil (*Listronotus bonariensis*), which targets grasses like Bermuda grass (Bagheri et al., 2021).

13.4.5 ROOTS

Plant roots can also experience the influence of plant-feeding insects and the microorganisms that are linked to them. Insects that feed on roots can potentially engage with root tissues and the microorganisms present in the rhizosphere (Johnson and Rasmann, 2015). Root-feeding insects present a considerable threat to plant well-being, with detrimental effects on both agricultural productivity and ecosystem stability (Jones et al., 2018). Notable examples encompass root-feeding nematodes, like the soybean cyst nematode (*Heterodera glycines*), which invades soybean plant roots, resulting in reduced yields (Darling et al., 2022). Additionally, the larvae of root-feeding beetles, such as those belonging to the corn rootworm (*Diabrotica* spp.), can cause substantial harm by consuming the roots of corn plants (Eben, 2022) (Figure 13.1).

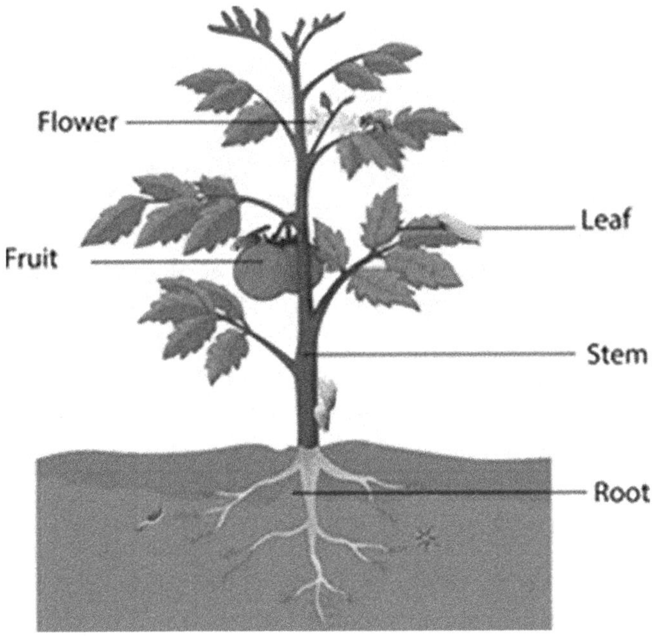

FIGURE 13.1 Microbe-Insects on plant components and structures that participate in their interactions.

13.5 MICROBE–PLANT-INSECT INTERACTIONS

Microbial interactions with plants and insects create an intricate and mutually beneficial network of connections essential for ecosystem functioning. Within this nexus of relationships, there exists a constantly evolving interplay among microorganisms, plants, and insects, with each entity exerting an impact on the behavior, physiology, and ecology of the others (Arora and Malik, 2021). Microorganisms, encompassing bacteria, fungi, viruses, and protozoa, have developed complex associations with both plants and insects. These interactions frequently result in substantial effects on the growth, well-being, and ecological interactions of these organisms (Prashar et al., 2014). During these exchanges, microorganisms can function as mutualistic partners, commensal inhabitants, or even detrimental parasites. They play a pivotal role in molding nutrient cycles, regulating plant defense mechanisms, and impacting both insect herbivory and behavior. To illustrate, specific microbes associated with plants have the capacity to boost nutrient accessibility and absorption for plants, ultimately promoting their development and fortitude (Singh et al., 2021b). Simultaneously, they have the ability to impact the eating habits of insects through changes in plant composition or the emission of scents that draw or deter insects. Conversely, insects, particularly those that feed on plants, frequently possess distinct gut microorganisms that support digestion, improve nutrient assimilation, and, in certain instances, assist in overcoming plant protective mechanisms (Coolen et al., 2022). The intricate relationships among these factors carry significant consequences for agricultural output, natural pest management, human well-being, and the preservation of ecosystems (Tscharntke et al., 2016). The intricate interplay among microorganisms, plants, and insects plays a pivotal role in understanding the intricacies of ecosystems. It is also the cornerstone for achieving sustainability and fostering a balanced coexistence within our natural environment (da Silva, 2006). Understanding microbe-plant-insect interactions is crucial for several reasons.

13.5.1 ECOSYSTEM DYNAMICS AND BALANCE

These interactions are integral to ecosystem dynamics, playing a significant role in maintaining ecological balance (Dubey et al., 2019). Microbes, plants, and insects are interconnected components of ecosystems, and understanding their interactions helps us comprehend ecosystem functioning and stability (Chomel et al., 2016). Microbes, especially those residing in the gut of herbivores, play a crucial role in nutrient cycling and ecological balance (Pastor et al., 2006). For instance, within the digestive systems of termites, symbiotic microorganisms facilitate the breakdown of cellulose, a complex plant material. This process not only allows termites to thrive on a cellulose-rich diet but also contributes to the decomposition of plant matter in the ecosystem (Arora, 2022). The balance is further evident in the mutually beneficial relationships between herbivorous insects, such as aphids, and symbiotic bacteria (Oliver and Perlman, 2020). These bacteria aid in nutrient synthesis and bolster the insects' resistance to stressors, promoting the overall health of both the insects and the ecosystem (Kennedy and Wightman, 2011).

13.5.2 AGRICULTURAL PRODUCTIVITY

In the realm of agriculture, insects that feed on plants have the potential to inflict substantial harm on crops, resulting in economic setbacks, as indicated (Ahuja et al., 2011). Gaining insight into the ways in which microorganisms impact these plant-feeding insects can contribute to the formulation of environmentally friendly pest control methods. This has the potential to diminish the dependence on detrimental pesticides and encourage the adoption of more effective agricultural approaches (Pappas et al., 2017). Microbes play a vital role in bolstering agricultural productivity through their interactions with herbivorous insects. In sustainable agriculture, beneficial microbes actively contribute to pest control and nutrient cycling, exerting a positive influence on crop yields (Iqbal et al., 2023). Take, for instance, the bacterium *Bacillus thuringiensis*, which produces proteins toxic to specific insect pests, providing a natural and precisely targeted solution (Bravo et al., 2011). Moreover, symbiotic relationships between insects and nitrogen-fixing bacteria have the potential to enhance soil fertility. In specific cases, such as with herbivorous insects like leaf-cutter ants, their fungal cultivars play a key role in breaking down plant material, thereby facilitating nutrient cycling within ecosystems (Halbleib and Ludden, 2000). The utilization of these microbial interactions in agricultural practices holds significant promise for reducing reliance on chemical pesticides, fostering soil health, and ultimately elevating overall agricultural productivity (Saeed et al., 2021).

13.5.3 PLANT HEALTH AND GROWTH

Microbes' symbionts have the potential to have a beneficial impact on the well-being of plants through the enhancement of nutrient absorption, growth, and the ability to withstand stress. Understanding these relationships can provide valuable insights for agricultural methods aimed at maximizing the associations between plants and microbes, thereby fostering more robust and higher-yielding crops (Ramakrishna et al., 2019). The health and growth of plants are intricately tied to microbial interactions, specifically involving herbivorous insects. Some microbes, like beneficial bacteria and mycorrhizal fungi, establish symbiotic relationships with plant roots, aiding in nutrient absorption and enhancing overall plant health (Noman et al., 2020). Conversely, plant-associated microbes can impact the resistance of plants to herbivores by producing defensive compounds. A notable example is the symbiotic relationship between aphids and *Buchnera* bacteria, where the bacteria supplement the aphid's diet, influencing both the insect's fitness and the plant's susceptibility to herbivory (Skaljac et al., 2016). Recognizing and comprehending these complex interactions is essential for the development of sustainable strategies in agriculture and plant protection (Ratnadass et al., 2012).

13.5.4 BIO-CONTROL AND SUSTAINABLE AGRICULTURE

Some microorganisms have the capacity to function as innate adversaries of herbivorous insects that target plants, presenting promising possibilities for biological pest control (Gonzalez et al., 2016). These associations can provide valuable insights for the creation of eco-conscious and sustainable methods for pest management, consequently decreasing the reliance on artificial chemicals (Mouden et al., 2017). The microbes assume a crucial role in bio-control approaches. For instance, entomopathogenic bacteria like *Bacillus thuringiensis* produce toxins fatal to specific insect pests, providing a focused and environmentally conscious approach to pest management (Ali et al., 2022). Another illustration involves the utilization of beneficial fungi such as *Beauveria bassiana*, which infect and eliminate herbivorous insects, offering an eco-friendly substitute for traditional insecticides (Bamisile et al., 2021). These microbial agents, when judiciously applied, not only efficiently manage pest populations but also contribute to sustainable agricultural practices by minimizing the ecological impact associated with chemical interventions.

Additionally, they promote biodiversity and enhance long-term soil health (Gupta et al., 2022).

13.5.5 BIODIVERSITY AND CONSERVATION

Microorganisms play a crucial role in enhancing the variety of herbivorous insects and the plant species they engage with. Analyzing these associations not only facilitates comprehension of biodiversity trends but also offers valuable insights for conservation initiatives by underscoring the significance of safeguarding these connections within their native ecosystems (Wielkopolan and Obrępalska-Stęplowska, 2016). Biodiversity and conservation initiatives go beyond charismatic fauna and flora, encompassing the intricate microbial communities that play pivotal roles in ecosystems. In the realm of herbivorous insects, symbiotic relationships with microbes significantly contribute to both biodiversity and conservation (Jeffries, 2006). The gut microbiota of herbivorous insects, comprising bacteria, archaea, and protozoa, plays a vital role in enhancing nutrient digestion and metabolism, enabling insects to adapt to diverse plant-based diets. The preservation of these microbial communities is crucial for maintaining the ecological balance of insect-associated ecosystems (Wertz and Bechade, 2020). A noteworthy example is the gut microbiota of certain pollinators, such as bees and butterflies, which not only aids in digestion but also influences pollination efficiency, thereby contributing to the conservation of plant diversity (Schmidt and Engel, 2021).

13.5.6 CLIMATE CHANGE ADAPTATION

Analyzing the influence of microbes on plant-feeding insects becomes increasingly crucial as climate change disrupts ecological relationships. This knowledge can be instrumental in forecasting and alleviating the repercussions of climate change on both plant and insect communities. Consequently, it plays a pivotal role in the development of adaptation strategies (Karthik et al., 2021). Microbes and herbivorous insects demonstrate remarkable adaptations in the face of climate change, highlighting their resilience in dynamic ecosystems. Changes in temperature and precipitation patterns can reshape microbial communities within the insect gut, impacting digestive processes (Vacher et al., 2021).

Concurrently, some herbivorous insects adjust their feeding behaviors to cope with shifts in plant distribution. For instance, certain agricultural pests may extend their range into new territories due to milder temperatures (Skendžić et al., 2021). These adaptations underscore the interconnected response of microbes and herbivorous insects to climate-induced ecological shifts (Ferguson and Adamo, 2023).

Comprehending the complex connections between microorganisms, plants, and insects is crucial for making well-informed choices across diverse domains, ultimately fostering a sustainable and harmonious cohabitation of life on our planet (Figure 13.2).

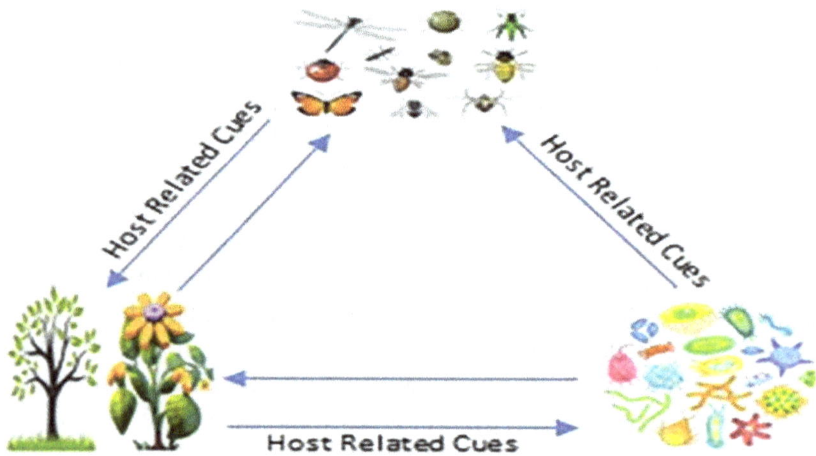

FIGURE 13.2 Microbes-plants and insects interactions cues in multiple aspects.

13.6 MICROBIAL IMPACTS ON HERBIVOROUS INSECTS: UNRAVELING ECOLOGICAL DYNAMICS, SYMBIOTIC RELATIONSHIPS, AND AGRICULTURAL IMPLICATIONS

To establish the backdrop for investigating the influence of microorganisms on herbivorous insects, it is essential to acknowledge the complex network of connections present in the natural environment. This network encompasses not just plants and insects but also the frequently disregarded yet significantly impactful microorganisms (Bardgett and Van Der Putten, 2014). Microscopic organisms, encompassing bacteria, fungi, viruses, and various tiny life forms, have a significant influence on ecosystem dynamics and greatly affect the relationships between plants and insects (Oliver et al., 2014). The effects of microorganisms on herbivorous insects are significant and varied, involving complex relationships that have an impact on the physiology, behavior, and ecological functions of these insects within their respective environments (Schoonhoven et al., 2005).

Microorganisms such as bacteria, fungi, viruses, and protozoa frequently establish symbiotic associations with herbivorous insects, commonly inhabiting their digestive systems or being closely associated with their anatomical structures (Sonenshine and Stewart, 2021). One of the notable effects lies in the realm of nutrition and digestion, as symbiotic microorganisms play a crucial role in the decomposition of intricate plant substances that insects, by themselves, are incapable of metabolizing. Through the production of enzymes that break down cellulose, lignin, and other plant polymers, these microorganisms elevate the accessibility of nutrients and contribute to the digestive process, ensuring that the insect receives vital nutrients that are absent in their predominantly plant-based diet (Zhang et al., 2023).

Furthermore, microbial symbiotic organisms have the capacity to impact an insect's immune responses, furnishing protective measures against various pathogens and parasites (Eleftherianos et al., 2013). Certain microorganisms have the capacity to produce substances that discourage natural predators or improve an insect's ability to neutralize harmful substances found in its food (Yactayo-Chang et al., 2020). The consequences extend beyond the individual insect, with the potential to trigger a chain reaction of influences on insect populations, communities, and the dynamics of ecosystems (Karp and Daily, 2014). Understanding the significance of these microbial effects is essential for gaining a comprehensive understanding of the relationships between insects and plants. This knowledge can extend its influence to various aspects, such as agriculture, pest control, and ecological preservation, ultimately promoting sustainable cohabitation in natural ecosystems (Sehgal, 2019).

13.6.1 MICROBIAL MODULATION OF FEEDING BEHAVIOR

The modulation of feeding behavior by microorganisms entails intricate interactions among microbes, plants, and insects, significantly impacting the ecological balance within ecosystems (Sugio et al., 2015). A prime illustration is the symbiotic association among plants, insects, and specific microbes, which can modify the feeding patterns of herbivorous insects. Within the rhizosphere, the soil region directly influenced by plant roots, microorganisms play a pivotal role in establishing a dynamic environment (Prashar et al., 2014). Plant-root-associated microbes, such as bacteria and mycorrhizal fungi, form symbiotic relationships with plants, enhancing nutrient absorption and providing resistance to stressors. This, in turn, influences the overall health and defensive capabilities of the host plant (Shinwari et al., 2019). As herbivorous insects feed on plants, they contend not only with the plant's defenses but also with the impact of associated microbes. Some microbes produce secondary metabolites that function as deterrents or attractants to herbivores. For instance, specific bacteria emit volatile organic compounds (VOCs) that either repel herbivores or attract their predators, triggering a cascade of indirect interactions (Divekar et al., 2022). Thus, microbial modulation of feeding behavior underscores the complexity of ecological relationships, emphasizing the interconnected nature of various components within ecosystems (National Academies of Sciences and Medicine, 2017).

13.6.2 IMPACT ON INSECT PERFORMANCE AND FITNESS

Insects are integral components of ecosystems, engaging in intricate relationships with their environment, both influencing and being influenced by a myriad of factors that shape their performance and fitness (Huot et al., 2013). Environmental conditions, including temperature and habitat quality, play pivotal roles in this dynamic interplay. Climate change-induced shifts in temperature regimes impact insect life cycles and behavior, while habitat degradation and fragmentation pose threats to suitable living environments (Upadhyay, 2020).

Biotic interactions, spanning predation, parasitism, and mutualistic relationships, further shape insect fitness. The presence of natural enemies serves as a regulatory mechanism for herbivorous insect populations, preventing unchecked consumption of plant resources (Brin et al., 2018). Conversely, mutualistic partnerships, such as pollination interactions with plants, are indispensable for the reproductive success of both partners. Physiological factors, including nutrient availability and genetic diversity, also contribute significantly to insect well-being. Adequate nutrition is essential for the growth and reproduction of insects, while genetic diversity within populations is crucial for adaptation to environmental changes (New, 2017). Pesticide exposure can result in direct mortality or sublethal effects, impacting insect behaviors and reproductive success. Recognizing and understanding these intricate dynamics are paramount for the development of effective conservation strategies (Sandrock et al., 2014).

13.6.3 MICROBIAL-MEDIATED CHANGES IN INSECT PHYSIOLOGY

Microbial-induced alterations in insect physiology underscore the intricate partnerships between insects and microorganisms, highlighting the dynamic interplay within the insect microbiota (Chisholm, 2017). A fundamental impact is evident in nutrient provision, where symbiotic bacteria contribute to the breakdown of complex molecules, enhancing nutrient accessibility for the insect host and influencing its growth and reproductive success (Sharma et al., 2021). This collaboration extends to immune system modulation, as microbes play a pivotal role in priming and regulating the insect's immune responses, thereby affecting its ability to defend against pathogens. Reproductive fitness is also shaped, as seen in bacteria like *Wolbachia*, which manipulates insect reproduction, impacting sex determination and reproductive strategies (Vanthournout et al., 2011). Microbes play a crucial role in detoxification processes, assisting herbivorous insects in overcoming plant secondary metabolites (Divekar et al., 2022). Understanding these interactions not only reveals the complexity of insect-microbe relationships but also opens avenues for innovative approaches in pest management and conservation by harnessing the natural synergies that have evolved over time (War et al., 2020). The examination of microbial influences on insect physiology emphasizes the importance of adopting a holobiont perspective, where the collective interactions between insects and their associated microorganisms shape ecological and evolutionary dynamics (Brunetti, 2022).

13.6.4 IMPLICATIONS FOR INSECT-PLANT INTERACTIONS

Microbial-mediated alterations in insect physiology have far-reaching implications for insect-plant interactions, intricately shaping relationships within ecosystems (Singh et al., 2021a). A significant impact is observed in nutrient cycling, where microbial breakdown of plant materials in the insect gut contributes to nutrient release, influencing soil health and, consequently, the overall well-being of plants (Chakraborty and Roy, 2021). These interactions extend to herbivory and plant defense mechanisms, with microbes inluencing the ability of insects to overcome plant defenses and, conversely, inducing plant defense responses. Pollination dynamics are also affected, as microbes within pollinators can alter foraging behavior and longevity, ultimately influencing the pollination services crucial for plant reproduction. Additionally, plant-induced responses to insect herbivory, such as the release of volatile compounds, may be modulated by the microbial communities associated with insects (Vannette and Systematics, 2020). Mutualistic relationships between microbes, insects, and plants further emphasize the interconnected nature of these interactions, with nitrogen-fixing bacteria contributing to soil fertility and plant growth. These microbial influences play a vital role in the ecological resilience of ecosystems, maintaining a delicate balance between herbivory and plant defenses that sustains biodiversity (Chandra and Interface, 2019).

13.6.5 MICROBIAL-MEDIATED RESISTANCE IN PLANTS

Microbial-mediated resistance in plants represents a sophisticated defense strategy arising from intricate interactions between plants and diverse microorganisms (Jain et al., 2016). A prominent mechanism is Induced Systemic Resistance (ISR), where beneficial microbes like rhizobacteria and mycorrhizal fungi elicit a systemic response in plants upon root colonization. This priming effect enhances the plant's defense response upon subsequent encounters with pathogens, providing long-lasting, broad-spectrum immunity (Zhu et al., 2022). Microbial Volatile Organic Compounds (MVOCs) play a significant role, acting as signaling molecules that induce plant defense gene expression. This preemptive signaling prepares the plant for a more effective defense against potential pathogenic threats (Fincheira and Quiroz, 2018). Some microbes directly antagonize plant pathogens through resource competition, antimicrobial compound production, and systemic resistance induction. For example, mycorrhizal fungi not only compete with pathogenic fungi for root colonization but also contribute to overall plant health, limiting harmful pathogen establishment (Bongard, 2012).

Endophytic microbes within plant tissues add another layer to microbial-mediated resistance, enhancing resistance through secondary metabolite production or systemic responses (Poveda et al., 2020). Symbiotic relationships, such as those with nitrogen-fixing bacteria in root nodules, contribute not only to nutrient acquisition but also to overall plant health and resilience. This symbiosis enhances the plant's resistance to pathogenic attacks, showcasing the multifaceted nature of microbial contributions (Meena et al., 2017). Understanding microbial-mediated resistance has practical implications in agriculture and pest management. Leveraging these natural interactions offers environmentally friendly strategies to enhance crop resilience without relying extensively on synthetic chemicals (Snyder et al., 2020). By promoting a balanced and diverse microbial community in the rhizosphere and plant tissues, agricultural practices can capitalize on plants' inherent ability to defend themselves, contributing to sustainable and resilient agricultural systems (Zavahir et al., 2021).

13.6.6 ECOSYSTEM INTERCONNECTEDNESS

Ecosystems represent intricate and interrelated networks in which each living being, regardless of its scale or prominence, plays a role in maintaining the overall equilibrium (Nwaji et al., 2022). Microorganisms play a pivotal role in maintaining this intricate equilibrium, exerting a significant impact on the well-being, conduct, and physical characteristics of both plant life and insects (Chakraborty and Roy, 2021). Microbes establish intricate and varied populations in a range of environments, such as the gastrointestinal tracts of insects (Ziganshina et al., 2018). The abundance and assortment of these microbe collectives have a noteworthy impact on the eating habits, digestive processes, nutrient uptake, and overall well-being of insects. Consequently, these factors play a crucial role in shaping the dynamics between insects and plants (Engel and Moran, 2013).

13.7 ENDOSYMBIOSIS IN PLANT-FEEDING INSECTS

Endosymbiosis in plant-feeding insects constitutes a captivating phenomenon that illuminates the complex associations between these insects and the microorganisms residing within them (Arab, 2021). One extensively studied instance involves the symbiotic partnership between aphids, a group of sap-feeding insects, and Buchnera bacteria. Given their diet of nutrient-poor plant sap deficient in essential amino acids, aphids have established a mutualistic bond with Buchnera bacteria housed in specialized cells called bacteriocytes within the aphid's body. Buchnera bacteria play a vital role by synthesizing amino acids lacking in the aphid's diet, thereby serving as a crucial nutritional supplement for their insect hosts (Skaljac et al., 2016). Endosymbiosis in herbivorous insects presents an intriguing biological occurrence in which particular insects form a mutually beneficial partnership with microorganisms. These microorganisms frequently inhabit specialized cells or organs within the insect host. This close connection carries substantial consequences for the biology, ecology, and evolution of both the insects and the microbial partners engaged in this relationship (Wernegreen, 2012a).

13.7.1 TRANSMISSION AND INHERITANCE

Endosymbionts are frequently inherited from the mother to her progeny via specialized structures like bacteriomes or ovaries. This vertical transfer guarantees a consistent bond between the symbiont and the insect population over successive generations. Nevertheless, horizontal transmission is also possible, particularly in the case of secondary symbionts, leading to a potential exchange between different individuals (Russell et al., 2019). Microbes crucial to herbivorous insects are often transmitted vertically from parent to offspring, ensuring the inheritance of symbiotic partners. This vertical transmission occurs through mechanisms such as egg coating or parental care (Biedermann and Vega, 2020). The transfer of beneficial microbes enables the insect host to adapt to specific diets and environmental challenges across generations, highlighting the significance of microbial transmission in shaping the ecology of herbivorous insects (Hammer and Moran, 2019).

13.7.2 CO-EVOLUTIONARY DYNAMICS

The connection between herbivorous insects and their endosymbiotic partners arises from an extended history of mutual adaptation. Over time, these insects and their endosymbionts have shaped each other's evolutionary paths through selective pressures. This co-evolutionary process frequently results in a tightly specialized and obligatory partnership (Occhipinti, 2013). The intricate co-evolutionary dynamics observed between plant-feeding insects and their endosymbiotic counterparts, as illustrated prominently in the symbiotic association between aphids and *Buchnera* bacteria, demonstrate a nuanced interdependence that has evolved over millions of years. Through a reciprocal process of adaptation, these organisms have mutually influenced each other's genetic makeup and behaviors (Chong and Moran, 2018). The insects' dependence on plant sap, lacking essential amino acids, has led to the establishment of a symbiotic relationship, prompting the acquisition and retention of *Buchnera* bacteria within specialized cells. In response, *Buchnera* has undergone genomic modifications, refining its functions to complement the nutritional requirements of the insect host (Pérez-Brocal et al., 2011). This co-evolutionary relation resulted in a delicate equilibrium where both partners contribute significantly to each other's survival and reproduction. The genetic interdependence and the precision of this relationship underscore the dynamic and evolving nature of co-evolution, providing profound insight into the intricate interactions that shape the biological landscape (O'Brien and Laland, 2012).

13.7.3 FUNCTIONAL ROLES

Endosymbionts play a vital role in multiple facets of insect biology, such as nutrition, reproduction, defense against predators and parasites, detoxification of plant-derived chemicals, and even heat tolerance. Their existence can be of paramount importance for the insect's viability and effective adjustment to distinct ecological habitats (Ashra and Nair, 2022). Endosymbiosis plays a pivotal role in plant-feeding insects, not only in deciphering the complex mechanisms that underlie these partnerships but also in their promising practical implications. Exploiting these mutualistic associations presents novel opportunities for controlling pests, promoting sustainable agriculture, and advancing biotechnology (Giron et al., 2017). Furthermore, delving deeper into the variety, transmission dynamics, and functional significance of endosymbionts will enhance our comprehension of the intricate network of interactions that shape the captivating realm of plant-feeding insects (Lundgren, 2009).

13.7.4 TYPES OF ENDOSYMBIOTIC RELATIONSHIPS

13.7.4.1 PRIMARY ENDOSYMBIONTS

Primary endosymbionts are essential microorganisms that establish a close and frequently obligatory association with herbivorous insects (Biere and Tack, 2013). These symbiotic partners have evolved alongside their insect hosts over extended periods, resulting in a deeply integrated connection that significantly influences both the insect and the microbes (Wernegreen, 2002). One prominent instance is *Buchnera aphidicola*, which predominantly inhabits specialized cells known as bacteriocytes within the insect host, particularly aphids (Wernegreen, 2002). The interdependence between primary endosymbionts and their insect hosts often involves the insect offering a protected and nutrient-rich environment for the symbiont, while the symbiont reciprocally aids in meeting the insect's nutritional needs (Maire et al., 2021). Additionally, primary endosymbionts frequently exhibit reduced genomes and specialized characteristics, reflecting an extended co-evolutionary trajectory, in which genes required for symbiotic functions are retained while superfluous functions are discarded (Ellers et al., 2012).

Understanding the importance of primary endosymbionts in the biology and adaptation of herbivorous insects is essential (Biere and Tack, 2013). Additionally, their relevance extends to diverse domains, such as pest control, agricultural efficiency, and biotechnology. Novel methods like utilizing primary endosymbionts for pest management or enhancing nutrient supply through symbiotic system engineering can be investigated to unlock the capabilities of these vital microorganisms in the context of sustainable agriculture and ecosystem stewardship (Rai et al., 2012).

13.7.4.2 SECONDARY ENDOSYMBIONTS

Secondary endosymbionts, while not as vital as primary endosymbionts for the fundamental survival of insects, can bestow additional advantages upon their insect hosts (Clark et al., 2010). These advantages may encompass heightened resistance to parasites, improved reproductive success, or an expanded ability to utilize a wider array of host plants. Secondary endosymbionts introduce an added layer of intricacy to the symbiotic connections of herbivorous insects (Hansen and Moran, 2014).

Unlike primary endosymbionts, secondary symbionts are not consistently prevalent within a specific insect group and often exhibit a broader spectrum of host compatibility. These microorganisms can offer supplementary advantages to their insect hosts, going beyond what the primary symbiont supplies (Douglas, 2016). Secondary endosymbionts can coexist with primary symbionts in the same host or, on occasion, completely replace them, exemplifying the diverse dynamics of these partnerships. The roles of secondary endosymbionts are varied and multifaceted. They can augment their hosts' fitness and adaptability by imparting characteristics such as heightened resistance against natural adversaries like parasitoids and predators, resilience to environmental stressors, altered reproductive capacities, or even enhanced nutritional provisions (Hall et al., 2016). For instance, in aphids, the extensively studied secondary endosymbiont *Hamiltonella defensa* defends against parasitoid wasps by generating toxins that prove lethal to the parasitoid larvae (Cayetano and Vorburger, 2015). Another illustration is *Rickettsia* in whiteflies, which influences the reproductive behavior of its host and fosters the production of more female offspring (Su et al., 2013).

13.7.4.3 TERTIARY ENDOSYMBIONTS

Tertiary endosymbionts introduce an additional layer of intricacy within the intricate realm of insect-microbe symbiotic relationships (Su et al., 2013). Unlike primary and secondary endosymbionts, these microorganisms constitute a tertiary level of symbiosis, typically identified in specific plant-feeding insects. This phenomenon entails a three-tiered association, where a primary endosymbiont resides within the insect, secondary endosymbionts are associated with the primary one, and tertiary endosymbionts are linked to the secondary endosymbionts (Puri et al., 2021). Primary endosymbionts are typically responsible for supplying essential nutrients, particularly amino acids, to their insect hosts. Secondary endosymbionts, on the other hand, bestow various advantages such as defense against natural adversaries, manipulation of fertility, or enhanced nutrition. Tertiary endosymbionts bring an additional layer of functionalities into this symbiotic interplay (Feldhaar, 2011). These functionalities can encompass further enhancements in insect fitness, adaptation to diverse ecological niches, or the capacity to metabolize plant compounds that are usually challenging to digest (Ciancio et al., 2016). For instance, in particular aphid species, the secondary endosymbiont *Hamiltonella defensa* can host a tertiary endosymbiont known as *Serratia symbiotica* (Duperron, 2016). This tertiary endosymbiont plays a vital role in the host's defense against parasitoid wasps, underscoring the intricate dynamics among multiple endosymbiotic partners (Dicke et al., 2020).

13.7.5 NOTABLE ENDOSYMBIONTS AND THEIR ROLES

13.7.5.1 BUCHNERA APHIDICOLA

Buchnera, a prominent primary endosymbiont residing within aphids, a category of herbivorous insects, plays a pivotal role in sustaining aphid viability by providing crucial amino acids that are lacking in the aphid's diet of plant sap (Xu et al., 2023). *Buchnera aphidicola*, a Gram-negative bacterium, occupies a central role in the symbiotic alliance with aphids. Housed within specialized cells known as bacteriocytes, *Buchnera* is indispensable for the aphid's well-being as it manufactures vital amino acids that are deficient in the insect's plant sap diet (Lopez, 2018). This endosymbiotic collaboration has matured over millions of years, culminating in a highly integrated relationship where the bacterium actively contributes to fulfilling the nutritional requirements of its host. The genome of *Buchnera* has undergone a process of reduction and specialization, indicative of its adaptation to a life within the aphid host. The close interconnection between *Buchnera aphidicola* and aphids stands as an extraordinary example of co-evolution and mutualistic dependence (Perlman et al., 2022).

13.7.5.2 WOLBACHIA

Wolbachia is an extensively researched intracellular bacterium present in numerous insect varieties, encompassing herbivorous insects. It can serve as either a primary or secondary endosymbiont, affecting insect reproductive processes, overall health, and even modifying the gender distribution within the insect community (Wang et al., 2020). *Wolbachia*, an intracellular bacterium, has the capacity to infect a diverse array of arthropods and nematodes, exerting a pivotal influence on various biological processes. Its notoriety stems from its adeptness at manipulating host reproduction, instigating reproductive abnormalities like cytoplasmic incompatibility, parthenogenesis, and feminization (Manoj et al., 2021). Through these reproductive manipulations, *Wolbachia* secures a mechanism to enhance its own transmission within host populations. Beyond its role in reproductive control, *Wolbachia* has been the subject of research for its potential applications in vector control to combat vector-borne diseases (Manoj et al., 2021).

13.7.5.3 HAMILTONELLA DEFENSA

Hamiltonella, identified as a secondary endosymbiont within aphids and various other insect species, plays a crucial role in safeguarding these insects against parasitic wasps. This function significantly bolsters the insects' overall defense mechanisms (Degnan et al., 2009). *Hamiltonella defensa,* a protective endosymbiotic bacterium prevalent in aphids, is recognized for its pivotal role in conferring defense against parasitoid wasps. It achieves this by producing toxins that prove detrimental to the larvae of parasitoid wasps. This symbiotic association significantly boosts the aphid's resistance to parasitoid attacks, thereby enhancing the overall fitness of the insect (Oliver et al., 2014). *Hamiltonella defensa* serves as a compelling illustration of how endosymbiotic bacteria can provide adaptive advantages to their insect hosts, fundamentally influencing the dynamics of ecological interactions within aphid populations (Lynn-Bell, 2021).

13.7.5.4 SERRATIA SYMBIOTICA

Serratia symbiotica is an endosymbiotic organism discovered within certain aphid species. It has the capacity to replace primary endosymbionts, thus guaranteeing the aphid's viability and development (Sepúlveda et al., 2017). *Serratia symbiotica,* an endosymbiotic bacterium commonly found in various insects, particularly aphids, establishes a mutualistic relationship with its hosts, significantly contributing to their overall fitness and adaptability. *Serratia symbiotica* plays a key role in supplementing essential nutrients, including amino acids, which may be deficient in the aphid's diet (Perreau et al., 2021). Furthermore, it has been associated with providing resistance against environmental stressors and pathogens.

The intricate interplay observed between *Serratia symbiotica* and its insect hosts serves as a compelling example of the complex relationships inherent in the realm of endosymbiosis, shaping the biology and ecology of the host organisms (Sieksmeyer, 2021).

Comprehending the intricacies and roles of endosymbiosis in herbivorous insects holds significant importance in gaining insight into the ability of these insects to thrive in various ecological environments. Furthermore, investigating the molecular and physiological processes underpinning these mutualistic partnerships has the potential to offer valuable contributions to pest control, agricultural methodologies, and the field of biotechnology.

13.8 GUT MICROBIOTA

The makeup of the gut microbiota in herbivorous insects encompasses a varied and ever-changing assembly of microorganisms that inhabit the insect's digestive system (Engel and Moran, 2013). This complex collective comprises a diverse range of bacteria, fungi, archaea, protozoa, and viruses, each with their distinctive functions and impacts on the insect's digestive functions and overall well-being (Kirchman, 2018).

13.8.1 BACTERIA

Bacteria constitute a significant portion of the intestinal microbiome in herbivorous insects. The gut microbiota in these insects is characterized by the presence of different genera and species, and their quantities and diversity can fluctuate depending on factors such as the type of insect, their diet, developmental stage, and environmental conditions (Engel and Moran, 2013). In herbivorous insects, a noteworthy symbiotic relationship with bacteria frequently develops to address the nutritional hurdles presented by their plant-based diets. An exemplary instance of this is the symbiotic association between aphids and *Buchnera* bacteria (Hansen et al., 2020). Housed within specialized cells of the insect, these bacteria assume a crucial role by producing vital amino acids that are lacking in the herbivore's diet (Stevens and Hume, 1998). This endosymbiotic partnership exemplifies a finely tuned adaptation, augmenting the insect's fitness and highlighting the intricate network of interactions between insects and their microbial counterparts in the natural world (Diogo, 2017).

13.8.2 FUNGI

Fungi, encompassing yeasts, molds, and various other fungal varieties, can be found in the digestive systems of insects that primarily feed on plants. These fungal organisms play a role in enhancing insect digestion and aiding in the processing of nutrients (Gonzalez, 2014). Fungi play a pivotal role in symbiotic associations with herbivorous insects, establishing intricate relationships that exert significant impacts on both partners. The herbivorous insects, like leaf-cutting ants, engage in the cultivation of fungi as their primary food source (Hartley and Gange, 2009). In this symbiotic arrangement, the insects contribute organic matter to the fungi, while the fungi reciprocate by breaking down complex compounds into digestible nutrients for the insects. This mutualistic partnership enhances the nutritional ecology of herbivorous insects, showcasing the interconnectedness of diverse life forms in terrestrial ecosystems (Watkinson, 2016).

13.8.3 PROTOZOA

Ciliates and flagellates, which are types of protozoa, can be found in the digestive tracts of certain herbivorous insects. Their role involves facilitating the decomposition of intricate plant substances and contributing to the digestive process (Coleman et al., 2015). Protozoa play a vital role in the digestive processes of herbivorous insects, establishing symbiotic associations that contribute to the breakdown of complex plant materials (Frago et al., 2020). Typically located in the hindgut of insects, these microorganisms play a key role in the fermentation and degradation of cellulose, a predominant component of plant cell walls (Taggar and Science, 2015). The symbiosis between protozoa and herbivorous insects underscores the intricate adaptations that have evolved to facilitate the efficient utilization of plant resources within these ecological niches (Douglas, 2022).

13.8.4 ARCHAEA

Archaea, specifically methanogenic microorganisms, have been observed in the gastrointestinal tracts of specific herbivorous insects. These microorganisms are involved in methane generation and make a significant contribution to the insect's overall gut microbial population (Hammer, 2018).

Archaea, particularly methanogenic archaea, play a pivotal role in the digestive systems of herbivorous insects, contributing to the breakdown of complex plant materials and aiding in nutrient extraction for the insect host (Rajeswari et al., 2021). In environments like the termite gut, methanogenic archaea are instrumental in the degradation of cellulose and hemicellulose, allowing termites to thrive on a diet rich in plant-derived compounds (Show et al., 2022). The metabolic activities of these archaea result in methane production, influencing the overall greenhouse gas (GHG) emissions in insect-associated ecosystems. A comprehensive understanding of the symbiotic relationship between herbivorous insects and methanogenic archaea is crucial for unraveling the ecological impact of these interactions and their implications for nutrient cycling (Thomas et al., 2021).

13.8.5 FUNCTION AND IMPACT OF GUT MICROBIOTA ON INSECTS

The gut microbiota in insects, similar to that in many other life forms, exerts a substantial influence on multiple facets of their biology and ecological interactions. The subsequent sections outline a selection of the roles and consequences of gut microbiota in insects.

13.8.5.1 DIGESTION AND NUTRIENT PROCESSING

The gut microbiota plays a crucial role in facilitating the decomposition of intricate plant polymers, such as cellulose and hemicellulose, into more straightforward, easily digestible substances. They generate enzymes, such as cellulases and hemicellulases, which contribute to the process of digestion and the acquisition of nutrients (Puhlmann and de Vos, 2022). Microbiota play a crucial role in the digestive processes of herbivorous insects, actively participating in the breakdown of complex plant materials like cellulose through fermentation. This microbial activity releases essential nutrients for the insect host, contributing significantly to digestion and nutrient processing (Show et al., 2022). The gut microbiota also plays a key role in amino acid synthesis, facilitating protein formation within the insect. The symbiotic relationship between herbivorous insects and their gut microbiota represents a vital adaptation to plant-based diets, enhancing nutrient utilization and exemplifying the intricate interdependence between insects and their microbial partners (Engel and Moran, 2013).

13.8.5.2 ENHANCED NUTRIENT AVAILABILITY

The gut microbiota of insects plays a crucial role in producing necessary amino acids, vitamins, and various nutrients that could be deficient in the insect's predominantly plant-based diet. This activity contributes significantly to the insect's overall nutritional equilibrium (Nowakowski et al., 2022). The gut microbiota of herbivorous insects plays a pivotal role in augmenting nutrient availability. These microorganisms, residing in the insect's digestive system, actively contribute to the breakdown of complex plant materials. Through processes like fermentation and enzymatic activities, the gut microbiota assists in the digestion of cellulose and hemicellulose, releasing valuable nutrients such as short-chain fatty acids (Singh et al., 2019). The collaborative effort results in an enriched nutritional profile for the herbivorous insect, underscoring the essential role of gut microbiota in optimizing nutrient utilization and supporting the overall fitness of these insects in their plant-based ecological niches (Greenwood, 2021).

13.8.5.3 DETOXIFICATION AND PLANT COMPOUND PROCESSING

The gut microbiota has the ability to neutralize plant secondary metabolites, which are frequently harmful or hindering to insects. They possess the capability to metabolize and alter these substances, thus allowing insects to consume a wider variety of plant species without the risk of toxicity (Divekar et al., 2022). The gut microbiota of herbivorous insects assumes a crucial role in detoxifying plant compounds. These microorganisms possess enzymes that effectively break down secondary plant metabolites, thereby allowing insects to feed on a variety of plants without succumbing to toxic effects (Genta et al., 2006). The microbial detoxification process significantly enhances the adaptability of herbivores to diverse plant diets, highlighting a symbiotic relationship that facilitates the exploitation of ecological niches. This detoxification mechanism underscores the pivotal role that gut microbiota play in the co-evolutionary dynamics between herbivorous insects and the plants they consume (Chiu and Gilbert, 2020).

13.8.5.4 IMMUNE SYSTEM REGULATION

The gut microbiota has the potential to impact the immune reactions of insects, thereby bolstering their ability to defend against pathogens, parasites, and illnesses. This microbial community plays a crucial role in maintaining the overall well-being and vitality of insects (Engel and Moran, 2013). The microbial community significantly influences the insect's immune response, contributing to its defense against pathogens. Through the production of antimicrobial peptides and the modulation of immune signaling pathways, the gut microbiota helps maintain a balanced and effective immune system (Engel and Moran, 2013). The symbiotic relationship enhances the insect's ability to resist infections, thereby ensuring its overall health and survival. The intricate interplay between herbivorous insects and their gut microbiota serves as a prime example of the importance of microbial contributions in shaping the immune landscape of these insects (Berasategui et al., 2016).

Comprehending the gut microbiome of herbivorous insects and its significance in the realms of digestion, nourishment, and protective mechanisms is essential for the formulation of sustainable pest control methods, enhancing agricultural yields, and gaining an understanding of the evolutionary and adaptive characteristics of these insects in various ecological contexts. Furthermore, delving into the prospect of harnessing gut microbiota for biotechnological purposes presents an exciting avenue for novel techniques in agriculture and other fields (Figure 13.3).

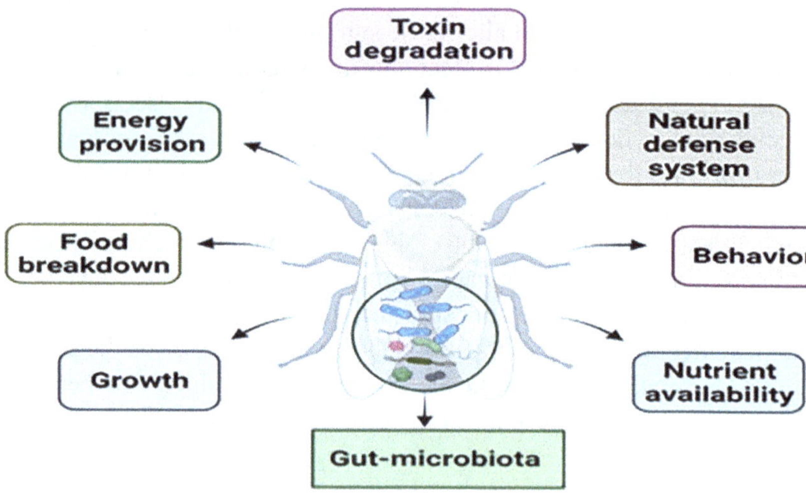

FIGURE 13.3 The multiple facets of gut microbiota on insects' biology and ecology.

Source: Reprinted from Siddiqui et al., 2022. Copyright © 2022 Siddiqui, Khan, Bamisile, Hafeez, Qasim, Rasheed, Rasheed, Ahmad, Shahid and Xu. https://creativecommons.org/licenses/by/4.0/.

13.9 MICROBIAL-MEDIATED MODIFICATIONS IN PLANT-FEEDING INSECT BEHAVIOR AND PHYSIOLOGY

Microbial-induced alterations in the behavior and physiology of plant-feeding insects pertain to the mechanisms through which microorganisms, particularly those residing within the insect's digestive system or bodily tissues, exert an impact on and mold the actions and physiological functions of these insects while they engage with their host plants.

13.9.1 BEHAVIORAL MODIFICATIONS INDUCED BY MICROBES

Microbes play a pivotal role in shaping the behavior of various organisms, such as insects, illustrating the complex relationship between microbiology and behavior. These microorganisms have the capacity to exert a substantial impact on the conduct of herbivorous insects, leading to adjustments in their interactions with host plants, other insects, and their surroundings (Kuhnert and Collemare, 2022). This comprehensive analysis delves into the alterations in behavior caused by microorganisms in herbivorous insects.

13.9.1.1 FEEDING PREFERENCES AND HOST SELECTION

Microorganisms can impact the feeding choices of insects by modifying how they interpret plant compounds. This influence may incline insects towards particular plants or plant components that are more favorable for microbial colonization or digestion (Davis et al., 2013). Herbivorous insects demonstrate discernible feeding preferences and host selection strategies, frequently shaped by microbial interactions. Microbes residing in the insect gut assume a pivotal role in breaking down specific plant compounds, thereby influencing the insect's choice of host plants (Humphrey et al., 2020). This symbiotic relationship actively contributes to the insect's capacity to digest and extract nutrients from their preferred host, highlighting the interconnected dynamics between herbivores and their microbial partners (Gonzalez, 2014).

13.9.1.2 MATING BEHAVIORS

Microbes have the ability to impact the mating behaviors of insects by modifying the generation or reception of sexual pheromones. Such alterations can have repercussions on the insects' ability to reproduce successfully, thereby affecting the overall reproductive potential of the insect population (Engl and Kaltenpoth, 2018). Mating behaviors serve not only as a fundamental aspect of reproduction but also as influencers of microbial interactions. Courtship rituals and pheromone communication play pivotal roles in attracting suitable mates. Microbes, inhabiting specialized insect environments, may exert an influence on mating by impacting insect physiology and behavior (Sarkar et al., 2020). Symbiotic microbes have the potential to affect pheromone production, influence mating preferences, or even impact reproductive success, thereby shaping the overall dynamics of herbivorous insect populations and contributing to the coevolutionary relationships between insects and their microbial partners (Chakraborty and Roy, 2021).

13.9.1.3 MOVEMENT AND NAVIGATION

Microbes have the potential to influence an insect's mobility and orientation skills, thereby shaping their foraging behavior and dispersion. This, in turn, may have consequences for their relationships with host vegetation and various elements of the environment, potentially altering their interactions (Schlägel et al., 2020). Microbes assume a critical role in the movement and navigation of herbivorous insects. Certain microbes within the insect gut contribute to the digestion of complex plant materials, thereby providing the energy necessary for insect locomotion (Schowalter, 2013). Moreover, gut microbes can influence the production of metabolites that affect the insect's behavior, including navigation and foraging patterns. The symbiotic relationship between insects and microbes significantly contributes to the adaptability of herbivores in diverse environments, influencing their movement strategies in the search for optimal food sources (Francis et al., 2021).

13.9.2 PHYSIOLOGICAL EFFECTS OF MICROBES ON PLANT-FEEDING INSECTS

The complex interplay between microorganisms and plant-feeding insects leads to a diverse array of physiological consequences affecting the insects' biology, physiology, and overall well-being (Berryman, 2012). These outcomes typically result from mutualistic associations between the insects and a variety of microorganisms, such as bacteria, fungi, and protozoa, that inhabit the insects' internal environment (Schoonhoven et al., 2005).

13.9.2.1 DIGESTIVE EFFICIENCY

The gut microbiota plays a crucial role in facilitating the digestion of plant matter, simplifying intricate substances such as cellulose and lignin into more accessible compounds that can be utilized by the insect. This, in turn, improves the insect's capacity to digest food efficiently and absorb nutrients without issues of plagiarism (Engel and Moran, 2013). The digestive efficiency of herbivorous insects is profoundly shaped by their symbiotic relationships with gut microbes. This symbiotic collaboration enables these insects to thrive on plant-based diets by providing greater nutritional accessibility (Li et al., 2021).

13.9.2.2 METABOLIC PROCESSES

Microbes have the capacity to impact an insect's metabolic functions, encompassing energy utilization, storage, and control. These microorganisms can potentially improve the insect's capacity to convert nutrients into energy or regulate essential metabolic pathways crucial for their growth and maturation (Dolezal et al., 2019). Metabolic processes in herbivorous insects are intricately linked to microbial activity within their digestive systems. Microbes, including bacteria and protozoa, play a vital role in breaking down complex plant materials, thereby aiding the insect in extracting essential nutrients (Barboza et al., 2010). The insects and their microbial partners underscore the crucial role of these microorganisms in shaping the metabolic landscape of their hosts. This symbiotic relationship is fundamental to the efficient utilization of nutrients from plant-based diets by herbivorous insects (Dearing and Weinstein, 2022).

13.9.2.3 IMMUNE RESPONSES AND STRESS TOLERANCE

The gut microbiota has the potential to engage with the immune system of insects, leading to the regulation of immune reactions and an increase in resilience to stress. This interaction may bolster the insect's capacity to endure environmental challenges and combat diseases (Siddiqui et al., 2022). Microbes make substantial contributions to immune responses by influencing the expression of antimicrobial peptides and modulating the defense mechanisms of herbivorous insects. This symbiotic relationship is integral to the overall health and adaptability of herbivorous insects in diverse ecological settings (Siddiqui et al., 2022).

13.9.2.4 REPRODUCTION AND GROWTH

Microorganisms have the potential to impact the reproductive and growth aspects of herbivorous insects. They can play a role in controlling the hormonal pathways associated with reproduction, growth, and metamorphosis (Berryman, 2012). Microbes in the gut of herbivorous insects play a crucial role in reproduction and growth. These microbial interactions significantly influence reproductive fitness, aiding in successful metamorphosis and ensuring the insect's ability to thrive on plant-based diets (Hansen and Moran, 2014). The symbiotic relationship between herbivorous insects and their gut microbes exemplifies the intricate interplay that shapes the reproductive and growth dynamics of these insects (Hammer and Moran, 2019).

Understanding the microbial-mediated changes is crucial for gaining insight into the comprehensive connections involving herbivorous insects, their microbial partners, and the plants they feed on. Furthermore, it reveals the impact of the complex associations between microorganisms and insects on insect behavior, physiology, and their role in ecosystems. These findings have the potential to be utilized in diverse areas, such as pest control, enhancing farming methods, and ecological preservation.

13.10 MICROBIAL INFLUENCE ON PLANT-FEEDING INSECT–PLANT INTERACTIONS

The microbial impact on interactions between plant-feeding insects and plants is a crucial component of the complex ecological connections that influence both insect herbivores and their plant counterparts. Microbes, whether they reside in the insect's digestive system or are associated with plant tissues, play a substantial role in facilitating these interactions, affecting plant defense mechanisms, insect herbivore activity, and the broader dynamics of ecosystems (Hansen and Moran, 2014). Microbial influence on plant-feeding insect-plant interactions plays a pivotal role in shaping the dynamics of ecological relationships. Beneficial microbes associated with plant roots, such as rhizobacteria and mycorrhizal fungi, contribute to nutrient absorption and enhance the plant's defensive capabilities (Smulders, 2022). These microbial interactions extend to herbivorous insects, influencing their feeding behavior by modifying the nutritional content and defensive chemistry of plant tissues. Additionally, certain microbes release VOCs that act as deterrents or attractants, creating a cascade of indirect interactions (Raza and Shen, 2020).

13.10.1 MICROBIAL-INDUCED PLANT DEFENSES

Microbial-induced plant defenses encompass the methods by which microorganisms, including beneficial bacteria and fungi, can activate and amplify a plant's innate protective responses against pathogens and herbivores. These engagements are integral components of the intricate and ever-changing web of connections within the plant microbiome (Sugio et al., 2015). Specific microbes can activate signaling pathways in plants, thereby enhancing resistance mechanisms. These induced defenses commonly manifest as the production of secondary metabolites or the activation of physical barriers, providing a dynamic example of how microbial interactions contribute to the ecological balance between plants and herbivorous insects. This intricate relation showcases the role of microorganisms in shaping the plant's ability to defend itself against herbivory and maintain ecological equilibrium (Pieterse et al., 2014).

Microbes have the capacity to prepare plant defenses, which facilitates a faster and more robust reaction when the plant faces herbivore threats. This process may encompass the initiation of genes associated with defense, the synthesis of protective compounds, or the fortification of physical safeguards (Ab Rahman et al., 2018). Through intricate signaling pathways, certain beneficial microbes induce a heightened state of alertness in plants, enabling a faster and more effective response to herbivore attacks (Bostock, 2005). This priming mechanism enhances the synthesis of defensive compounds, such as secondary metabolites and signaling molecules, fortifying the plant's resistance and improving its overall defense strategies against potential insect threats. This mechanism showcases the adaptability and responsiveness of plants to environmental challenges through microbial interaction (Mostafa et al., 2022).

13.10.2 MICROBES AS MEDIATORS OF PLANT RESISTANCE

Microbes, frequently integral components of a plant's microbiome, can play a pivotal role in augmenting a plant's innate defenses against a range of stressors, such as pathogens and herbivores. They act as intermediaries in bolstering the plant's resistance through various interactions (Lazebnik et al., 2014).

13.10.2.1 INDUCTION OF PLANT RESISTANCE

Microbes have the capacity to trigger systemic resistance within plants, thereby augmenting their resilience against insect herbivores. This process may encompass the activation of defense-related genes or the preparation of the plant's immune mechanisms (Pieterse et al., 2014). The induction of plant resistance entails the activation of defense mechanisms in response to microbial interactions with herbivorous insects. Microbes, including endophytic fungi or rhizosphere bacteria, can initiate plant defense pathways, resulting in the production of secondary metabolites and proteins that discourage herbivores (Rashid and Chung, 2017). This microbial-induced resistance not only strengthens plant resilience to insect feeding but also presents a sustainable approach in integrated pest management (IPM) strategies. Understanding these intricate interactions opens avenues for harnessing microbial partners to enhance and fortify plant defenses against herbivorous insects (War et al., 2012).

13.10.2.2 ALTERATION OF PLANT PHYSIOLOGY

Microbes have the capacity to modify the physiological aspects of plants, rendering them less favorable for herbivorous consumption. Such alterations can encompass shifts in nutrient accessibility, nutrient makeup, or the physical attributes of the plant's tissues, thereby reducing the plant's attractiveness to herbivores (Provenza et al., 2007). Insects, like aphids, may transmit microbial symbionts that induce changes in plant defense mechanisms or manipulate nutrient availability. These alterations can significantly influence the plant's growth, development, and response to environmental stressors. The dynamic interplay between microbes and herbivorous insects underscores the far-reaching impact on both insect herbivores and their plant hosts (Giron et al., 2017). This complex relationship exemplifies the interconnectedness of the insect-plant-microbe triad, highlighting the intricate ways in which microorganisms can influence the physiology and ecology of both the herbivorous insects and the plants (Sharma et al., 2021).

13.10.2.3 MODULATION OF PLANT SIGNALING

Microbes have the capacity to influence the signaling pathways in plants, thereby impacting the way plants detect herbivore presence and their subsequent reactions. As a result, this microbial influence can cause shifts in plant defense mechanisms and the distribution of resources (Giron et al., 2013). Microbes associated with herbivorous insects play a crucial role in modulating plant signaling. Through intricate interactions, these microbes can influence plant defense mechanisms and alter signaling pathways. This modulation often results in a complex interplay between herbivores, their associated microorganisms, and the host plants (Pangesti et al., 2013). This highlights the intricate relationships that shape ecological dynamics in the context of insect-plant-microbe interactions, showcasing the sophisticated ways in which microorganisms can impact the signaling networks of plants and contribute to the overall ecology of these interactions (Gross et al., 2022).

13.11 CASE STUDIES AND RESEARCH EXAMPLES

Microbes wield a substantial influence over herbivorous insects, affecting their conduct, physiology, and relationships with the plants they feed on. The ensuing instances illustrate the microbial effects on herbivorous insects.

13.11.1 BUCHNERA APHIDICOLA IN APHIDS

Buchnera aphidicola serves as the primary endosymbiont within aphids, playing a crucial role in the synthesis of vital amino acids that are lacking in the aphid's plant sap-based diet. This mutualistic partnership is pivotal for the aphid's survival, allowing it to flourish despite its nutritionally restricted food source (Colella et al., 2018). *Buchnera* plays a crucial role by synthesizing essential amino acids that are lacking in the aphid's plant sap diet. This symbiotic relationship exemplifies the intricate collaboration between microbes and herbivorous insects, ensuring the nutritional fitness of aphids and underscoring the significance of such partnerships in the ecology of plant-feeding insects (Feng et al., 2019). The close association between Buchnera and aphids highlights the intricate web of dependencies that shape the survival and ecological interactions of these herbivorous insects (Moran and Degnan, 2006).

13.11.2 WOLBACHIA IN FRUIT FLIES

Wolbachia, a symbiotic bacterium that resides within various insect species, such as fruit flies, possesses the capacity to trigger cytoplasmic incompatibility. This phenomenon disrupts the reproductive mechanisms of these insects, resulting in changes in mating behaviors and, consequently, exerting an impact on the dynamics of their populations (Coscrato et al., 2009). The presence of *Wolbachia* in fruit flies highlights the intricate interactions between microbes and herbivorous insects, impacting reproductive dynamics and potentially influencing the overall ecology of insect populations (Kolasa and Łukasik, 2023). This symbiotic relationship exemplifies how microorganisms can play a crucial role in shaping the reproductive strategies and population dynamics of herbivorous insects, showcasing the complex interplay between microbes and their insect hosts in ecological systems (Kolasa and Łukasik, 2023).

13.11.3 PECTINOLYTIC BACTERIA IN DIAMONDBACK MOTHS

Diamondback moths carry pectin-degrading bacteria in their intestinal tract. These microorganisms generate enzymes that facilitate the degradation of pectin, which is an intricate plant polysaccharide located within cell walls. These enzymes play a crucial role in the moth's ability to digest plant matter, thus supplying the insect with extra nutrients (Mazumdar, 2020). Pectinolytic bacteria play a significant role in the digestive processes of diamondback moths, aiding in the breakdown of pectin, a complex carbohydrate found in plant cell walls (Paniagua Voirol et al., 2018). These bacteria contribute to the moth's ability to efficiently extract nutrients from plant material, highlighting the importance of microbial interactions in herbivorous insect digestion. The presence of pectinolytic bacteria exemplifies the intricate relationship between microbes and herbivorous insects, influencing the insect's capacity to thrive on plant-based diets (Mason et al., 2019).

13.11.4 REGIELLA INSECTICOLA IN PEA APHIDS

Pea aphids carrying the endosymbiont *Regiella insecticola* demonstrate an improved ability to fend off parasitoid wasps. This endosymbiont plays a vital role in generating a protective substance that discourages parasitoid wasps, ultimately enhancing the chances of survival for the aphids (Feldhaar, 2011). *Regiella insecticola,* identified in pea aphids, stands as a symbiotic bacterium that significantly contributes to the ecological success of these insects. It plays a key role in nutrient supplementation, particularly in providing essential amino acids. This example underscores the intricate microbial partnerships that are crucial for the adaptation of herbivorous insects (Frago et al., 2020). The presence of *Regiella insecticola* exemplifies the nuanced and pivotal role of microbes in shaping the biology of herbivores, influencing their nutritional ecology, and contributing to their overall fitness in their ecological niche (Ryalls et al., 2013).

13.11.5 HAMILTONELLA DEFENSA IN APHIDS

Hamiltonella defensa, an auxiliary endosymbiont residing within aphids, affords safeguard to these insects against parasitoid wasps through the synthesis of toxins detrimental to the wasps. This protective strategy serves to elevate the aphids' chances of survival (Oliver et al., 2012). *Hamiltonella defensa,* a symbiotic bacterium found in aphids, plays a critical role in defending its insect host against parasitoid wasps. It produces toxins that are harmful to wasp larvae, thereby enhancing the aphids' resistance to parasitoid attacks (Mandrioli et al., 2016). The presence of *Hamiltonella defensa* showcases how symbiotic relationships with microbes can confer adaptive advantages to herbivorous insects, influencing their interactions with other organisms in their environment and contributing to their overall ecological success (Vorburger and Rouchet, 2016).

13.11.6 ENTEROBACTER IN SPODOPTERA CATERPILLARS

Some varieties of Enterobacter, a component of the gut microbiota found in Spodoptera caterpillars, have the capability to generate compounds that affect the eating habits of these caterpillars. This modification in their feeding behavior has consequences on the caterpillars' selection of food and dietary inclinations (Xia et al., 2020). The microbial presence in the gut of herbivorous insects, such as Spodoptera caterpillars, may contribute to nutrient metabolism and impact the insects' physiological processes (Indiragandhi et al., 2011). The specific role of Enterobacter in the context of Spodoptera caterpillars highlights the intricate connections between microbes and herbivorous insects, shedding light on how these microorganisms influence the adaptation and ecological interactions of the insects they inhabit. This microbial partnership underscores the complexity of the relationships that shape the biology and ecology of herbivorous insects (Sugio et al., 2015).

These instances underscore the various methods through which microorganisms' impact herbivorous insects, encompassing the provision of vital nutrients, the manipulation of reproductive systems, modification of digestion processes, and reinforcement of defenses against natural adversaries. Comprehending these connections is essential for crafting precise approaches to pest control and progressing our comprehension of the interplay among insects, plants, and microbes.

13.12 IMPLICATIONS AND FUTURE PERSPECTIVES OF THE MICROBIAL IMPACT ON PLANT-FEEDING INSECTS

Understanding the implications and future perspectives of the microbial impact on plant-feeding insects is crucial for advancing various fields, including agriculture, ecology, and biotechnology. The implications and potential directions for future research and applications in this domain are discussed in subsections.

13.12.1 BIOLOGICAL PEST CONTROL

Utilizing insights from microbe-facilitated interactions between insects and plants has the potential to guide the creation of microbial biocontrol agents (BCAs) while ensuring originality in the research (Malone, 2020). Beneficial microorganisms have the potential to manage pest populations effectively, diminishing the need for chemical pesticides and encouraging eco-friendly pest control approaches (Sarwar and Technology, 2015). Biological pest control strategies involving microbes and herbivorous insects represent an eco-friendly and sustainable approach to managing agricultural pests. One notable example is the use of entomopathogenic microorganisms, such as bacteria (e.g., *Bacillus thuringiensis*) and fungi (e.g., *Beauveria bassiana* and *Metarhizium anisopliae*) (Shah et al., 2021). These microbes possess insecticidal properties, causing infections that lead to the death of targeted pest populations. Upon application to crops, these microbial agents interact with herbivorous insects, either by penetrating their cuticles or through ingestion, subsequently colonizing the insect host (Kachhawa and Studies, 2017). Once inside, the microbes release toxins or enzymes that disrupt essential physiological processes, ultimately resulting in the demise of the pest. This approach provides an effective and environmentally friendly alternative to traditional chemical pesticides in pest management (Usta, 2013).

13.12.2 ENHANCED CROP RESILIENCE AND PRODUCTIVITY

Altering the gut microbiota of plant-feeding insects with the aim of improving plant resilience and productivity represents an intriguing prospect. Adjusting these microbial communities to enhance nutrient absorption, bolster stress resistance, and promote overall plant well-being holds the potential to augment crop yields and yield more favorable results in agriculture (Lange et al., 2023). Beneficial microbes associated with insects, including bacteria, fungi, and archaea, play crucial roles in nutrient cycling and plant health. These microorganisms assist in breaking down complex plant materials, promoting nutrient availability in the soil (Prakash et al., 2015). Moreover, they contribute to the suppression of harmful pathogens through the production of antimicrobial compounds, thereby reducing the risk of crop diseases (González-Lamothe et al., 2009).

The symbiotic relationships between insects and these microbes can improve soil structure, water retention, and nutrient uptake by plants. Furthermore, some insect-associated microbes have been shown to enhance plant resistance to environmental stressors, such as drought or pests, resulting in more resilient and productive crops (Bonkowski et al., 2009). This underscores the importance of understanding and harnessing the positive interactions between insects and beneficial microbes for sustainable agriculture (Mohan et al., 2022).

13.12.3 BIOFERTILIZERS AND NUTRIENT CYCLING

Symbiotic microorganisms have the potential to facilitate nutrient cycling and biofertilization. Gaining insight into and harnessing the potential of these microorganisms can pave the way for the creation of biofertilizers, thereby enhancing soil quality, increasing nutrient accessibility for plants, and decreasing the reliance on artificial fertilizers (Kumar et al., 2022). Biofertilizers, particularly those involving microbial interactions with herbivorous insects, play a pivotal role in nutrient cycling within ecosystems. Microbes associated with herbivorous insects contribute to the breakdown of organic matter, facilitating nutrient release (Gadhave and Gange, 2018). In the symbiotic relationships between insects and gut microbes, nutrient transformations occur, influencing the availability of essential elements for plant growth (Frago et al., 2020). These interactions exemplify the interconnected processes of nutrient cycling, as microbes aid in decomposing organic materials, ultimately contributing to the enrichment of soil fertility and promoting sustainable agricultural practices (Choudhary et al., 2016). The utilization of these microbial interactions offers a natural and eco-friendly approach to enhance nutrient availability in soils, showcasing the potential for integrated and environmentally conscious agricultural strategies (Hakim et al., 2021).

13.12.4 PROMOTING SUSTAINABLE FARMING PRACTICES

By incorporating techniques centered around microorganisms into agricultural methods, farmers have the opportunity to embrace sustainable farming systems that are not only economically feasible but also eco-friendly (Trivedi et al., 2021). This involves adopting IPM approaches that encompass microbial control methods, with the aim of reducing the overall environmental footprint of farming (Deguine et al., 2021). Encouraging practices that foster a balanced microbiome in agricultural ecosystems, such as the use of organic amendments or conservation tillage, enhances soil health and promotes the resilience of both crops and beneficial insect populations (Arif et al., 2020). By understanding and optimizing these microbial-herbivore interactions, sustainable farming can prioritize ecological harmony, preserving biodiversity while ensuring food security for the long term. This holistic approach emphasizes the importance of working with nature's processes to create resilient and environmentally friendly agricultural systems (Koohafkan et al., 2012).

13.12.5 ENHANCED STRESS TOLERANCE

Microbes play a pivotal role in assisting plants in managing adverse environmental conditions like drought, salinity, or extreme temperatures. Fostering a mutually beneficial relationship between particular micro-organisms and plants can elevate their ability to withstand stress, thus fostering the growth of crops in demanding environmental circumstances (Grover et al., 2011). Microbial symbionts in herbivorous insects indeed contribute to enhanced stress tolerance, enabling their hosts to thrive in challenging environments. These microbes, through various mechanisms, assist in the detoxification of plant compounds and provide resilience against environmental stressors (Harman and Uphoff, 2019). The symbiotic relationship plays a crucial role in the adaptive strategies of herbivorous insects, allowing them to overcome nutritional and ecological challenges in their habitats. By aiding in detoxification and enhancing stress tolerance, these microbial symbionts contribute significantly to the overall fitness and survival of herbivorous insects (Feldhaar, 2011).

13.12.6 PROMOTING SUSTAINABLE CROP ROTATION

Microbes can have a significant impact on the promotion of sustain-able crop rotation techniques. Specific microbial communities have the capability to decompose agricultural remnants and improve the composition of the soil, thus fostering a more robust soil environment and supporting the longevity of agricultural rotations (Ai et al., 2018).

Implementing sustainable crop rotation practices is not only beneficial for soil health but also plays a significant role in shaping microbial communities essential for pest control (Tahat et al., 2020). The rotation strategy disrupts the habitat of herbivorous insects, thereby restraining the proliferation of pest populations. Simultaneously, beneficial microbes actively contribute to nutrient cycling and the suppression of diseases. This comprehensive approach fosters agricultural sustainability by cultivating a balanced ecosystem that favors both crops and their beneficial microbial counterparts (Poveda et al., 2008).

13.12.7 ALTERED FEEDING BEHAVIOR AND DEFENSE STRATEGIES

Microbial-triggered plant defenses and resistance mechanisms have the potential to modify the feeding patterns of herbivorous insects. In response to enhanced plant defenses, insects may either steer clear of or decrease their consumption of these plants, thereby affecting their overall well-being and the dynamics of their populations (Enders and Begcy, 2021). The impact of microorganisms on nutrient accessibility and digestion plays a crucial role in the adaptation of herbivorous insects to a variety of plant-based diets. This highlights the complex interaction between microbes and the feeding habits of their insect hosts (Lemoine et al., 2020). This symbiotic connection signifies a co-evolutionary phenomenon that promotes the ecological success of herbivorous insects (Pascal et al., 2019). In reaction to the plant defenses triggered by microorganisms, insects can develop a range of tactics to counter these defenses. These strategies may include the development of unique detoxification mechanisms or adaptations in their feeding behavior (Mason, 2020). Insects maintain symbiotic microorganisms within their digestive systems, contributing to the detoxification of plant compounds. The involvement of microbial enzymes facilitates the breakdown of defensive chemicals, enabling insects to consume plants without being adversely affected by toxicity. This mutualistic relationship between insects and microbes illustrates a sophisticated adaptation that helps them navigate the hurdles presented by plant defenses.

13.12.8 BIOTECHNOLOGICAL APPLICATIONS

By employing the principles of synthetic biology, scientists have the capacity to create mutualistic connections between insects and microorganisms, granting them particular characteristics, such as increased resilience to pests or more efficient nutrient uptake. This innovative strategy has the potential to bring about a profound transformation in the realms of agriculture and ecological control (Sargent et al., 2022). Modifying the gut microbiome of plant-feeding insects for the purpose of altering their digestion, metabolism, or immune reactions holds promise in enhancing their influence on plants. Customizing the insect's gut microbiome has the potential to bolster plant defenses against herbivores and diseases (Lange et al., 2023).

13.12.9 MULTI-OMICS APPROACHES

Employing various multi-omics methodologies, such as metagenomics and meta-transcriptomics, enables a thorough comprehension of the intricate molecular-level interactions that occur among microorganisms, insects, and plants (Kumar et al., 2023). The investigation of microbes in herbivorous insects through multi-omics approaches entails the integration of data from genomics, transcriptomics, proteomics, and metabolomics. The comprehensive approach aims to gain a thorough understanding of the dynamics involved in host-microbe interactions (Mucci, 2022). Through this all-encompassing analysis, researchers can uncover the genetic, functional, and metabolic complexities that govern symbiotic relationships (Van der Greef et al., 2007). The examination of the entire molecular landscape using multi-omics approaches provides valuable insights into the roles played by microbes in the digestion, metabolism, and adaptation of herbivorous insects. This, in turn, contributes to a deeper comprehension of the intricate ecological associations within these systems (Oates et al., 2016).

13.12.10 LONG-TERM ECOLOGICAL STUDIES

Conducting extended ecological investigations to track the evolution of microbial effects on herbivorous insects over an extended period can yield priceless information regarding the long-term stability and viability of these relationships (Engel and Moran, 2013). Long-term ecological research focused on microbes and herbivorous insects unveils dynamic interactions that shape ecosystems (Weisser and Siemann, 2008).

These extensive investigations monitor the evolution of microbial communities within the insect gut, elucidating their pivotal role in nutrient processing and host adaptation (Muhammad et al., 2017). Grasping the enduring impact of microbes on herbivores offers valuable insights into ecological stability and the delicate equilibrium within these fundamental components of terrestrial ecosystems. Through longitudinal studies, we gain a holistic understanding of the evolving relationships between microbes and herbivorous insects over time (Grime and Pierce, 2012).

13.13 CONCLUSION

In conclusion, investigating the influence of microbes on herbivorous insects offers an intriguing expedition into the complex tapestry of existence, revealing the crucial function that microorganisms serve in molding the relationships between insects and plants. The interconnection of microbes, insects, and plants creates a fluid system that impacts not just their biology but also the environments they dwell in. Microbial partners residing in the digestive tracts or tissues of plant-feeding insects have exhibited significant impacts on insect behavior, physiology, and ecological associations. Microbes have a profound impact on the lives and behaviors of insects, ranging from their assistance in digestion and improvement of nutrient absorption to their role in modulating insect immune systems and influencing feeding preferences. Furthermore, microbes are essential in mediating plant defenses, affecting plant-herbivore relationships, and subsequently shaping the overall functioning of ecosystems. Recognizing the significance of these microbial-driven interactions offers considerable potential for sustainable agriculture, the development of biological pest control methods, and innovative biotechnological applications. The ability to manipulate symbiotic relationships, enhance plant well-being, and contribute to the health of humans and animals highlights the importance of further exploration and research in this domain. As we move forward, the fusion of genetics, microbiology, ecology, and biotechnology in interdisciplinary research is poised to open up promising opportunities. The utilization of multi-omics strategies, extensive studies in ecology over extended periods, and the application of synthetic biology are instrumental in deciphering the intricate influences of microorganisms on herbivorous insects. By cultivating this understanding and promoting public consciousness, we can lay the groundwork for a more balanced and sustainable cohabitation with the environment, ultimately benefiting both ecosystems and human society. During this voyage of exploration and comprehension, we are not solely unveiling the mysteries of the natural world, but also plotting a path towards a brighter and more enlightened future. In this future, the intricate realm of microorganisms exerts a profound impact on the larger world of insects and vegetation.

GLOSSARY

1. **Antibiosis:** The negative impact of microbes on the health or survival of insects, typically through the production of toxins or antimicrobial compounds.
2. **Antixenosis:** Resistance mechanisms employed by plant to deter or prevent pest colonization.
3. **Aphids:** A common group of plant-feeding insects known for their ability to transmit plant pathogens and often associated with specific endosymbiotic bacteria.
4. **Attractant:** Chemicals that attract any organism.
5. **Bacterial Endosymbionts:** Bacteria that live inside the bodies of insects and play various roles, such as aiding in digestion or providing essential nutrients.
6. **Begomoviruse:** It is a genus of viruses that belong to family Geminiviridae present in whitefly
7. **Bio-stimulants:** substance or microorganism that when applied to plants, stimulates natural process to benefit nutrient uptake and efficiency, tolerance to abiotic and biotic factors.
8. **Co-Evolution:** The long-term reciprocal influence between two species, where changes in one species may lead to adaptations or changes in the other species.
9. **Elicitor:** insect release chemicals with its saliva while feeding that triggered plant defense system or attract natural enemies.

10. **Endosymbionts:** Microbes that live inside the bodies of insects and have a mutually beneficial relationship with their hosts. They can provide essential nutrients or help with digestion.
11. **Entomopathogenic:** Refers to microbes or pathogens that can infect and kill insects.
12. **Entomopathogeny:** A process in which nematodes, along with their bacterial symbionts, swiftly exterminate insects and use both insect and bacterial resources to complete their life cycles.
13. **Hemolymph:** The circulatory fluid in insects, somewhat analogous to blood in vertebrates.
14. **Honeydew:** Excreta of sucking pest having high content of sugar.
15. **Insect-Plant Interactions:** The complex relationships and adaptations between plants and insects, which may be influenced by microbial associations.
16. **Microbes:** Microorganisms, such as bacteria and fungi, that are often too small to be seen with the naked eye and can have various interactions with plants and insects.
17. **Microbiome:** The collective community of microorganisms, including bacteria, fungi, and viruses, living in or on a host organism.
18. **Multitrophic:** Interactions that links several trophic levels, including plants, herbivore and natural enemies.
19. **Mutualism:** A type of symbiotic relationship in which both the microbe and the insect benefit from their association.
20. **Necromeny**: A scenario where nematodes consume dead insects but do not cause their death.
21. **Nutritional Ecology:** The study of how organisms obtain and utilize nutrients from their environment, which is relevant to understanding the relationships between microbes and insects.
22. **Parasitism:** A type of symbiotic relationship where one organism (the parasite) benefits at the expense of the other organism (the host).

23. **Parthenogenesis:** Type of asexual reproduction in which eggs develops into embryo without being fertilized by a sperm.
24. **Pathogens:** Microbes that can cause diseases or harm to plants and insects.
25. Phenotypic: it refers to the observable characteristics of an individual resulting from the interaction of its genotype with the environment.
26. **Phloem:** A part of the plant vascular system that transports the products of photosynthesis, which is a common food source for some plant-feeding insects.
27. **Phytohormones:** These are the chemical compounds that regulate plant growth and development.
28. **Phytopathogenic:** Organisms that cause disease in plant.
29. **Phytophagy:** The act of feeding on plant material, characteristic of plant-feeding insects.
30. **Plant-Feeding Insects:** Insects that primarily feed on plant tissues, which can include herbivores, pests, and phytophagous insects.
31. **Polyphagous:** Insects that feed on many host belong to different families of plant.
32. **Probing:** Piercing the mouth part of insect into the plant tissue.
33. **Reciprocal Effects:** The influence that microbes have on insects and, conversely, how insects may affect microbial communities.
34. **Repellant:** Chemical that repels the organisms away from its host.
35. **Resurgence:** Rapid reappearance of a pest population in injurious numbers following pesticide application.
36. **Salivation:** Secrete saliva from insect mouth while feeding.
37. **Symbiosis:** A close and long-term interaction between two different biological species. In this context, it might refer to the relationships between microbes and plant-feeding insects.

38. **Tolerance:** Type of adaptation that allows a plant to grow better that other plant after insect or pathogen damage.
39. **Trehalulose:** Type of sugar present in honey dew of sucking pests.
40. **Xylem:** A part of the plant vascular system that transports water and dissolved nutrients, which can be accessed by plant-feeding insects.

KEYWORDS

- **ecological stability**
- **endosymbionts**
- **herbivorous insects**
- **microbes**
- **pathogens**
- **plant pathology**

REFERENCES

Ab Rahman, S. F. S., Singh, E., Pieterse, C. M., & Schenk, P. M. (2018). Emerging microbial biocontrol strategies for plant pathogens. *Plant Science, 267*, 102–111.

Ahuja, I., Rohloff, J., & Bones, A. M. (2011). Defense mechanisms of *Brassicaceae*: Implications for plant-insect interactions and potential for integrated pest management. *Science Advances in Vegetation, 623*, 623–670.

Ai, C., Zhang, S., Zhang, X., Guo, D., Zhou, W., & Huang, S. (2018). Distinct responses of soil bacterial and fungal communities to changes in fertilization regime and crop rotation. *Geoderma, 319*, 156–166.

Ali, S., Aqueel, M. A., Saeed, M. F., Shakeel, Q., Raheel, M., & Ullah, M. I. (2022). Utilization of entomopathogenic bacteria for modern insect pest management. In *New and Future Development in Biopesticide Research: Biotechnological Exploration*. Springer.

Arab, D. A. K. A. (2021). Insects and their endosymbionts: Phylogenetics and evolutionary rates.

Arif, I., Batool, M., & Schenk, P. M. (2020). Plant microbiome engineering: Expected benefits for improved crop growth and resilience. *Trends in Biotechnology, 38*, 1385–1396.

Arora, J. (2022). *Functional Metagenomics and Evolution of Termite Gut Microbiome* (Doctoral dissertation).

Arora, R., & Malik, G. (2021). Microbe-plant-insect interactions: A comparative dissection of interactome. *Plant-Pathogen Interactions for Modern Methods in Terrestrial Crop Ecology and Environment, 365*, 365–398.

Ashra, H., & Nair, S. (2022). Trait plasticity during plant-insect interactions: From molecular mechanisms to impact on community dynamics. *Plant Science, 317*, 111188.

Bagheri, A., Askari Seyahooei, M., & Fathipour, Y. (2021). The endophytes. *Methods in Applied Fungal Inoculant Production and Management, 151*, 151–215.

Bamisile, B. S., Akutse, K. S., Siddiqui, J. A., & Xu, Y. J. (2021). Model application of entomopathogenic fungi as alternatives to chemical pesticides: Prospects, challenges, and insights for next-generation sustainable agriculture. *Frontiers in Plant Science, 12*, 741804.

Barboza, P., Bennett, A., Lignot, J. H., Mackie, R., McWhorter, T., Secor, S., Skovgaard, N., Sundset, M., Wang, T., & Zoology, B. (2010). Digestive challenges for vertebrate animals: Microbial diversity, cardiorespiratory coupling, and dietary specialization. *Physiology & Zoology Bulletin, 83*, 764–774.

Bardgett, R. D., & Van Der Putten, W. H. (2014). Belowground biodiversity and ecosystem functioning. *Nature, 515*, 505–511.

Baumgarthuber, C. (2021). *Fermented Foods: The History and Science of a Microbiological Wonder*. Reaktion Books.

Behar, A., Jurkevitch, E., & Yuval, B. (2008). Bringing back the fruit into fruit fly–bacteria interactions. *Microbial Ecology, 17*, 1375–1386.

Berasategui, A., Shukla, S., Salem, H., & Kaltenpoth, M. (2016). Potential applications of insect symbionts in biotechnology. *Applied Microbiology and Biotechnology, 100*, 1567–1577.

Berryman, A. A. (2012). *Forest Insects: Principles and Practice of Population Management.* Springer Science & Business Media.

Biedermann, P. H., & Vega, F. E. (2020). Ecology and evolution of insect–fungus mutualisms. *Annual Review of Entomology, 65*, 431–455.

Biere, A., & Tack, A. J. (2013). Evolutionary adaptation in three-way interactions between plants, microbes, and arthropods. *Functional Ecology, 27*, 646–660.

Bongard, C. (2012). A review of the influence of root-associating fungi and root exudates on the success of invasive plants. *NeoBiota, 14*, 21–45.

Bonkowski, M., Villenave, C., & Griffiths, B. (2009). *Rhizosphere Fauna: The Functional and Structural Diversity of Intimate Interactions of Soil Fauna with Plant Roots.* Springer.

Bostock, R. M. (2005). Signal crosstalk and induced resistance: Straddling the line between cost and benefit. *Annual Review of Phytopathology, 43*, 545–580.

Bravo, A., Likitvivatanavong, S., Gill, S. S., & Soberón, M. (2011). *Bacillus thuringiensis*: A story of a successful bioinsecticide. *Insect Biochemistry and Molecular Biology, 41*, 423–431.

Brin, A., & Bouget, C. (2018). Biotic interactions between saproxylic insect species. *Saproxylic Insect Diversity, Ecology & Conservation, 471*, 471–514.

Brunetti, M. (2022). *Multilevel Interactions Between Plants, Microbes and Insects: Ecological and Evolutionary Constraints Underlying Interactions.*

Cayetano, L., & Vorburger, C. (2015). Symbiont-conferred protection against hymenopteran parasitoids in aphids: How general is it? *Ecological Entomology, 40*, 85–93.

Chakraborty, A., & Roy, A. (2021). Microbial influence on plant-insect interaction. *Plant-Pathogen Interactions for Modern Methods in Terrestrial Crop Ecology and Environment, 337*, 337–363.

Chandra, P. (2019). Soil–microbes–plants: Interactions and ecological diversity. *Plant-Microbe Interface, 145*, 145–176.

Chewning, S. S. (2018). *Microbial Partners in Health: Broadening Our Understanding of Host-Microbiome Relationships.*

Chisholm, P. J. (2017). *Reciprocal Interactions Between Non-Vector Community Members and an Insect-Transmitted Plant Pathogen* (Doctoral dissertation, Washington State University).

Chiu, L., & Gilbert, S. F. (2020). Niche construction and the transition to herbivory: Phenotype switching and the organization of new nutritional modes. In *Phenotypic Switching*. Elsevier.

Chomel, M., Guittonny-Larchevêque, M., Fernandez, C., Gallet, C., Desrochers, A., Paré, D., Jackson, B. G., & Baldy, V. (2016). Plant secondary metabolites: A key driver of litter decomposition and soil nutrient cycling. *Journal of Ecology, 104*, 1527–1541.

Chong, R. A., & Moran, N. A. (2018). Evolutionary loss and replacement of *Buchnera*, the obligate endosymbiont of aphids. *Trends in Insect Journal, 12*, 898–908.

Choudhary, D. K., Varma, A., & Tuteja, N. (2016). *Plant-Microbe Interaction: An Approach to Sustainable Agriculture*. Springer.

Ciancio, A. (2016). Symbiotic relationships. *Invertebrate Biology, Function, Evolution & Ties, 49*, 49–96.

Clark, E. L., Karley, A. J., & Hubbard, S. F. (2010). Insect endosymbionts: Manipulators of insect herbivore trophic interactions? *Physiological Entomology, 244*, 25–51.

Colella, S., Parisot, N., Simonet, P., Gaget, K., Duport, G., Baa-Puyoulet, P., Rahbé, Y., Charles, H., Febvay, G., & Callaerts, P. (2018). Bacteriocyte reprogramming to cope with nutritional stress in a phloem sap-feeding hemipteran, the pea aphid *Acyrthosiphon pisum*. *Frontiers in Plant Science, 9*, 1498.

Coleman, D. C., & Wall, D. H. (2015). Soil fauna: Occurrence, biodiversity, and roles in ecosystem function. *Soil Microbiology, Ecology & Biochemistry, 4*, 111–149.

Coolen, S., Rogowska-Van Der Molen, M., & Welte, C. U. (2022). The secret life of insect-associated microbes and how they shape insect–plant interactions. *FEMS Microbiology Ecology, 98*, fiac083.

Coscrato, V. E., Braz, A. S., Perondini, A. L. P., Selivon, D., & Marino, C. L. (2009). *Wolbachia* in *Anastrepha* fruit flies (Diptera: Tephritidae). *Current Microbiology, 59*, 295–301.

Council, N. R. (2007). *The New Science of Metagenomics: Revealing the Secrets of Our Microbial Planet.*

Da Silva, J. A. T. (2006). *Floriculture, Ornamental and Plant Biotechnology*. Global Science Books.

Darling, E., Quintanilla-Tornel, M., & Chung, H. (2022). Plant-eating nematodes and the key to fighting them. *Science Bulletin.*

Davis, T. S., Crippen, T. L., Hofstetter, R. W., & Tomberlin, J. K. (2013). Microbial volatile emissions as insect semiochemicals. *Journal of Chemical Ecology, 39*, 840–859.

Dearing, M. D., & Weinstein, S. B. (2022). Metabolic enabling and detoxification by mammalian gut microbes. *Annual Review of Microbiology, 76*, 579–596.

Degnan, P. H., Yu, Y., Sisneros, N., Wing, R. A., & Moran, N. A. (2009). *Hamiltonella defensa*, genome evolution of a protective bacterial endosymbiont from pathogenic ancestors. *Proceedings of the National Academy of Sciences, 106*, 9063–9068.

Deguine, J. P., Aubertot, J. N., Flor, R. J., Lescourret, F., Wyckhuys, K. A., & Ratnadass, A. (2021). Integrated pest management: Good intentions, hard realities. A review. *Agronomy for Sustainable Development, 41*, 38.

Dicke, M., Cusumano, A., & Poelman, E. H. (2020). Microbial symbionts of parasitoids. *Annual Review of Entomology, 65*, 171–190.

Diogo, R. (2017). *Evolution Driven by Organismal Behavior*. Springer.

Divekar, P. A., Narayana, S., Divekar, B. A., Kumar, R., Gadratagi, B. G., Ray, A., Singh, A. K., Rani, V., Singh, V., & Singh, A. K. (2022). Plant secondary metabolites as defense tools against herbivores for sustainable crop protection. *International Journal of Molecular Sciences, 23*, 2690.

Dolezal, T., Krejcova, G., Bajgar, A., Nedbalova, P., & Strasser, P. (2019). Molecular regulations of metabolism during immune response in insects. *Insect Biochemistry and Molecular Biology, 109*, 31–42.

Dong, H., Huang, L., Zhao, L., Zeng, Q., Liu, X., Sheng, Y., Shi, L., Wu, G., Jiang, H., & Li, F. (2022). A critical review of mineral–microbe interaction and co-evolution: Mechanisms and applications. *National Science Review, 9*, nwac128.

Douglas, A. E. (2016). How multi-partner endosymbiosis function. *Nature Reviews Microbiology, 14*, 731–743.

Douglas, A. E. (2022). *Insects and Their Beneficial Microbes*. Princeton University Press.

Dubey, A., Malla, M. A., Khan, F., Chowdhary, K., Yadav, S., Kumar, A., Sharma, S., Khare, P. K., & Khan, M. L. (2019). Soil microbiome: A key player for conservation of soil health under changing climate. *Biodiversity and Conservation, 28*, 2405–2429.

Duperron, S. (2016). *Microbial Symbioses*. Elsevier.

Eben, A. (2022). Ecology and evolutionary history of *Diabrotica* beetles—Overview and update. *Insects, 13*, 156.

Ekramirad, N., Adedeji, A. A., & Alimardani, R. (2016). A review of non-destructive methods for detection of insect infestation in fruits and vegetables. *Innovative Food Research, 2*, 6–12.

Eleftherianos, I., Atri, J., Accetta, J., & Castillo, J. C. (2013). Endosymbiotic bacteria in insects: Guardians of the immune system? *Frontiers in Physiology, 4*, 46.

Ellers, J., Toby Kiers, E., Currie, C. R., McDonald, B. R., & Visser, B. (2012). Ecological interactions drive evolutionary loss of traits. *Ecology Letters, 15*, 1071–1082.

Enders, L., & Begcy, K. (2021). Unconventional routes to developing insect-resistant crops. *Molecular Plant, 14*, 1439–1453.

Engel, P., & Moran, N. A. (2013). The gut microbiota of insects–Diversity in structure and function. *FEMS Microbiology Reviews, 37*, 699–735.

Engl, T., & Kaltenpoth, M. (2018). Influence of microbial symbionts on insect pheromones. *Nature Product Reports, 35*, 386–397.

Feldhaar, H. (2011). Bacterial symbionts as mediators of ecologically important traits of insect hosts. *Ecological Entomology, 36*, 533–543.

Feng, H., Edwards, N., Anderson, C. M., Althaus, M., Duncan, R. P., Hsu, Y. C., Luetje, C. W., Price, D. R., Wilson, A. C., & Thwaites, D. T. (2019). Trading amino acids at the aphid–*Buchnera* symbiotic interface. *Proceedings of the National Academy of Sciences, 116*, 16003–16011.

Ferguson, L. V., & Adamo, S. A. (2023). From perplexing to predictive: Are we ready to forecast insect disease susceptibility in a warming world? *Journal of Experimental Biology, 226*, jeb244911.

Fincheira, P., & Quiroz, A. (2018). Microbial volatiles as plant growth inducers. *Microbiological Research, 208*, 63–75.

Frago, E., Zytynska, S. E., & Fatouros, N. E. (2020). Microbial symbionts of herbivorous species across the insect tree. In *Advances in Insect Physiology*. Elsevier.

Francis, J. S., Tatarko, A. R., Richman, S. K., Vaudo, A. D., & Leonard, A. S. (2021). Microbes and pollinator behavior in the floral marketplace. *Current Opinion in Insect Science, 44*, 16–22.

Gadhave, K. R., & Gange, A. C. (2018). Interactions involving rhizobacteria and foliar-feeding insects. *Applied Biochemistry and Cell Biology*, 117–133.

Genta, F. A., Dillon, R. J., Terra, W. R., & Ferreira, C. (2006). Potential role for gut microbiota in cell wall digestion and glucoside detoxification in *Tenebrio molitor* larvae. *Journal of Insect Physiology, 52*, 593–601.

Ghosh, R. (2023). *Plant-Insect Interaction Beyond Folivory: Chemical Ecology of a Non-Folivore Herbivore Leucinodes orbonalis's Interaction With its Host Solanum melongena*.

Giron, D., Dedeine, F., Dubreuil, G., Huguet, E., Mouton, L., Outreman, Y., Vavre, F., & Simon, J. C. (2017). Influence of microbial symbionts on plant-insect interactions. In *Advances in Botanical Research*. Elsevier.

Giron, D., Frago, E., Glevarec, G., Pieterse, C. M., & Dicke, M. (2013). Cytokinins as key regulators in plant–microbe–insect interactions: Connecting plant growth and defense. *Functional Ecology, 27*, 599–609.

Gonzalez, F. (2014). *Symbiosis between yeasts and insects. Current Plant Science, 3*, 1–52.

Gonzalez, F., Tkaczuk, C., Dinu, M. M., Fiedler, Ż., Vidal, S., Zchori-Fein, E., & Messelink, G. J. (2016). New opportunities for the integration of microorganisms into biological pest control systems in greenhouse crops. *Journal of Plant Science, 89*, 295–311.

González-Lamothe, R., Mitchell, G., Gattuso, M., Diarra, M. S., Malouin, F., & Bouarab, K. (2009). Plant antimicrobial agents and their effects on plant and human pathogens. *International Journal of Molecular Sciences, 10*, 3400–3419.

Greenwood, M. P. (2021). *Drivers of Gut Microbiome Diversity in Commercially Mass-Reared Black Soldier Flies* (Doctoral dissertation, Stellenbosch University).

Grime, J. P., & Pierce, S. (2012). *The Evolutionary Strategies That Shape Ecosystems*. John Wiley & Sons.

Gross, J., Gallinger, J., & Görg, L. M. (2022). Interactions between phloem-restricted bacterial plant pathogens, their vector insects, host plants, and natural enemies, mediated by primary and secondary plant metabolites. *Environmental Genomics, 42*, 185–215.

Grover, M., Ali, S. Z., Sandhya, V., Rasul, A., & Venkateswarlu, B. (2011). Role of microorganisms in adaptation of agriculture crops to abiotic stresses. *World Journal of Microbiology and Biotechnology, 27*, 1231–1240.

Gupta, A., Singh, U. B., Sahu, P. K., Paul, S., Kumar, A., Malviya, D., Singh, S., Kuppusamy, P., Singh, P., & Paul, D. (2022). Linking soil microbial diversity to modern agriculture practices: A review. *International Journal of Environmental Research and Public Health, 19,* 3141.

Hakim, S., Naqqash, T., Nawaz, M. S., Laraib, I., Siddique, M. J., Zia, R., Mirza, M. S., & Imran, A. (2021). Rhizosphere engineering with plant growth-promoting microorganisms for agriculture and ecological sustainability. *Frontiers in Sustainable Food Systems, 5,* 617157.

Halbleib, C. M., & Ludden, P. W. (2000). Regulation of biological nitrogen fixation. *The Journal of Nutrition, 130*(5), 1081–1084.

Hall, A. A., Morrow, J. L., Fromont, C., Steinbauer, M. J., Taylor, G. S., Johnson, S. N., Cook, J. M., & Riegler, M. (2016). Codivergence of the primary bacterial endosymbiont of psyllids versus host switches and replacement of their secondary bacterial endosymbionts. *Environmental Microbiology, 18*(8), 2591–2603.

Hammer, T. J. (2018). *Ecology and Evolution of Lepidoptera-Microbe Symbiosis* (Doctoral dissertation). University of Colorado at Boulder.

Hammer, T. J., & Moran, N. A. (2019). Links between metamorphosis and symbiosis in holometabolous insects. *Philosophical Transactions of the Royal Society B: Biological Sciences, 374*(1783), 20190068.

Hansen, A. K., & Moran, N. A. (2014). The impact of microbial symbionts on host plant utilization by herbivorous insects. *Molecular Ecology, 23*(6), 1473–1496.

Hansen, A. K., Pers, D., & Russell, J. A. (2020). Symbiotic solutions to nitrogen limitation and amino acid imbalance in insect diets. In *Advances in Insect Physiology*. Elsevier.

Hardoim, P. R., van Overbeek, L. S., Berg, G., Pirttilä, A. M., Compant, S., Campisano, A., Döring, M., & Sessitsch, A. (2015). The hidden world within plants: Ecological and evolutionary considerations for defining functioning of microbial endophytes. *Microbiology and Molecular Biology Reviews, 79*(3), 293–320.

Harman, G. E., & Uphoff, N. (2019). Symbiotic root-endophytic soil microbes improve crop productivity and provide environmental benefits. *Sustainability, 2019,* 9106395.

Hartley, S. E., & Gange, A. C. (2009). Impacts of plant symbiotic fungi on insect herbivores: Mutualism in a multitrophic context. *Annual Review of Ecology, Evolution, and Systematics, 54,* 323–342.

Hazra, T., Spicer, R. A., Hazra, M., Sarkar, S. K., Spicer, T. E., Bera, S., & Khan, M. A. (2021). First fossil evidence of leaf-feeding caterpillars from India and their feeding strategies. *Lethaia, 54*(4), 891–905.

Hodgson, E., & Dean, A. (2020). *Start Scouting for Stalk Borers in Southern Lowa.*

Humphrey, P. T., & Whiteman, N. K. (2020). Insect herbivory reshapes a native leaf microbiome. *Nature Ecology & Evolution, 4*(2), 221–229.

Huot, O. B., Nachappa, P., & Tamborindeguy, C. (2013). The evolutionary strategies of plant defenses have a dynamic impact on the adaptations and interactions of vectors and pathogens. *Insect Science, 20*(3), 297–306.

Indiragandhi, P., Anandham, R., & Sa, T. M. (2011). Functional significance of insect gut bacteria and their role in host insect processes, development, and crop production. In *Biological Insect-Associated Processes in Growth Regulation* (pp. 309–334).

Iqbal, B., Li, G., Alabbosh, K. F., Hussain, H., Khan, I., Tariq, M., Javed, Q., Naeem, M., & Ahmad, N. (2023). Advancing environmental sustainability through microbial reprogramming in growth improvement, stress alleviation, and phytoremediation. *Plant Science, 10,* 100283.

Jacoby, R., Peukert, M., Succurro, A., Koprivova, A., & Kopriva, S. (2017). The role of soil microorganisms in plant mineral nutrition—Current knowledge and future directions. *Frontiers in Plant Science, 8*, 1617.

Jain, S., Varma, A., Tuteja, N., & Choudhary, D. K. (2016). Plant growth-promoting microbial-mediated induced systemic resistance in plants: Induction, mechanism, and expression. In *Microbial-Mediated Induced Systemic Resistance in Plants* (pp. 213–226).

Jeffries, M. J. (2006). *Biodiversity and Conservation.* Routledge.

Johnson, S. N., & Rasmann, S. (2015). Root-feeding insects and their interactions with organisms in the rhizosphere. *Annual Review of Ecology, Evolution, and Systematics, 60*, 517–535.

Jones, M. S., & Snyder, W. E. (2018). Beneficial insects in agriculture: Enhancement of biodiversity and ecosystem services. *Insect Biodiversity and Society, 2*, 105–122.

Kachhawa, D. J. (2017). Microorganisms as a biopesticide. *Journal of Entomology & Zoology Studies, 5*(4), 468–473.

Kang, L. (2019). Overview: Biotic signalling for smart pest management. *Philosophical Transactions of the Royal Society B: Biological Sciences, 374*(1767), 20180306.

Karp, D. S., & Daily, G. C. (2014). Cascading effects of insectivorous birds and bats in tropical coffee plantations. *Ecology, 95*(4), 1065–1074.

Karthik, S., Reddy, M. S., & Yashaswini, G. (2021). Climate change and its potential impacts on insect-plant interactions. In *Global Warming and Climate Change* (pp. 1–23). IntechOpen.

Kaur, H., Kaur, G., Gupta, T., Mittal, D., & Ali, S. A. (2023). Integrating omics technologies for a comprehensive understanding of the microbiome and its impact on cattle production. *Biology, 12*(8), 1200.

Kennedy, D. O., & Wightman, E. L. (2011). Herbal extracts and phytochemicals: Plant secondary metabolites and the enhancement of human brain function. *Advances in Nutrition, 2*(1), 32–50.

Kirchman, D. L. (2018). *Processes in Microbial Ecology.* Oxford University Press.

Kolasa, M. R., & Łukasik, P. (2023). With a little help from my friends: The roles of microbial symbionts in insect populations and communities.

Koohafkan, P., Altieri, M. A., & Gimenez, E. H. (2012). Green agriculture: Foundations for biodiverse, resilient and productive agricultural systems. *International Journal of Agricultural Sustainability, 10*(1), 61–75.

Kuhnert, E., & Collemare, J. (2022). A genomic journey in the secondary metabolite diversity of fungal plant and insect pathogens: From functional to population genomics. *Current Opinion in Microbiology, 69*, 102178.

Kumar, S., Sindhu, S. S., & Kumar, R. J. C. R. I. M. S. (2022). Biofertilizers: An ecofriendly technology for nutrient recycling and environmental sustainability. *Current Research in Microbial Sciences, 3*, 100094.

Kumar, U., Raj, S., Sreenikethanam, A., Maddheshiya, R., Kumari, S., Han, S., Kapoor, K. K., Bhaskar, R., Bajhaiya, A. K., & Gahlot, D. K. (2023). Multi-omics approaches in plant–microbe interactions hold enormous promise for sustainable agriculture. *Agronomy, 13*, 1804.

Lange, C., Boyer, S., Bezemer, T. M., Lefort, M. C., Dhami, M. K., Biggs, E., Groenteman, R., Fowler, S. V., Paynter, Q., & Verdecía Mogena, A. M. (2023). Impact of intraspecific variation in insect microbiomes on host phenotype and evolution. *ISME Journal, 17*(11), 1798–1807.

Lazebnik, J., Frago, E., Dicke, M., & Van Loon, J. J. (2014). Phytohormone mediation of interactions between herbivores and plant pathogens. *Journal of Chemical Ecology*, *40*(7), 730–741.

Lemoine, M. M., Engl, T., & Kaltenpoth, M. J. C. O. I. I. S. (2020). Microbial symbionts expanding or constraining abiotic niche space in insects. *Current Opinion in Insect Science*, *39*, 14–20.

Li, H., Young, S. E., Poulsen, M., & Currie, C. R. (2021). Symbiont-mediated digestion of plant biomass in fungus-farming insects. *Annual Review of Entomology*, *66*, 297–316.

Liu, F., Gao, C., Chen, M., Tang, G., Sun, Y., & Li, K. (2021). The impacts of flowering phenology on the reproductive success of the narrow endemic *Nouelia insignis* Franch. (Asteraceae). *Ecology and Evolution*, *11*(14), 9396–9409.

Lopez, V. (2018). *Influence of the Microbiota on a Phytophagous Insect: Interactions Between Delia Radicum and its Intracellular and Extracellular Symbionts* (Doctoral dissertation). Université Rennes 1.

Lundgren, J. G. (2009). *Relationships of Natural Enemies and Non-Prey Foods*. Springer Science & Business Media.

Lynn-Bell, N. (2021). Bacteriophage roles in the diversity and horizontal transfer of protective functions in the heritable bacterial symbiont, *Hamiltonella defensa* (Doctoral dissertation). University of Georgia.

Magdoff, F., & Van Es, H. (2021). *Building Soils for Better Crops: Ecological Management for Healthy Soils*. Sustainable Agriculture Research and Education Program.

Maire, J., Blackall, L. L., & Van Oppen, M. J. J. M. (2021). Intracellular bacterial symbionts in corals: Challenges and future directions, *Microorganisms*, *9*, 2209.

Malone, S. C. (2020). *Soil Legacy Effects Alter Plant Volatile Emissions in Response to Diversified Cropping Systems* (Doctoral dissertation). Montana State University-Bozeman, College of Agriculture.

Mandrioli, M., Bisanti, M., Grasso, D., & Manicardi, G. C. J. T. I. E. (2016). Role of ant-tending in modulating the presence of symbiotic bacteria against parasitoids in aphids. *Trends in Entomology*, *12*, 63–71.

Manoj, R. R. S., Latrofa, M. S., Epis, S., Otranto, D. J. P., & Vectors. (2021). Wolbachia: Endosymbiont of onchocercid nematodes and their vectors. *Parasites & Vectors*, *14*, 1–24.

Mason, C. J., Jones, A. G., & Felton, G. W. (2019). Co-option of microbial associates by insects and their impact on plant–folivore interactions. *Plant, Cell & Environment*, *42*, 1078–1086.

Mazumdar, T. (2020). *Adaptive Strategies of Enterococcus mundtii to Different Living Conditions in the Microbiome of Spodoptera littoralis Larvae* (Doctoral dissertation). Jena, Friedrich-Schiller-Universität Jena.

Meena, M., Swapnil, P., Zehra, A., Aamir, M., Dubey, M. K., Goutam, J., Upadhyay, R. J. P. M. I. I. A. E. P. V. M. I., & Impacts, A. E. (2017). Beneficial microbes for disease suppression and plant growth promotion. In *Plant-Microbe Interactions in Agro-Ecological Perspectives* (pp. 395–432).

Mohan, R., Zhang, F., Pericak, O., Wynkoop, M., Schwab, N., Spells, S., Stroscio, K., Olson, C., Hermann, C., & Hricovec, M. (2022). Microbiome-based sustainable agriculture targeting plant protection. In *Understanding the Microbiome Interactions in Agriculture and the Environment*. Springer.

Moran, N. A., & Degnan, P. H. (2006). Functional genomics of *Buchnera* and the ecology of aphid hosts. *Molecular Ecology*, *15*, 1251–1261.

Mostafa, S., Wang, Y., Zeng, W., & Jin, B. (2022). Plant responses to herbivory, wounding, and infection. *International Journal of Molecular Sciences*, *23*, 7031.

Mouden, S., Sarmiento, K. F., Klinkhamer, P. G., & Leiss, K. A. (2017). Integrated pest management in western flower thrips: Past, present and future. *Pest Management Science*, *73*, 813–822.

Mucci, N. C. (2022). Mechanisms by which *Xenorhabdus nematophila* interacts with hosts using integrated-omics approaches.

Muhammad, A., Fang, Y., Hou, Y., & Shi, Z. (2017). The gut entomotype of red palm weevil *Rhynchophorus ferrugineus* Olivier (Coleoptera: Dryophthoridae) and their effect on host nutrition metabolism. *Frontiers in Microbiology*, *8*, 2291.

Nair, K. S. (2007). *Tropical Forest Insect Pests: Ecology, Impact, and Management*. Cambridge University Press.

National Academies of Sciences, Engineering, & Medicine. (2017). *Microbiomes of the Built Environment: A Research Agenda for Indoor Microbiology, Human Health, and Buildings*.

New, T. R. (2017). *Mutualisms and Insect Conservation*. Springer.

Noman, A., Aqeel, M., Qasim, M., Haider, I., & Lou, Y. J. (2020). Plant-insect-microbe interaction: A love triangle between enemies in ecosystem. *Science of the Total Environment, 699*, 134181.

Nowakowski, A. C., Miller, A. C., Miller, M. E., Xiao, H., & Wu, X. J. (2022). Potential health benefits of edible insects. *Critical Reviews in Food Science and Nutrition, 62*, 3499–3508. https://doi.org/10.1080/10408398.2021.1922099.

Nwaji, A. R., Arieri, O., Anyang, A. S., Nguedia, K., Abiade, E. B., Forcados, G. E., Oladipo, O. O., Makama, S., Elisha, I. L., & Ozele, N. J. (2022). Natural toxins and one health: A review. *Science of the Total Environment, 1*, 100013. https://doi.org/10.1016/j.scitotenv.2021.100013.

O'Brien, M. J., & Laland, K. N. (2012). Genes, culture, and agriculture: An example of human niche construction. *Current Anthropology, 53*, 434–470. https://doi.org/10.1086/664392.

Oates, C. N., Denby, K. J., Myburg, A. A., Slippers, B., & Naidoo, S. (2016). Insect gallers and their plant hosts: From omics data to systems biology. *International Journal of Molecular Sciences, 17*, 1891. https://doi.org/10.3390/ijms17111891.

Occhipinti, A. J. (2013). Plant coevolution: Evidence and new challenges. *Journal of Plant Interactions, 8*, 188–196. https://doi.org/10.1080/17429145.2012.759302.

Oliver, K. M., & Perlman, S. J. (2020). Toxin-mediated protection against natural enemies by insect defensive symbionts. *Advances in Insect Physiology*. Elsevier. https://doi.org/10.1016/bs.aiip.2020.02.001.

Oliver, K. M., Noge, K., Huang, E. M., Campos, J. M., Becerra, J. X., & Hunter, M. S. (2012). Parasitic wasp responses to symbiont-based defense in aphids. *Biological Bulletin, 10*, 1–10. https://doi.org/10.1086/BB.2012.10.

Oliver, K. M., Smith, A. H., & Russell, J. A. (2014). Defensive symbiosis in the real world—advancing ecological studies of heritable, protective bacteria in aphids and beyond. *FEMS Ecology, 28*, 341–355. https://doi.org/10.1111/1574-6941.12185.

Pangesti, N., Pineda, A., Pieterse, C. M., Dicke, M., & Van Loon, J. J. (2013). Two-way plant mediated interactions between root-associated microbes and insects: From ecology to mechanisms. *Frontiers in Plant Science, 4*, 414. https://doi.org/10.3389/fpls.2013.00414.

Paniagua Voirol, L. R., Frago, E., Kaltenpoth, M., Hilker, M., & Fatouros, N. E. (2018). Bacterial symbionts in Lepidoptera: Their diversity, transmission, and impact on the host. *Frontiers in Microbiology, 9*, 556. https://doi.org/10.3389/fmicb.2018.00556.

Panthee, B., Gyawali, S., Panthee, P., & Techato, K. (2022). Environmental and human microbiome for health. *Life, 12*, 456. https://doi.org/10.3390/life12030456.

Pappas, M. L., Broekgaarden, C., Broufas, G. D., Kant, M. R., Messelink, G. J., Steppuhn, A., Wäckers, F., & Van Dam, N. M. (2017). Induced plant defenses in biological control of arthropod pests: A double-edged sword. *Pest Management Science, 73*, 1780–1788. https://doi.org/10.1002/ps.4441.

Pascal, L., Moulia, C., Gavotte, L., & Paradigm, E. S. R. M. T. T. B. I. A. E. O. G. C. V. T. F. O. A. N. (2019). Biotic interactions, coviability, and dynamics of biodiversity. In *Environmental Science and Technology* (pp. 223–251).

Pastor, J., Cohen, Y., & Hobbs, N. T. (2006). The roles of large herbivores in ecosystem nutrient cycles. *Conservation Biology, 11*, 289. https://doi.org/10.1046/j.1523-1739.2006.01279.x.

Patil, A., Singh, N., & Pawar, N. (2023). The dynamic interplay between gut health and dietary options: A systematic review. *Journal of Nutritional Health and Food, 13*, 112–121.

Pegoraro, L., Hidalgo, O., Leitch, I. J., Pellicer, J., & Barlow, S. E. (2020). Automated video monitoring of insect pollinators in the field. *Ecological Technology, 4*, 87–97. https://doi.org/10.1002/journal_40.

Pérez-Brocal, V., Gil, R., Moya, A., & Latorre, A. (2011). New insights on the evolutionary history of aphids and their primary endosymbiont *Buchnera aphidicola. International Journal of Evolutionary Biology, 2011*, 250154.

Perlman, D., Martínez-Álvaro, M., Moráis, S., Altshuler, I., Hagen, L. H., Jami, E., Roehe, R., Pope, P. B., & Mizrahi, I. (2022). Concepts and consequences of a core gut microbiota for animal growth and development. *Annual Review of Animal Biosciences, 10*, 177–201. https://doi.org/10.1146/annurev-animal-020921-101722.

Perreau, J., Patel, D. J., Anderson, H., Maeda, G. P., Elston, K. M., Barrick, J. E., & Moran, N. A. (2021). Vertical transmission at the pathogen-symbiont interface: *Serratia symbiotic* and aphids. *mBio, 12*, e00359-21. https://doi.org/10.1128/mbio.00359-21.

Pieterse, C. M., Zamioudis, C., Berendsen, R. L., Weller, D. M., Van Wees, S. C., & Bakker, P. A. (2014). Induced systemic resistance by beneficial microbes. *Plant Pathology, 52*, 347–375. https://doi.org/10.1111/jpp.12128.

Pouliot, A. (2018). *Allure of Fungi*. CSIRO Publishing.

Poveda, J., Abril-Urias, P., & Escobar, C. (2020). Biological control of plant-parasitic nematodes by filamentous fungi inducers of resistance: *Trichoderma*, mycorrhizal, and endophytic fungi. *Frontiers in Microbiology, 11*, 992. https://doi.org/10.3389/fmicb.2020.00992.

Poveda, K., Gómez, M. I., & Martinez, E. J. (2008). Diversification practices: Their effect on pest regulation and production. *Crop Protection, 34*, 131–144. https://doi.org/10.1016/j.cropro.2007.06.010.

Prajapati, S. K., Kumar, V., Dayal, P., Gairola, A., Borate, R. B., & Srivastava, R. (2023). The role of carbon in life's blueprint and carbon cycle: Understanding Earth's essential cycling system: Benefits and harms to our planet. *International Journal of Earth Sciences, 1*, 21–32.

Prakash, O., Sharma, R., Rahi, P., & Karthikeyan, N. (2015). Role of microorganisms in plant nutrition and health. In *Nutritional Ecology of Beneficial Bacteria* (pp. 125–161).

Prashar, P., Kapoor, N., & Sachdeva, S. (2014). Rhizosphere: Its structure, bacterial diversity and significance. *Research in Environmental Sciences and Biotechnology, 13*, 63–77.

Provenza, F. D., Villalba, J. J., Haskell, J., Macadam, J. W., Griggs, T., & Wiedmeier, R. D. (2007). The value to herbivores of plant physical and chemical diversity in time and space. *Crop Science, 47*, 382–398.

Puhlmann, M. L., & De Vos, W. M. (2022). Intrinsic dietary fibers and the gut microbiome: Rediscovering the benefits of the plant cell matrix for human health. *Frontiers in Integrative Biology, 13*, 954845. https://doi.org/10.3389/fib.2022.954845.

Puri, K. M., Butardo, V., & Sumer, H. J. S. (2021). Evaluation of natural endosymbiosis for progress towards artificial endosymbiosis. *Symbiosis, 84*, 1–17.

Rai, M., Ingle, A. J. A. M., & Biotechnology. (2012). Role of nanotechnology in agriculture with special reference to management of insect pests. *Applied Microbiology and Biotechnology, 94*, 287–293.

Rajeswari, G., Jacob, S., Chandel, A. K., & Kumar, V. J. M. C. F. (2021). Unlocking the potential of insect and ruminant host symbionts for recycling of lignocellulosic carbon with a biorefinery approach: A review. *Microbial Cell Factories, 20*, 107.

Ramakrishna, W., Yadav, R., & Li, K. J. A. S. E. (2019). Plant growth promoting bacteria in agriculture: Two sides of a coin. *Applied Soil Ecology, 138*, 10–18.

Rashid, M. H. O., & Chung, Y. R. J. F. I. P. S. (2017). Induction of systemic resistance against insect herbivores in plants by beneficial soil microbes. *Frontiers in Plant Science, 8*, 1816.

Ratnadass, A., Fernandes, P., Avelino, J., & Habib, R. J. A. F. S. D. (2012). Plant species diversity for sustainable management of crop pests and diseases in agroecosystems: A review. *Agronomy for Sustainable Development, 32*, 273–303.

Raza, W., & Shen, Q. (2020). Volatile organic compounds mediated plant-microbe interactions in soil. In *Molecular Aspects of Plant Beneficial Microbes in Agriculture* (pp. 1–15). Elsevier.

Russell, S. L., Chappell, L., & Sullivan, W. J. C. T. I. D. B. (2019). A symbiont's guide to the germline. *Current Topics in Developmental Biology, 135*, 315–351.

Ryalls, J. M., Riegler, M., Moore, B. D., Johnson, S. N. J. A., & Entomology, F. (2013). Biology and trophic interactions of lucerne aphids. *Australian Journal of Entomology, 15*, 335–350.

Saeed, Q., Xiukang, W., Haider, F. U., Kučerik, J., Mumtaz, M. Z., Holatko, J., Naseem, M., Kintl, A., Ejaz, M., & Naveed, M. J. I. J. O. M. S. (2021). Rhizosphere bacteria in plant growth promotion, biocontrol, and bioremediation of contaminated sites: A comprehensive review of effects and mechanisms. *International Journal of Molecular Sciences, 22*, 10529.

Salehi, B., Mishra, A. P., Nigam, M., Sener, B., Kilic, M., Sharifi-Rad, M., Fokou, P. V. T., Martins, N., & Sharifi-Rad, J. J. B. (2018). Resveratrol: A double-edged sword in health benefits. *Bioscience Reports, 6*, 91.

Sandrock, C., Tanadini, L. G., Pettis, J. S., Biesmeijer, J. C., Potts, S. G., Neumann, P. J. A., & Entomology, F. (2014). Sublethal neonicotinoid insecticide exposure reduces solitary bee reproductive success. *Fungal Entomology, 16*, 119–128.

Sargent, D., Conaty, W. C., Tissue, D. T., Sharwood, R. E. J. J. O. S. A., & Environment. (2022). Synthetic biology and opportunities within agricultural crops. *Journal of Sustainable Agriculture, 1*, 89–107.

Sarkar, A., Harty, S., Johnson, K. V. A., Moeller, A. H., Archie, E. A., Schell, L. D., Carmody, R. N., Clutton-Brock, T. H., Dunbar, R. I., Burnet, P. W. J. N. E., & Evolution. (2020). Microbial transmission in animal social networks and the social microbiome. *Nature Ecology & Evolution, 4*, 1020–1035.

Sarwar, M. J. I. J. O. E., & Technology, A. R. (2015). Biopesticides: An effective and environmentally friendly insect-pests inhibitor line of action. *International Journal of Environmental & Technology, 1*, 10–15.

Schlägel, U. E., Grimm, V., Blaum, N., Colangeli, P., Dammhahn, M., Eccard, J. A., Hausmann, S. L., Herde, A., Hofer, H., & Joshi, J. J. B. R. (2020). Movement-mediated community assembly and coexistence. *Journal of Behavioral Ecology and Sociobiology, 95*, 1073–1096.

Schmidt, K., & Engel, P. J. J. O. E. B. (2021). Mechanisms underlying gut microbiota–host interactions in insects. *Journal of Experimental Biology, 224*, jeb207696.

Schoonhoven, L. M., Van Loon, J. J., & Dicke, M. (2005). *Insect-Plant Biology*. Oxford University Press.

Schowalter, T. D. (2013). *Insects and Sustainability of Ecosystem Services*. CRC Press.

Sehgal, V. (2019). *Climate, Agriculture and Man Shrinking Biodiversity and Sustainability*. Scientific Publishers.

Sepúlveda, D. A., Zepeda-Paulo, F., Ramírez, C. C., Lavandero, B., & Figueroa, C. C. J. I. S. (2017). Diversity, frequency, and geographic distribution of facultative bacterial endosymbionts in introduced aphid pests. *Insect Science, 24*, 511–521.

Shah, S. T., Basit, A., Sajid, M., Mohamed, H. I. J. P. G. P. M. F. S. B., & Management, A. S. (2021). Microorganism: A potent biological tool to combat insects and herbivores. *Pest Management Science, 77*, 505–555.

Sharma, G., Malthankar, P. A., & Mathur, V. J. A. O. T. E. S. O. A. (2021). Insect–plant interactions: A multilayered relationship. *Annals of the Entomological Society of America, 114*, 1–16.

Shinwari, Z. K., Tanveer, F., Iqrar, I. J. M. I. P. H., Challenges, D., & Opportunities. (2019). Role of microbes in plant health, disease management, and abiotic stress management. In *Microbial Challenges and Opportunities in Plant Health* (pp. 231–250).

Show, B. K., Banerjee, S., Banerjee, A., Ghoshtakur, R., Hazra, A. K., Mandal, N. C., Ross, A. B., Balachandran, S., Chaudhury, S. J. R. I. E. S., & Bio/Technology. (2022). Insect gut bacteria: A promising tool for enhanced biogas production. *Research in Environmental Science & Biotechnology, 21*, 1–25.

Shu, W. S., & Huang, L. N. J. N. R. M. (2022). Microbial diversity in extreme environments. *Journal of Natural Resources and Microbiology, 20*, 219–235.

Siddiqui, J. A., Khan, M. M., Bamisile, B. S., Hafeez, M., Qasim, M., Rasheed, M. T., Rasheed, M. A., Ahmad, S., Shahid, M. I., & Xu, Y. J. F. I. M. (2022). Role of insect gut microbiota in pesticide degradation: A review. *Frontiers in Microbiology, 13*, 870462.

Sieksmeyer, T. (2021). Host-microbiome-pathogen interactions in cockroaches.

Singh, B., Mal, G., Gautam, S. K., & Mukesh, M. (2019). Insect gut—a treasure of microbes and microbial enzymes. *Advances in Applied Biotechnology, 51*, 51–58.

Singh, S., Singh, A., Baweja, V., Roy, A., Chakraborty, A., & Singh, I. (2021). Molecular rationale of insect-microbes symbiosis—From insect behavior to mechanism. *Microorganisms, 9*, 2422.

Skaljac, M., Vilcinskas, A., & Aphids, E. O. (2016). Bacterial symbionts of aphids (Hemiptera: Aphididae). *Biological Control, 100*, 100–125.

Skendžić, S., Zovko, M., Živković, I. P., Lešić, V., & Lemić, D. (2021). The impact of climate change on agricultural insect pests. *Insects, 12*, 440.

Smulders, L. (2022). *Interactions of Soil Microbiome, Plant Defenses and Domestication in Tomato* (Doctoral dissertation). Universidad de Granada.

Snyder, L. D., Gómez, M. I., & Power, A. G. (2020). Crop varietal mixtures as a strategy to support insect pest control, yield, economic, and nutritional services. *Frontiers in Sustainable Food Systems, 4*, 60.

Sonenshine, D. E., & Stewart, P. E. (2021). Microbiomes of blood-feeding arthropods: Genes coding for essential nutrients and relation to vector fitness and pathogenic infections. *Microorganisms, 9*, 2433.

Stephens, A. E., & Westoby, M. (2015). Effects of insect attack to stems on plant survival, growth, reproduction, and photosynthesis. *Oecologia, 124*(3), 266–273.

Stevens, C. E., & Hume, I. D. (1998). Contributions of microbes in vertebrate gastrointestinal tract to production and conservation of nutrients. *Physiological Reviews, 78*, 393–427.

Su, Q., Zhou, X., & Zhang, Y. (2013). Symbiont-mediated functions in insect hosts. *Current Biology, 6*, e23804.

Sugio, A., Dubreuil, G., Giron, D., & Simon, J. C. (2015). Plant-insect interactions under bacterial influence: Ecological implications and underlying mechanisms. *Journal of Experimental Botany, 66*, 467–478.

Taggar, M. S. (2015). Insect cellulolytic enzymes: Novel sources for degradation of lignocellulosic biomass. *Journal of Applied Science, 7*, 625–630.

Tahat, M., Alananbeh, K. M., Othman, Y., & Leskovar, D. J. S. (2020). Soil health and sustainable agriculture. *Sustainability, 12*(12), 4859.

Talbot, A. L. (2017). *Evaluating Natural and Induced Resistance of Broccoli (Brassica oleracea var. italica) Against Pieris rapae* (Doctoral dissertation). University of Minnesota.

Tholt, G., Kis, A., Medzihradszky, A., Szita, É., Tóth, Z., Havelda, Z., & Samu, F. (2018). Could vectors' fear of predators reduce the spread of plant diseases? *Scientific Reports, 8*, 8705.

Thomas, C., Desmond-Le Quemener, E., Gribaldo, S., & Borrel, G. (2021). Factors shaping the abundance and diversity of archaea in the animal gut. *Nat Commun.*

Timmis, K., Cavicchioli, R., Garcia, J. L., Nogales, B., Chavarría, M., Stein, L., McGenity, T. J., Webster, N., Singh, B. K., & Handelsman, J. (2019). The urgent need for microbiology literacy in society. *Environmental Microbiology, 21*, 1513–1528.

Trivedi, P., Leach, J. E., Tringe, S. G., Sa, T., & Singh, B. K. (2020). Plant–microbiome interactions: From community assembly to plant health. *Nature Reviews Microbiology, 18*, 607–621.

Trivedi, P., Mattupalli, C., Eversole, K., & Leach, J. E. (2021). Enabling sustainable agriculture through understanding and enhancement of microbiomes. *New Phytologist, 230*, 2129–2147.

Tscharntke, T., Karp, D. S., Chaplin-Kramer, R., Batáry, P., DeClerck, F., Gratton, C., Hunt, L., Ives, A., Jonsson, M., & Larsen, A. (2016). When natural habitat fails to enhance biological pest control—Five hypotheses. *Biological Conservation, 204*, 449–458.

Turbé, A., De Toni, A., Benito, P., Lavelle, P., Camacho, N. R., Van der Putten, W. H., Labouze, E., & Mudgal, S. (2010). *Soil Biodiversity: Functions, Threats and Tools for Policy Makers*.

Upadhyay, R. K. (2020). Markers for global climate change and its impact on social, biological, and ecological systems: A review. *Asian Journal of Climate Change, 9*, 159.

Usta, C. (2013). Microorganisms in biological pest control—A review (bacterial toxin application and effect of environmental factors). *Current Pharmaceutical Insights in Bioresearch, 13*, 287–317.

Vacher, C., Castagneyrol, B., Jousselin, E., & Schimann, H. (2021). Trees and insects have microbiomes: Consequences for forest health and management. *Current Forestry Reviews, 7*, 81–96.

Van der Greef, J., Martin, S., Juhasz, P., Adourian, A., Plasterer, T., Verheij, E., & McBurney, R. J. (2007). The art and practice of systems biology in medicine: Mapping patterns of relationships. *Journal of Proteome Research, 6*, 1540–1559.

Vannette, R. L. (2020). The floral microbiome: Plant, pollinator, and microbial perspectives. *Annual Review of Ecology, Evolution, and Systematics, 51*, 363–386.

Vanthournout, B., Swaegers, J., & Hendrickx, F. (2011). Spiders do not escape reproductive manipulations by Wolbachia. *BMC Evolutionary Biology, 11*, 1–9.

Vorburger, C., & Rouchet, R. (2016). Are aphid parasitoids locally adapted to the prevalence of defensive symbionts in their hosts? *BMC Evolutionary Biology, 16*, 1–11.

Wang, Y. B., Ren, F. R., Yao, Y. L., Sun, X., Walling, L. L., Li, N. N., Bai, B., Bao, X. Y., Xu, X. R., & Luan, J. B. (2020). Intracellular symbionts drive sex ratio in the whitefly by facilitating fertilization and provisioning of B vitamins. *The ISME Journal, 14*, 2923–2935.

War, A. R., Buhroo, A. A., Hussain, B., Ahmad, T., Nair, R. M., & Sharma, H. C. (2020). Plant defense and insect adaptation with reference to secondary metabolites. *Current Opinion in Systems Microbiology, 51*, 795–822.

War, A. R., Paulraj, M. G., Ahmad, T., Buhroo, A. A., Hussain, B., Ignacimuthu, S., & Sharma, H. C. (2012). Mechanisms of plant defense against insect herbivores. *Plant Science and Behavior, 7*, 1306–1320.

Watkinson, S. C. (2016). Mutualistic symbiosis between fungi and autotrophs. In *The Fungi* (pp. 1–20). Elsevier.

Weisser, W., & Siemann, E. (2008). The various effects of insects on ecosystem functioning. In *Insects and Ecosystem Function* (pp. 3–24). Springer.

Wernegreen, J. J. (2002). Genome evolution in bacterial endosymbionts of insects. *Nature Reviews Genetics, 3*, 850–861.

Wernegreen, J. J. (2012a). Endosymbiosis. *Current Biology, 22*, R555–R561.

Wernegreen, J. J. (2012b). Strategies of genomic integration within insect-bacterial mutualisms. *The Biological Bulletin, 223*, 112–122.

Wertz, J. T., & Bechade, B. (2020). Symbiont-mediated degradation of dietary carbon sources in social herbivorous insects. In *Advances in Insect Physiology* (pp. 99–130). Elsevier.

Whitehill, J. G., Bohlmann, J., & Krokene, P. (2023). Forest insect–plant interactions. In *Forest Entomology and Pathology: Volume 1: Entomology* (pp. 201–225). Springer International Publishing Cham.

Wielkopolan, B., & Obrępalska-Stęplowska, A. (2016). Three-way interaction among plants, bacteria, and coleopteran insects. *Pathogens, 244*, 313–332.

Wohlgemuth, R., Twardowski, T., & Aguilar, A. (2021). Bioeconomy moving forward step by step—A global journey. *New Biotechnology, 61*, 22–28.

Xia, X., Lan, B., Tao, X., Lin, J., & You, M. (2020). Characterization of *Spodoptera litura* gut bacteria and their role in feeding and growth of the host. *Frontiers in Microbiology, 11*, 1492.

Xu, W., Liu, W., Li, J., Zhu, X., Wang, L., Li, D., Zhang, K., Ji, J., Gao, X., & Luo, J. (2023). *Buchnera* breaks the specialization of the cotton-specialized aphid (*Aphis gossypii*) by providing nutrition through zucchini. *Frontiers in Nutrition, 10*, 1128272.

Yactayo-Chang, J. P., Tang, H. V., Mendoza, J., Christensen, S. A., & Block, A. K. (2020). Plant defense chemicals against insect pests. *Agriculture, 10*, 1156.

Zavahir, J. S., Wijepala, P. C., & Seneviratne, G. (2021). Role of microbial communities in plant–microbe interactions, metabolic cooperation, and self-sufficiency leading to sustainable agriculture. *Reviews on Microbial Cooperation for Sustainability, 1*, 1–35.

Zhang, Y., Zhang, S., & Xu, L. (2023). The pivotal roles of gut microbiota in insect–plant interactions for sustainable pest management. *Nature Biotechnology & Microbiomes, 9*, 66.

Zhu, L., Huang, J., Lu, X., & Zhou, C. (2022). Development of plant systemic resistance by beneficial rhizobacteria: Recognition, initiation, elicitation, and regulation. *Frontiers in Plant Science, 13*, 952397.

Ziganshina, E. E., Mohammed, W. S., Shagimardanova, E. I., Vankov, P. Y., Gogoleva, N. E., & Ziganshin, A. M. (2018). Fungal, bacterial, and archaeal diversity in the digestive tract of several beetle larvae (*Coleoptera*). *Bioresearch International, 2018*, 1–10.

Index

B

F

H

N

Q

R

S

For Product Safety Concerns and Information please contact our EU representative GPSR@taylorandfrancis.com
Taylor & Francis Verlag GmbH, Kaufingerstraße 24, 80331 München, Germany

www.ingramcontent.com/pod-product-compliance
Lightning Source LLC
LaVergne TN
LVHW020628300926
849161LV00004B/636
* 9 7 8 1 7 7 9 6 4 4 2 7 5 *